Handbook of
OFFSHORE OIL AND GAS
OPERATIONS

Handbook of
OFFSHORE OIL AND GAS OPERATIONS

JAMES G. SPEIGHT
CD & W Inc.,
Laramie, Wyoming, USA

Amsterdam • Boston • Heidelberg • London
New York • Oxford • Paris • San Diego
San Francisco • Singapore • Sydney • Tokyo
Gulf Professional Publishing is an Imprint of Elsevier

Gulf Professional Publishing is an imprint of Elsevier
225 Wyman Street, Waltham, MA 02451, USA
The Boulevard, Langford Lane, Kidlington, Oxford, OX5 1GB, UK

First edition 2015

Notice
No responsibility is assumed by the publisher for any injury and/or damage to persons or
property as a matter of products liability, negligence or otherwise, or from any use or operation
of any methods, products, instructions or ideas contained in the material herein. Because of rapid
advances in the medical sciences, in particular, independent verification of diagnoses and drug
dosages should be made.

Library of Congress Cataloging-in-Publication Data

Speight, James G.
  Handbook of offshore oil and gas operations / by James G. Speight. — First edition.
    pages cm
  Includes bibliographical references and index.
  ISBN 978-1-85617-558-6
  1. Offshore oil well drilling. 2. Petroleum—Transportation. 3. Petroleum pipelines—Maintenance
and repair. 4. Oil spills—Prevention. 5. Gas well drilling. 6. Gas well drilling—Environmental aspects.
7. Natural gas—Transportation. 8. Natural gas pipelines—Maintenance and repair. I. Title.
  TN871.3.S692 2015
  363.738'27—dc23

                                                                                            2014034659

British Library Cataloguing in Publication Data
A catalogue record for this book is available from the British Library

ISBN: 978-1-85617-558-6

For information on all Gulf Professional publications
visit our website at http://store.elsevier.com/

This book has been manufactured using Print On Demand technology. Each copy is produced to order
and is limited to black ink. The online version of this book will show color figures where appropriate.

Working together
to grow libraries in
developing countries

ELSEVIER | Book Aid International

www.elsevier.com • www.bookaid.org

TABLE OF CONTENTS

ABOUT THE AUTHOR

DR JAMES G. SPEIGHT

Dr. James G. Speight, who has doctorate degrees in Chemistry, Geological Sciences, and Petroleum Engineering, is the author of more than 60 books in petroleum science, petroleum engineering, and environmental sciences. He has served as Adjunct Professor in the Department of Chemical and Fuels Engineering at the University of Utah and in the Departments of Chemistry and Chemical and Petroleum Engineering at the University of Wyoming. In addition, he has been a Visiting Professor in Chemical Engineering at the following universities: the University of Missouri-Columbia, the Technical University of Denmark, and the University of Trinidad and Tobago.

As a result of his work, Dr. Speight has been honored as the recipient of the following awards:

- Diploma of Honor, United States National Petroleum Engineering Society. *For Outstanding Contributions to the Petroleum Industry, 1995.*
- Gold Medal of the Russian Academy of Sciences. *For Outstanding Work in the Area of Petroleum Science, 1996.*
- Einstein Medal of the Russian Academy of Sciences. *In recognition of Outstanding Contributions and Service in the field of Geologic Sciences, 2001.*
- Gold Medal - Scientists without Frontiers, Russian Academy of Sciences. *In recognition of His Continuous Encouragement of Scientists to Work Together across International Borders, 2005.*
- Methanex Distinguished Professor, University of Trinidad and Tobago. *In Recognition of Excellence in Research, 2006.*
- Gold Medal – Giants of Science and Engineering, Russian Academy of Sciences. *In recognition of Continued Excellence in Science and Engineering, 2006.*

PREFACE

Offshore oil and gas fields have received much attention since the first out-of-sight-land offshore platform that produced oil went on stream in 1947. Since that time, the design and construction of offshore pipelines and platforms have continued and the technology capable of finding and producing oil in deepwater and in harsh environments around the world has seen remarkable developments. As the industry enters a new era in technology, the offshore industry will continue to provide current and future supplies of oil and gas.

This text deals with the science and technology of offshore oil and gas production. The book contains an extensive coverage of offshore operations, including a historic background of offshore operations. The authors compare and contrast offshore operations and gas and oil properties to onshore operations throughout the text. In addition, this book provides coverage and discussion on environmental impact of the offshore oil and gas industry. Specifically, worldwide advances in studies, control, and prevention of the impact of the offshore crude oil and natural gas on the marine environment are addressed. Based on currently available information, environmental requirements for discharges and seawater quality are also presented.

This book can be used as a classroom text as well as a reference book for information on subsea exploration, drilling and completing wells, development systems, fixed structures, production systems, transportation, and related environmental issues.

Dr. James G. Speight, PhD DSc
Laramie, Wyoming, USA

CHAPTER ONE

Occurrence and Formation of Crude Oil and Natural Gas

Outline

1.1 INTRODUCTION

Crude oil is the most commonly used source of energy and liquid fuels to such an extent that the use of crude oil is projected to continue in the current amounts for several decades (Speight and Ozum, 2002; Parkash, 2003; Hsu and Robinson, 2006; Gary et al., 2007; Speight, 2011a, 2014b). Geologically, crude oil is scattered throughout the earth's crust, which is divided into chronological strata that are based on the distinctive systems of organic debris (as well as fossils, minerals, and other characteristics) (Table 1.1). Carbonaceous materials natural products such as coal, crude oil, and natural gas (fossil fuels) occur in many of these geological strata—the actual origin of fossil fuels within these formations is a question that has been debated for a century or more and still remains open to conjecture and speculation (Fig. 1.1) (Gold, 2013; Speight, 2013d, 2014a).

Handbook of Offshore Oil and Gas Operations. http://dx.doi.org/10.1016/B978-1-85617-558-6.00001-5
1

Table 1.1 The Geologic Timescale

Era	Period	Epoch	Approximate duration (millions of years)	Approximate number of years ago (millions of years)
Cenozoic	Quaternary	Holocene	10,000 years ago to the present	
		Pleistocene	2	0.01
	Tertiary	Pliocene	11	2
		Miocene	12	13
		Oligocene	11	25
		Eocene	22	36
		Paleocene	71	58
Mesozoic	Cretaceous		71	65
	Jurassic		54	136
	Triassic		35	190
Paleozoic	Permian		55	225
	Carboniferous		65	280
	Devonian		60	345
	Silurian		20	405
	Ordovician		75	425
	Cambrian		100	500
Precambrian			3,380	600

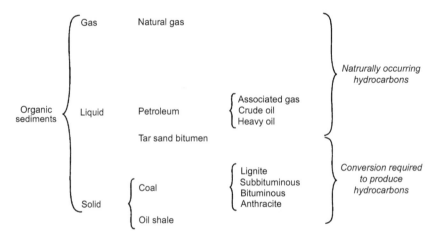

Figure 1.1 Relationship of crude oil to other fossil fuels.

However, in recent years, the quality of crude oil shipped to refineries has deteriorated insofar as there are larger quantities of heavy oil and tar sand bitumen (oil sand bitumen) in the crude mix (Speight, 2013a,b,c). This is reflected in a progressive decrease in API gravity (i.e., increase in density or an increase in specific gravity), which is usually accompanied by a rise in sulfur content (Speight, 2014a). Furthermore, the increasing need for crude oil products (such as liquid fuels) has resulted in an increasing need to develop and recover crude oil from other than the conventional land-based operations. Outside the United States, deposits are now being tapped in Siberia, Australia, India, and China, with additional accumulations being developed on a world-wide scale. In fact, the distribution of crude oil occurs in such diverse locations as polar regions, temperate regions, and tropical regions as well as in land-based sedimentary basins to offshore (under-sea) sedimentary basins.

Crude oil use and the associated technologies, in one form or another, is viable for (at least) the next 50 years until suitable alternative forms of energy are readily available and developed on a sufficient scale to (at first) comple-ment crude oil use and then (eventually) to replace crude oil as the dominant source of liquid fuels (Giampietro and Mayumi, 2009; EREC, 2010; Lan-geveld et al., 2010; Nersesian, 2010; Seifried and Witzel, 2010; Speight, 2011b; Lee and Shah, 2013). Therefore, a thorough understanding of the benefits and limitations of crude oil recovery, especially from offshore locations, is necessary and will be introduced within the pages of this book. For this particular chapter, the focus is on the various types of nomenclature used by the oil and gas industry as well as the types of crude oil that exist and could be recovered from offshore locations in which the reservoirs are under large bodies of water (specifically oceans), which is the subject matter of this book.

1.2 OFFSHORE OIL AND GAS

The reservoir rocks (onshore and offshore) that yield crude oil and natural gas range in age from Precambrian to Recent geologic time but rocks deposited during the Tertiary, Cretaceous, Permian, Pennsylvanian, Mississippian, Devonian, and Ordovician periods are particularly productive. In contrast, rocks of Jurassic, Triassic, Silurian, and Cambrian age are less pro-ductive and rocks of Precambrian age yield crude oil only under exceptional circumstances. Most of the crude oil currently recovered is produced from underground reservoirs (Magoon and Dow, 1994). However, surface seepage of crude oil and natural gas are common in many regions (Abraham, 1945).

In fact, it is the surface seepage of oil that led to the first use of the high boiling material (bitumen, sometime incorrectly called natural asphalt) in the area known as the Fertile Crescent (in ancient Mesopotamia–modern-day Iraq) (Forbes, 1958a,b, 1959, 1964).

The reserves of onshore of oil and gas have been sought and developed since the second half of the nineteenth century but the exploitation of offshore oil, on the other hand, dates back little more than 40 years [except for the shallow water exploitation of fields, which were essentially the continuation of onshore activities: Lake Maracaibo (Venezuela), the Caspian Sea, and the Gulf of Mexico]. Deep-water (water having a depth in excess of 160 ft) activities were initiated in the 1960s principally in the United States, whereas elsewhere offshore activities seeking reserves of oil and gas not intimately connected of onshore deposits were initiated in the 1970s by changes in the organization and structure of the international oil industry, which reduced and then eliminated access by the major international oil companies to most of the oil reserves that had been discovered in the Middle East and other OPEC countries. While these oil companies maintained business relationships with various oil-producing countries, the companies also pursued expansion of offshore activities in the US Gulf of Mexico, other major searches for offshore reserves thus began in the North Sea, the South China Sea, Australia, the Gulf of Suez, eastern Canada, Southern Brazil, the Mexican Gulf of Mexico, and West Africa.

To date, significant reserves of offshore oil and gas have been discovered of and it is predicted that proven (identified) reserves of oil and gas in discovered fields will expand considerably over the next 30 years (Humphries et al., 2010). In addition, the development of offshore fields is much more recent than the development of onshore fields and, therefore, less fully evaluated and, as a result, a greater percentage of the additional reserves expected from field extensions and re-evaluations will come from the offshore locales. This positive aspect for the relatively greater appreciation of the offshore fields may be partially offset by the greater scope onshore for securing improved rates of recovery from the oil and gas in place, including the oil and gas found in tight shale formations (Speight, 2013d, 2014a) but the development plans for exploiting offshore fields have generally been structured so as to achieve higher recovery rates than those typically observed and reported (at the time of writing).

In addition, there is a perceived strong likelihood that new onshore and offshore fields will continue to be discovered, which will provide additional oil and gas reserves—many numbers for such discoveries have been bandied about but are generally uncertain (unproven) and, if presented here, would very likely have changed by the time of publication of this book.

Progress in discovering such fields will be orientated somewhat more heavily to the offshore following on the transition of exploration efforts to deep water environments (Larue and Yue, 2003). The deep waters related to many of the modern offshore activities generally offer more attractive development prospects for the companies. Such prospects have been identified in the Atlantic Margin (an area lying within a plate at the boundary between continental and oceanic crust), the Gulf of Mexico, the south Atlantic, off the coasts of West and Southern Africa, the Timor Sea, and the South China Sea. On the other hand, if there is the possibility of the development and use of more of newly discovered and lower cost onshore oil and gas reserves, this may compromise the development of deep-water reserves. The size and speed of development and exploitation of oil and gas discoveries in deep water depend on sustainable contemporary international oil price levels.

In terms of reserves and although it might seem like the ocean is the border of many countries with a coastline, this is not the case. For example, using the United States as an example, the border is actually 200 miles out from the land. This 200-mile-wide band around the country is called the Exclusive Economic Zone (EEZ) and set a landmark for the offshore industry when it was claimed by President Reagan in 1983 by signing Proclamation 5030, which established the EEZ of the United States, claiming all rights to 200 miles off the coastline for the United States. In 1994, as a follow-up to the Reagan pronouncement, all countries with a coastline were granted an EEZ of 200 miles from their coastline according to the International Law of the Sea. The United States Minerals Management Service (MMS) leases the land under the ocean to producers, who pay rental fees and royalties to the MMS on all of the minerals (crude oil and natural gas are included in the definition of minerals) that are extracted from the ocean floor.

Physically, the beach extends from the shore into the ocean on a *continental shelf* that gradually descends to a sharp drop (the *continental slope*). The continental shelf drops off at the continental slope, ending in *abyssal plains* that are 1–3 miles below sea level.

By definition, the *continental shelf* is that part of the continental margin that is between the shoreline and the continental slope or, when there is no noticeable continental slope, the continental shelf is shallow, rarely more than 500–650 ft deep. On the other hand, the *continental slope* is the seaward border of the continental shelf, which leads to the beginning of the ocean basins that are located at depths of 330–10,500 ft. Continental slopes are indented by numerous submarine canyons and mounds. The predominant sediments of continental slopes are various types of mud and there are smaller amounts of sand-based sediments and gravel-based sediment—over geologic time, the continental slopes are temporary depositional sites for sediments, which build up until the mass becomes unstable and sloughs off to the lower slope and the continental rise. During periods of high sea level, these processes slow down as the coastline retreats landward across the continental shelf and more of the sediments delivered to the coast are trapped in estuaries and lagoons.

An *abyssal plain* is an underwater plain on the deep ocean floor, usually found at depths of 10,000–20,000 ft, which (typically) lie between the foot of a *continental rise* and a *mid-ocean ridge*. Abyssal plains result from the blanketing of an originally uneven surface of oceanic crust by fine-grained sediments, mainly clay and silt. Much of this sediment is deposited by turbidity currents that have been channeled from the continental margins—the zone of the ocean floor that separates the oceanic crust from thick continental crust and which includes the continental shelf, the continental slope, and continental rise—along submarine canyons down into deeper water. The remainder of the sediment is composed chiefly of *pelagic* sediments, which forms a sea or lake or that is neither close to the bottom nor near the shore (*pelagic zone*). The *continental rise* is an underwater feature found between the continental slope and the abyssal plain and represents the final stage in the boundary between continents and the deepest part of the ocean.

The *pelagic zone* can be contrasted with the *benthic zone* and the *demersal zone* at the bottom of the sea. The *benthic zone* is the lowest level of a body of water such as an ocean or a lake, including the sediment surface and some sub-surface layers and the *demersal zone* in that part of the sea or ocean or deep lake comprising the water column that is near to and is significantly affected by the seabed and the benthic zone.

An *abyssal plain* is typically a flat and smooth region and is a major geologic element of an ocean basin (the other elements of interest are an elevated mid-ocean ridge and flanking abyssal hills). In addition to these elements, *active* oceanic basins (i.e., basins associated with a moving plate tectonic boundary) also typically include an oceanic trench and a subduction zone. While an abyssal plain is typically flat, some exhibit mountain ridges, deep canyons, and valleys—the peaks of some of the undersea mountain ridges form islands where they extend above the water.

Owing to the vast size, abyssal plains are currently believed to be a major site of bioactivities and exert significant influence upon ocean carbon cycling, dissolution of calcium carbonate, and atmospheric carbon dioxide concentrations over timescales of 100–1000 years. The structure and function of abyssal ecosystems are strongly influenced by the rate of deposition of carbonaceous material on to the seafloor and the composition of the material that eventually settles.

Activities in deeper water (a depth in excess of 150 ft) were initiated in the 1960s (principally in the United States), whereas elsewhere offshore activities seeking reserves of crude oil not intimately connected to the on-shore deposits were eventually very powerfully stimulated in the 1970s by the radical changes in the organization and structure of the international oil industry (Speight, 2011a,b). These changes removed the control exerted by the oil international oil companies and gave control of most of the oil reserves to the countries in which the oil was found. Apart from the further expansion of offshore activities in the Gulf of Mexico, other major searches for offshore oil reserves began in the North Sea, the South China Sea, Australia, the Gulf of Suez, eastern Canada, Southern Brazil, the Mexican Gulf of Mexico, and West Africa.

As an extra note, Brazil is the largest producer of liquid fuels in South America and is the ninth-largest energy consumer in the world. The majority of Brazil's crude oil reserves are contained in the offshore pre-salt fields of the Campos, Espírito Santo, and Santos basins, which are located off the country's southeast coast and hold estimated reserves of approximately 100 billion barrels (100×10^9 bbls) of oil equivalent. However, other crude oil and natural gas producers should take note that Brazil lacks the installed refining capacity to meet its surging domestic demand for transportation fuels. At present, the Brazilian refining industry operates 13 refineries with a throughput capacity of just over 2 million barrels per day, which is between supply and demand, and plans are afoot to expand the refining capacity (Nichols, 2013).

However, given the inherently higher costs involved in exploiting this crude oil, the impact of the development and use of more of the much lower cost reserves of the Middle East (and of Venezuela and, possibly, Russia and Kazakhstan) would compromise the ability of the various companies to tackle the deep-water frontier. However, the discovery and development of offshore fields can (for the United States and other coastal crude oil importers) reduce the dependence of the oil-consuming nations on oil from politically uncertain areas or dependence on governments that might be hostile to the United States (Speight, 2011b; Hamilton, 2013; Luciani, 2013). Oil from tight shale formations (like gas from tight shale) might offer some relief to the United States and reduce dependence upon imported oil—the price of oil independence is an issue that various levels of government have faced over the past four decades and will continue to face unless realistic decisions are made (Speight, 2011a; Hamilton, 2013; Speight, 2013d, Speight, 2014a).

Currently (at the time of writing—2014), many governments will not face the prospect of high costs for crude oil and are satisfied to continue to import high levels of crude oil. In the 1960s, the Government of Canada decided to move toward a level of energy independence (the level of energy independence was not defined) when various Federal and Provincial levels of government formed government–industry consortia for the purposes of developing the tar sand deposits found in the north-east sector of the Province of Alberta. Currently, production levels of synthetic crude oil have surpassed the 1-million-barrel-per-day market and are projected to move to the 5-million-barrel-per-day mark within the next decade (2013a).

Furthermore, there has been, and will continue to be, much debate about opening the Outer Continental Shelf of the United States to oil exploration and drilling programs. According to an assessment by the United States Department of the Interior's Minerals Management Service (MMS), the Outer Continental Shelf in the Lower 48 States currently under leasing moratorium holds a mean estimate of 19 billion barrels (19×10^9 bbls) of technically recoverable oil. This has led some to claim that opening the Outer Continental Shelf will not significantly improve the energy situation; the recoverable reserves of crude oil may sustain only approximately 2 years of current consumption in the United States. However, a more appropriate (or realistic) way to consider the issue of the development of the Outer Continental Shelf is that measured and controlled additional production of crude oil (instead of the as-much-as-possible-now syndrome) would reduce crude oil imports from oil-producing countries and prolong the active life of the Outer Continental Shelf resources for another 50 years (Speight, 2011a,b, 2014a).

Finally, there is the projection (or hope) that new fields will continue to be discovered, both onshore and offshore, to create additional reserves. Progress in achieving such additional reserves through discoveries will be orientated toward the contribution of offshore reserves though the development of fields in deeper waters, which generally offer more attractive prospects and is the next *frontier* for the crude oil industry. In fact, in addition to the possibility of fields in the Atlantic Margin, such prospects also occur in the deeper waters of the Gulf of Mexico, in the south Atlantic, off the coasts of West and Southern Africa, in the Timor Sea, the South China Sea, and off the east coast of Trinidad and Tobago. With regard to Trinidad and Tobago, exploration to date has occurred on the western side (Caribbean side) of the islands but the deeper water on the eastern side (Atlantic side) may prove to be the *mother lode*—because of continental drift, there is the possibility that the geology of the eastern deep-water sea bed may have similar geological features to those of the Niger delta area where oil has been produced for the past decades.

1.3 FORMATION AND ACCUMULATION

The occurrence of crude oil in subterranean formations and sub-sea sediments is now well known and there are discussion about the formation and quality of crude oil in the different locales. The quality of crude oil is variable, whatever the current location of the reservoir, and it must be remembered that a reservoir that is now sub-sea may once have been sub-land and *vice versa*. The quality of the oil is subject to various aspects of the maturation process that may be influenced by depth below the surface and the accompanying physical parameters of pressure and temperatures as well as the character and type of the precursors that were first deposited in the ancient sediment.

1.3.1 Formation

There are two theories for the origin of carbon fuels: (1) the *abiogenic theory* or *abiotic theory* and (2) the *biogenic theory* or *biotic theory* (Table 1.2) (Herndon, 2006; Speight, 2014a, and references cited therein) and both theories have been intensely debated for approximately 150 years since the discovery of widespread occurrence of crude oil. It is not the intent of this section to sway the reader in his/her views of the origin of crude oil and natural gas, but the intent is to place before the reader both points of view from which the reader can decide that may ultimately assist the reader to understand the most appraised places to search for crude oil.

The *abiogenic theory* proposes that large amounts of hydrocarbons are formed without the aid of bio-organisms or any biotic activity. However, thermodynamic calculations and experimental studies confirm that n-alkanes (common crude oil components) do not spontaneously evolve from methane at pressures typically found in sedimentary basins, and so the theory of an abiogenic origin of hydrocarbons suggests deep generation (Holm and Charlou, 2001; Sherwood Lollar et al., 2002; Kieft et al., 2005; Glasby, 2006;

Table 1.2 Differences Between the Biogenic and Abiogenic Theories

Raw material
- Abiogenic theory: deep carbon deposits from when the planet formed or subducted material.
- Biogenic theory: remnants of buried plant and animal life.

Events before conversion
- Abiogenic theory: at depths of hundreds of kilometers, carbon deposits are a mixture of hydrocarbon molecules that leak upward through the crust. Much of the material becomes methane.
- Biogenic theory: large quantities of plant and animal life were buried. Sediments accumulating over the material slowly compressed it and covered it. At a depth of several hundred meters, conversion commences.

Conversion to crude oil and methane
- Abiogenic theory: when the material passes through temperatures at which *extremophile* microbes can survive; some of it will be consumed and converted to heavier hydrocarbons.
- Biogenic theory: conversion of the buried source material occurs as the depth of burial increases and the heat and pressure are instrumental in the formation of crude oil.

Kvenvolden, 2006; Speight, 2014a, and references cited therein). Furthermore, alternative theories of crude oil formation should not be dismissed until it can be conclusively established that crude oil formation is due to one particular aspect of geochemistry.

1.3.2 Accumulation

In addition to these readily appreciated requirements, there must also be a source rock in which the oil and gas are formed. Thus, whether the accumulation is land-based or sea-based accumulation, crude oil and natural gas require the existence of a reservoir rock to store the fluids and a cap to prevent escape (Tissot and Welte, 1978; Bjorøy et al., 1987; Speight, 2014a). For convenience, the term *reservoir fluid* includes gases, liquids, semi-solid, solids, and water that may be found in the reservoir.

Typically, crude oil and natural gas accumulations are generally found in relatively coarse-grained porous and permeable rocks that contain little or no insoluble organic matter. However, the recent discovery of oil and gas in tight shale formations offers almost contradictory statement to the typical definition of reservoir rock (Speight, 2013d, Speight, 2014a). Nevertheless, the reservoir rock must possess fluid-holding capacity (*porosity*) and also fluid-transmitting capacity (*permeability*); a variety of different types of openings in rocks are responsible for these properties in reservoir rocks. The most common are the pores between the grains of which the rocks are made or the cavities inside fossils, openings formed by solution or fractures, and joints that have been created in various ways. The relative proportion of the different kinds of openings varies with the rock type, but pores usually account for the bulk of the storage space. The effective porosity for oil storage results from continuously connected openings. These openings alone provide the property of permeability, and although a rock must be porous to be permeable, there is no simple quantitative relationship between the two.

When sediments such as mud and clay are deposited, they contain water—as might occur during the deposition of sea-bed sediments. As the layers of sediment accumulate, the increasing load of material causes compaction of the sediment and part of the water is expelled from the pores. Migration probably begins at this stage and may involve a substantial horizontal component. Initially, the oil hydrocarbons may be present in the water either suspended as tiny globules or dissolved, hydrocarbons being slightly soluble in water.

Crude oil constituents may migrate through one or more formations that have permeability and porosity similar to those of the ultimate reservoir rock. It is the occurrence of an impermeable, or a very low permeability, barrier that causes the formation of oil and gas accumulations. As practically all pores in the subsurface are water saturated, the movement of crude oil constituents within the network of capillaries and pores must take place in the presence of the aqueous pore fluid. Such movement may be due to active water flow or may occur independently of the aqueous phase, either by displacement or by diffusion. There may be a single phase (oil and gas dissolved in water) or a multiphase (separate water and hydrocarbon phases) fluid system. The loss of hydrocarbons from a trap is referred to as *dismigration*.

The specific gravity of gas and crude oil, the latter generally has a specific gravity (at 60 °F, 15.6 °C) that varies from approximately 0.75–1.00 (57–10° API), with the specific gravity of most crude oil types falling in the range 0.80–0.95 (45–17° API) are considerably lower than those of saline pore waters (specific gravity: 1.0–1.2). Thus, crude oil accumulations are usually found in geological structures where reservoir rocks of suitable porosity and permeability are covered by a dense, relatively impermeable cap rock, such as evaporite (salt, anhydrite, and gypsum) or shale (a fine-grained, sedimentary rock composed of mud and clay minerals and silt-sized particles of other minerals) as well as a basement rock of similar geological character—that is, far lower permeability than the reservoir rock—and impermeable to oil and gas.

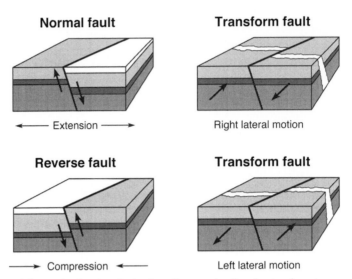

Figure 1.2 Schematic of various faults that allow crude oil and natural gas to accumulate.

A reservoir rock sealed by a cap rock caused by a fault in the strata (Fig. 1.2) or in the position of a geological high, such as an *anticline*, is known as a *structural crude oil trap* (Fig. 1.3). Other types of traps, such as *sand lenses, reefs, pinch-outs* of more permeable and porous rock units, as well as salt domes, are also known and occur in various fields (Speight, 2014a). While it is convenient to show the segregation of gas, oil, and water in such traps, it must be recognized that the straight-line relationship between the phases does not exist. Each phase (gas, oil, or water) is separated from the other phases by an intermediate transition zone (Fig. 1.4). It is the recovery of material

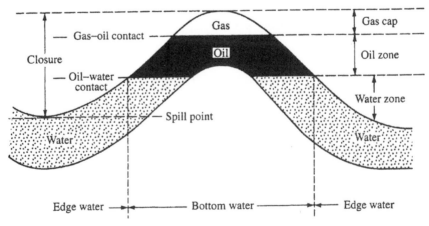

Figure 1.3 Typical anticlinal crude oil trap.

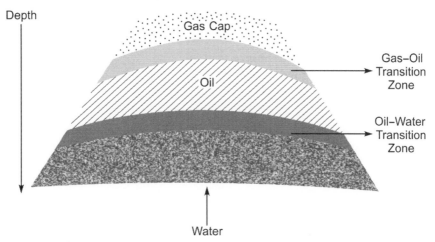

Figure 1.4 The various gas–oil–water zones in a reservoir.

from the transition zones that can give rise to difficult-to-break emulsions and foamy oil. Hence, there is a need to have an intimate knowledge of the anatomy of the reservoir to determine the most appropriate point(s) to penetrate through drilling.

In all these situations, the changes in permeability and porosity determine the location of crude oil and/or a gas accumulation. The predominant theory assumes that as the sedimentary layers superimposed in the source bed became thicker, the pressure increased and the compression of the source bed caused liquid organic matter to migrate to sediments with a higher permeability, which, as a rule, is sand or porous limestone. In fact, there are several mechanisms that have been postulated for the migration of crude oil from the source rock to the reservoir rock and there is differentiation between the mechanisms of *primary migration* and *secondary migration* (Table 1.3).

Table 1.3 General Mechanisms for Crude oil Migration

Geological event	Migration effect
Basin development	Downward fluid flow
Mature basin	Sediments move downward
Hydrocarbon generation	Thermal effects on source sediment
Hydrocarbons dissolve	Pore fluid become saturated
Geothermal gradient changes	Isotherms depressed
Pore fluids cool	Hydrocarbons form separate phase
Hydrocarbons separate	Move to top of carrier fluid (water)
Updip migration	Buoyancy effect
Intermittent faulting	Hydrocarbon migration to traps

Primary migration is the movement of hydrocarbons (oil and natural gas) from mature, organic-rich source rocks to a point where the oil and gas can collect as droplets or as a continuous phase of liquid hydrocarbon. *Secondary migration* is often attributed to various aspects of buoyancy and hydrodynamics (Barker, 1980) and is the movement of the hydrocarbons as a single, continuous fluid phase through water-saturated rocks, fractures, or faults followed by accumulation of the oil and gas in sediments (*traps*) from which further migration is prevented. However, these migration mechanisms do not necessarily compete with each other but are in fact complementary to each other. The prevailing conditions in the subterranean strata (or in subsea strata) may dictate that a particular migration mechanism is favored and this is therefore the dominant mode of migration, with alternative migration mechanisms playing lesser roles. In another given situation, the role assigned to one mechanism as the dominant mode of migration mechanism may be reversed in a different reservoir.

The distribution of the fluids crude oil, natural gas, and water in a reservoir rock is dependent on the density of each fluid as well as on the properties of the rock. If the pores are of uniform size and evenly distributed, there is: (1) an upper zone where the pores are filled mainly by gas—the gas cap, (2) a middle zone in which the pores are occupied principally by oil with gas in solution, and (3) a lower zone with its pores filled by water—interstitial water. In addition, there is a transition zone from the pores occupied entirely by water to pores occupied mainly by oil in the reservoir rock, and the thickness of this zone depends on the densities and interfacial tension of the oil and water as well as on the sizes of the pores. Similarly, there is some water in the pores in the upper gas zone that has at its base a transition zone from pores occupied largely by gas to pores filled mainly by oil. Thus, in reality (as opposed to the three recognizable fluids—water, crude oil, and natural gas), there are five distribution zones within a reservoir, each zone being dependent upon the character of the reservoir and the liquids contained therein.

1.3.3 Composition

The composition of crude oil and natural gas is influenced not only by the nature of the precursors that eventually form crude oil but also by the relative amounts of these precursors (that are dependent upon the local flora and fauna) that occur in the source material (Speight, 2014a). Hence, it is not surprising that composition varies with the location and age of the field in addition to any variations that occur with the depth of the individual well. Two adjacent wells are more than likely to produce crude oil with very different characteristics—the same applies to natural gas from two different wells.

Of all the properties, specific gravity, or American Crude oil Institute gravity, is the variable usually observed. The changes may simply reflect compositional differences, such as the gasoline content or asphalt content, but analysis may also show significant differences in sulfur content or even in the proportions of the various hydrocarbon types.

Elemental sulfur is a common component of sediments and, if present in the reservoir rock, dissolves in the crude oil and reacts slowly with it to produce various sulfur compounds and/or hydrogen sulfide, which may react further with certain components of the oil. These reactions are probably much the same as those that occur in the source bed and are presumed to be largely responsible for the sulfur content of a crude oil; these reactions are accompanied by a darkening of the oil and a significant rise in specific gravity and viscosity.

The variation in the character of crude oil and natural as with depth of burial of the reservoir is also of interest. An increase in the lower-boiling constituents and a decrease in the amount of higher-boiling constituents—reflected in a decrease in density (increase in API gravity)—has been noted (Speight, 2014a).

The processes involved in crude oil maturation are extremely complex—the determining factor is *site specificity*—but several trends are notable in the maturation process: (1) mature crude oils are low in asphaltic fraction (asphaltenes plus resins) and, consequently, lower in nitrogen, oxygen, and sulfur; (2) asphaltene constituents from mature crude oils are more aromatic (lower atomic H/C ratio) than asphaltenes from immature crude oils; and (3) a concentration effect (the extent of which depends upon the reactions involved in the maturation process) causes the majority of the heteroelements (nitrogen, oxygen, and sulfur) to be located in the higher-molecular-weight fractions such as the asphaltene and resin fraction (Fig. 1.5). Other constituents, such as the arbitrarily named the *carbene fraction* and the *carboid fraction* (Fig. 1.5), are regarded as being the products of thermal or even oxidative degradation of the asphaltene and resin constituents and, as a result, are not classified as naturally occurring constituents of crude oil. In addition to maturation by thermal means, crude oil also undergoes non-thermal transformations after accumulation in the reservoir. These changes are usually restricted to shallow oil deposits located near surface outcrops or near faults (Fig. 1.2), which provide access to surface waters containing dissolved oxygen (Speight, 2014a).

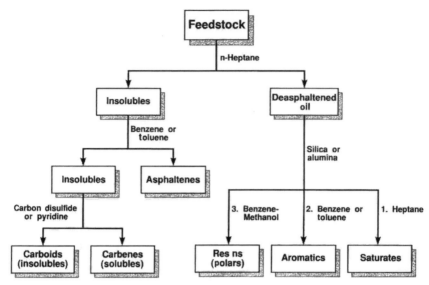

Figure 1.5 Separation scheme for crude oil.

In summary, the mechanism of crude oil formation and the resulting quality of the oil is *site specific*, even to the point where variations in quality can occur from well to well in the same reservoir due to geologic variations within the reservoir.

1.4 DEFINITION OF OIL AND GAS

Crude oil and natural gas currently are known by a variety of names and the purpose of this section is to clear up some of the confusion that exists. Therefore, formal statements of the meaning and significance of the nomenclature are personated here because resources such as crude oil and natural gas are necessary in order to stay up to date with the ever-expanding vocabulary.

1.4.1 Conventional Crude Oil

When crude oil occurs in a reservoir that allows the crude material to be recovered by pumping operations as a free-flowing dark- to light-colored liquid, it is often referred to as *conventional crude oil*.

At the present time, several countries are recognized as producers of crude oil and have available reserves. In particular, the United States is an oil-based culture (Speight, 2011b) and one of the largest importers of crude oil, and the imports of crude oil into the United States continue to rise. The United States now imports approximately 60%–65% (v/v) of its daily crude oil (and crude oil product) requirements. In addition, the crude oils available today to the refinery are substantially different in composition and properties from those available some 50 years ago, and rather than processing one crude oil, many refineries process blends of different crude oils and will continue to do so for the foreseeable future (Speight, 2011a,b, 2014a).

The current crude oils typically have higher proportions of nonvolatile (asphaltic) constituents—reflected in the higher density and lower API gravity (as well as, in many cases, higher sulfur content). In fact, by the standards of yester-year, many of the crude oils currently in use would have been classified by refiners as heavy feedstocks, bearing in mind that such crude oils may not approach the more modern definitions of heavy crude oil. Changes in feedstock character, such as this tendency to greater amounts of higher boiling constituents, require adjustments to recovery (and refinery) operations to adapt to the heavier crude oils by reducing the amount of coke formed during refining and to balance the overall product slate to market demand (Speight, 2011a, 2013c, 2014a).

Typically, conventional crude oil is a brownish green to black liquid of specific gravity ranging from about 0.810 to 0.985 and having a boiling range (excluding the C_1 to C_4 gases) from approximately 20 C (68 F) to above 350 C (660 F), above which active decomposition ensues when distillation is attempted without reducing the pressure. The oils contain from 0% to 35% or more of naphtha (the precursor to gasoline) as well as varying proportions of kerosene, gas oil, and higher boiling constituents up to the viscous and nonvolatile compounds present in lubricant oil and in asphalt (Speight and Ozum, 2002; Parkash, 2003; Hsu and Robinson, 2006; Gary et al., 2007; Speight, 2014a). The composition of the crude oil obtained from the well is variable and depends not only on the original composition of the oil *in situ* but also on the manner of production and the stage reached in the life of the well or reservoir.

1.4.2 Natural Gas

Natural gas is the gaseous mixture associated with crude oil reservoirs and is predominantly methane, but does contain other combustible hydrocarbon compounds as well as nonhydrocarbon compounds (Speight, 2007, 2014a). Natural gas has no distinct odor and the main use is for fuel, but it can also be used to make chemicals and liquefied crude oil gas. The gas occurs in the porous rock of the earth's crust either alone or with accumulations of crude oil (Fig. 1.3, Fig. 1.4). In the latter case, the gas forms the gas cap, which is the mass of gas trapped between the liquid crude oil and the impervious (very low permeability) cap rock. When the pressure in the reservoir is sufficiently high, the natural gas may be dissolved in the crude oil and is released upon release of pressure (by penetration of the reservoir) as a result of drilling operations.

Initially, natural gas is characterized on the basis of the mode of the natural gas found in reservoirs where there is no or, at best only minimal amounts of, crude oil. Thus, there is *nonassociated* natural gas, which is found in reservoirs in which there is no, or at best only minimal amounts of, crude oil—the only liquid hydrocarbons might be small amounts of *gas condensate*. Nonassociated gas is usually richer in methane but is markedly leaner in terms of the higher-molecular-weight hydrocarbons and condensate. Conversely, there is also *associated* natural gas (*dissolved* natural gas) that occurs either as free gas or as gas in solution in the crude oil. Gas that occurs as a solution with the crude crude oil is *dissolved gas*, whereas the gas that exists in contact with the crude crude oil (*gas cap*) is *associated gas*. Associated gas is usually leaner in methane than the nonassociated gas but is richer in the higher-molecular-weight constituents.

There are several general definitions that have been applied to natural gas and need to be considered here: (1) *lean* gas is gas in which methane is the major constituent and (2) *wet* gas contains considerable amounts of higher-molecular-weight hydrocarbons such as ethane (C_2) propane (C_3), butane (C_4), and hydrocarbons as high as octane (C_8). To further define the terms *dry* and *wet* in quantitative measures, the term *dry natural gas* indicates that there is less than 0.1 gallon (1 US gallon = 264.2 m^3) of gasoline vapor (higher molecular weight paraffins) per 1000 ft^3 (1 ft^3 = 0.028 m^3). The term *wet natural gas* indicates that there are such paraffins present in the gas, in fact more than 0.1 gal/1000 ft^3 gasoline vapor (higher-molecular-weight paraffins).

On the other hand, *sour gas* contains hydrogen sulfide whereas *sweet gas* contains very little, if any, hydrogen sulfide. *Residue gas* is natural gas from which the higher-molecular-weight hydrocarbons have been extracted and *casing head gas* (*casinghead gas*) is derived from crude oil but is separated at the separation facility at the wellhead. In addition, *associated* or *dissolved* natural gas occurs either as free gas or as gas in solution in the crude oil. Gas that occurs as a solution in the crude oil is *dissolved* gas, whereas the gas that exists in contact with the crude oil (*gas cap*) is *associated* gas.

Other components such as carbon dioxide (CO_2), hydrogen sulfide (H_2S), and mercaptans (thiols) (RSH, where R is a hydrocarbon moiety), as well as trace amounts of other constituents, may also be present in natural gas. Furthermore, there is no single composition of the components that might be representative of a so-called *typical* natural gas.

As with crude oil, natural gas from different wells varies widely in composition and analyses and the proportion of nonhydrocarbon constituents can vary over a very wide range (Mokhatab et al., 2006; Speight, 2007, 2014a). Thus, a particular natural gas field could require production, processing, and handling protocols different from those used for gas from another field. *Liquefied crude oil gas (LPG)* is composed of propane (C_3H_8), butane (C_4H_{10}), and/ or mixtures thereof; small amounts of ethane (C_2H_6) and pentane (C_5H_{12}) may also be present as impurities. On the other hand, natural gasoline (like refinery gasoline) consists mostly of pentane (C_5H_{12}) and higher-molecular-weight hydrocarbons. The term *natural gasoline* has als,o on occasion in the gas industry, been applied to mixtures of liquefied crude oil gas, pentanes, and higher-molecular-weight hydrocarbons. Caution should be taken not to confuse *natural gasoline* with the term *straight-run gasoline* (often also incorrectly referred to as natural gasoline), which is the gasoline distilled unchanged from crude oil.

1.4.3 Gas Condensate

Gas condensate is a hydrocarbon liquid stream separated from natural gas and consists of higher-molecular-weight hydrocarbons that exist in the reservoir as constituents of natural gas but which are recovered as liquids in separators, field facilities, or gas-processing plants. Typically, gas condensate contains hydrocarbons boiling up to C_8 (Fig. 1.6).

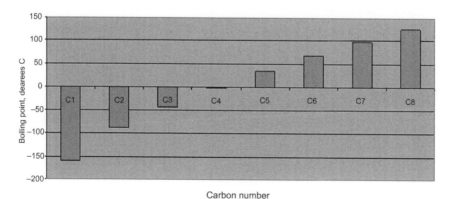

Carbon number

Figure 1.6 Boiling points of hydrocarbons from C1 to C8, which are representative of the boiling range of gas condensate.

After the natural gas is brought to the surface, separation is achieved by use of a tank battery at or near the production lease into a hydrocarbon liquid stream (crude oil or *gas condensate*), a produced water stream (brine or salty water), and a gaseous stream. The gaseous stream is traditionally very rich (*rich gas*) in *natural gas liquids* (NGLs). NGLs include ethane, propane, butanes, and pentanes and higher-molecular-weight hydrocarbons (C_6+). The higher-molecular-weight hydrocarbons product is commonly referred to as *natural gasoline*.

At the top of the well, the crude oil and gas mixture passes into a separation plant that drops the pressure down to nearly atmospheric in two stages. The higher boiling constituents and water exit the bottom of the lower-pressure separator, from where it is pumped to tanks for separation of the condensate and water. The gas produced in the separators is recompressed and any NGLs are treated in a gas plant to provide propane and butane or a mixture of the two (liquefied crude oil gas, LPG). The higher boiling fraction, after removal of propane and butane, is condensate, which is mixed with the crude oil or exported as a separate product (Mokhatab et al., 2006; Speight, 2007, 2014a).

1.4.4 Gas Hydrates

In crude oil and gas production technology, gas hydrates are undesirable compounds that may be formed during production of transportation of natural gas (Fink, 2007; Speight, 2011c; Marcelle-De Silva et al., 2012). Large accumulations of gas hydrates are, however, known in nature, and these are now considered as potential nonconventional energy sources. Natural gas hydrates are an unconventional source of energy and occur abundantly in nature, both in Arctic regions and in marine sediments (Bishnoi and Clarke, 2006). The formation of gas hydrate occurs when water and natural gas are present at low temperature and high pressure. Such conditions often exist in oil and gas wells and pipelines. Gas hydrates offer a source of energy as well as a source of hydrocarbons for the future.

Gas hydrates are crystalline inclusion compounds of gas molecules in water, which form at low temperature and high pressures—typical of the conditions in deep water. The water molecules form ice-cage structures around the appropriate gas molecules (Table 1.4 and Fig. 1.7). The lattice structure formed from the water molecules is thermodynamically unstable and always stabilized by the incorporation of gas molecules. Depending on pressure and gas composition, these ice-like compounds can exist above the freezing point of water, at up to 25 °C (77 °F). In the crude oil and natural

Table 1.4 Composition of Gas Hydrates

Hydrate	Chemical composition
Methane	$CH_4 \times 6H_2O$
Ethane	$C_2H_6 \times 8H_2O$
Propane	$C_3H_8 \times 17H_2O$
Isobutane	$C_4H_{10} \times 17H_2O$
Nitrogen	$N_2 \times 6H_2O$
Carbon dioxide	$CO_2 \times 6H_2O$
Hydrogen sulfide	$H_2S \times 6H_2O$

gas industry, great significance attaches to the hydrates, which form from water and methane, ethane, propane, isobutane, n-butane, nitrogen, carbon dioxide, or hydrogen sulfide. They pose a great problem in modern natural gas extraction, especially in conditions where wet gas or multiphasic mixtures of water, gas, and alkane mixtures are subjected to low temperatures under high pressure. Because of their insolubility and crystalline structure, the formation of gas hydrates leads to the blockage of extraction equipment such as pipelines, valves, or production equipment in which wet gas or multiphase mixtures are transported over long distances, as occurs in colder regions or on the seabed (Dahlmann and Feustel, 2008). Gas hydrate formation can also lead to problems in drilling operations.

Gas hydrate formation can be suppressed by using relatively large amounts of smaller alcohols such as methanol, glycol, or diethylene glycol. These additives cause the thermodynamic limit of gas hydrate formation to shift to lower temperatures and higher pressures (thermodynamic inhibition), but

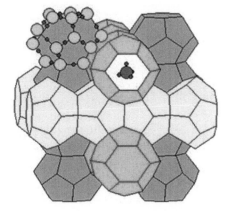

Figure 1.7 Representation of a gas hydrate showing methane in the ice cage.

they can cause serious safety problems (flash point and toxicity of the alcohols), logistic problems (large storage tanks, recycling of these solvents), and accordingly high costs, especially in offshore extraction.

Attempts are therefore being made to replace these inhibitors by using additives in amounts of <2% in temperature and pressure ranges in which gas hydrates can form. These additives either delay gas hydrate formation (kinetic inhibitors) or keep the gas hydrate agglomerates small and therefore pumpable, so that they can be transported through the pipeline. The inhibitors either prevent nucleation and the growth of the gas hydrate particles, or they modify the hydrate growth in such a way that relatively small hydrate particles result.

The main problem with hydrate formation arises in pipelines that transport natural gas because they are solids and deposit in the pipeline, which reduces the effective diameter of the pipe and can therefore restrict or even clog the flow. Another issue—often ignored or unacknowledged—is that gas hydrates are highly flammable and can decompose explosively.

1.4.5 Opportunity Crudes

Opportunity crude oils are either new crude oils with unknown or poorly understood processing issues or are existing crude oils with well-known processing concerns. Opportunity crude oils are often, but not always, heavy crude oils but in either case are more difficult to desalt, most commonly due to high solids content, high levels of acidity, viscosity, electrical conductivity, or contaminants. They may also be oils that are incompatible, causing excessive equipment fouling when processed either in blends or separately.

Typically, opportunity crude oils often need cleaning before refining by removal of undesirable constituents such as high-sulfur, high-nitrogen, and high-aromatics (such as polynuclear aromatic) components (Speight, 2014a,b). A controlled visbreaking treatment would *clean up* such crude oils by removing these undesirable constituents (which, if not removed, would cause problems further down the refinery sequence) as coke or sediment (Radovanovic´ and Speight, 2011). In addition to taking preventative measure for the refinery to process these feedstocks without serious deleterious effects on the equipment, refiners will need to develop programs for detailed and immediate feedstock evaluation so that they can understand the qualities of a crude oil very quickly and it can be valued appropriately and management of the crude processing can be planned meticulously.

Compatibility of opportunity crudes with other opportunity crudes and with conventional crude oil and heavy oil is a very important property to consider when making decisions regarding which crude to purchase. Blending crudes that are incompatible can lead to extensive fouling and processing difficulties due to unstable asphaltene constituents (Speight, 2014a). These problems can quickly reduce the benefits of purchasing the opportunity crude in the first place. For example, extensive fouling in the crude preheat train may occur resulting in decreased energy efficiency, increased emissions of carbon dioxide, and increased frequency at which heat exchangers need to be cleaned. In a worst-case scenario, crude throughput may be reduced leading to significant financial losses.

Before refining—in addition to initial economic advantages and because of the potential problems at the refinery, harming infrastructure, yield, and profitability—there is the need for comprehensive evaluations of opportunity crudes to give the buyer and seller the needed data to make informed decisions regarding the suitability of a particular opportunity crude oil for a refinery. This will assist the refiner to manage the ever-changing crude oil quality input to a refinery—including quality and quantity requirements and situations, crude oil variations, contractual specifications, and risks associated with such opportunity crudes.

1.4.6 High Acid Crudes

Acidity in crude oils is typically caused by the presence of naphthenic acids that are natural constituents of crude oil, where they evolve through the oxidation of naphthenes (cycloalkanes) (Speight, 2014a,b). Initially, the presence of these acidic species was suggested to be due to process artifacts formed during refining processes—and this may still be the case in some instances. However, it was shown that only a small quantity of acids was produced during these processes (Costantinides and Arich, 1967). Currently, it is generally assumed that acids may have been incorporated into the oil from three different sources: (1) acidic compounds found in source rocks, derived from the original organic matter that created the crude oil (plants and animals); (2) neo-formed acids during biodegradation (although the high acid concentration in biodegraded oils is believed to be related principally to the removal of non-acidic compounds, leading to a relative increase in the acid concentration levels); and (3) acids that are derived from the bacteria themselves, for example, from cell walls that the organisms leave behind when their life cycle is completed (Mackenzie et al., 1981; Behar and Albrecht, 1984; Thorn and Aiken, 1998; Meredith et al., 2000; Tomczyk et al., 2001;

Watson et al., 2002; Wilkes et al., 2003; Barth et al., 2004; Kim et al., 2005; Fafet et al., 2008).

The naphthenic acid sub-class of the oxygen-containing species is known as *naphthenic acids* and the term *naphthenic acids* is commonly used to describe an isomeric mixture of carboxylic acids (*predominantly* mono carboxylic acids) containing one or several saturated fused alicyclic rings (Hell and Medinger, 1874; Lochte, 1952; Ney et al., 1943; Tomczyk et al., 2001; Rodgers et al., 2002; Barrow et al., 2003; Clemente et al., 2003a,b; Zhao et al., 2012). However, in crude oil terminology, it has become customary to use this term to describe the whole range of organic acids found in crude oils; species such as phenols and other acidic species are often included in the naphthenic acid category (Speight, 2014b).

High acid crude oils (Speight, 2014a,b) cause corrosion in the refinery—corrosion is predominant at temperatures in excess of 180 °C (355 °F) (Speight, 2014c)—and occurs particularly in the atmospheric distillation unit (the first point of entry of the high-acid crude oil) and also in the vacuum distillation units. In addition, overhead corrosion is caused by the mineral salts, magnesium, calcium, and sodium chloride, which are hydrolyzed to produce volatile hydrochloric acid, causing a highly corrosive condition in the overhead exchangers. Therefore, these salts present a significant contamination in opportunity crude oils. Other contaminants in opportunity crude oils that are shown to accelerate the hydrolysis reactions are inorganic clays and organic acids.

1.4.7 Foamy Oil

Foamy oil is oil-continuous foam that contains dispersed gas bubbles produced at the well head from heavy oil reservoirs under solution gas drive and often during the production of heavy oil (Maini, 1996). The nature of the gas dispersions in oil distinguishes foamy oil behavior from conventional heavy oil. The gas that comes out of solution in the reservoir does not coalesce into large gas bubbles or into a continuous flowing gas phase. Instead it remains as small bubbles entrained in the crude oil, keeping the effective oil viscosity low while providing expansive energy that helps drive the oil toward the producing. Foamy oil accounts for unusually high production in heavy oil reservoirs under solution-gas drive (Mastmann et al., 2001; Bennion et al., 2003; Kumar and Mahadevan, 2012).

Thus, foamy oil is formed in solution gas drive reservoirs when gas is released from solution as reservoir pressure declines. It has been noted that the oil at the wellhead of these heavy-oil reservoirs resembles the form of foam, hence the term *foamy oil*. The gas initially exists in the form of small bubbles within individual pores in the rock. As time passes and pressure continues to decline, the bubbles grow to fill the pores. With further declines in pressure, the bubbles created in different locations become large enough to coalesce into a continuous gas phase. Once the gas phase becomes continuous (i.e., when gas saturation exceeds the critical level)—the minimum saturation at which a continuous gas phase exists in porous media—traditional two-phase (oil and gas) flow with classical relative permeability occurs. As a result, the production gas–oil ratio increases rapidly after the critical level.

Before analysis, the foam should be dismissed either by use of an appropriate separator vessel or by use of anti-foaming agents. However, modification of the separator design may not always be feasible because of the limited space at many wellhead facilities, especially offshore platforms. Therefore, chemical additives (antifoaming agents, foam inhibitors) are employed to prevent or break up the foam. In the case of antifoaming agents, analysts must (1) determine the chemical nature of the agents, (2) remove the agents prior to commencing analysis, and (3) if item 2 is not possible or difficult, ensure that the presence of these agents during does not interfere with the test method result.

1.4.8 Heavy Oil

Heavy oil is a type of crude oil that is different from conventional crude oil insofar as they are much more difficult to recover from the subsurface reservoir (Meyer and Attanasi, 2003; Speight, 2009; Speight, 2013b,c, 2014a). These materials have a much higher viscosity (and lower API gravity) than conventional crude oil, and primary recovery of these crude oil types usually requires thermal stimulation of the reservoir.

The definition of *heavy oil* is arbitrarily based on the API gravity or viscosity, and the definition is quite arbitrary, although there have been attempts to rationalize the definition based upon viscosity, API gravity, and density. Thus, the generic term *heavy oil* is often applied inconsistently to crude oil that has an API gravity of less than 20°. Other definitions classify heavy oil as heavy oil having an API gravity less than 22° API, or less than 25° API and usually, but not always, a sulfur content higher than 2% by weight (Speight, 2000; Ancheyta and Speight, 2007; Speight, 2007, 2014a). In contrast to conventional crude oils, heavy oils are darker in color and may even be black. The term *heavy oil* has also been arbitrarily used to describe both the heavy oils that require thermal stimulation of recovery from the reservoir and the bitumen in bituminous sand (tar sand) formations from which the heavy bituminous material is recovered by a mining operation.

1.4.9 Extra Heavy Oil

Extra heavy oil is a material that occurs in the solid or near-solid state and generally has mobility under reservoir conditions (Speight 2013b,c, 2014a). The term *extra heavy oil* is a recently evolved term (related to viscosity) of little scientific meaning. While this type of oil may resemble tar sand bitumen and does not flow easily, extra heavy oil is generally recognized as having mobility in the reservoir compared to tar sand bitumen, which is typically incapable of mobility (free flow) under reservoir conditions. For example, the tar sand bitumen located in Alberta Canada is not mobile in the deposit and requires extreme methods of recovery to recover the bitumen. On the other hand, much of the extra heavy oil located in the Orinoco belt of Venezuela requires recovery methods that are less extreme because of the mobility of the material in the reservoir.

Whether the mobility of extra heavy oil is due to a high reservoir temperature (that is higher than the pour point of the extra heavy oil) or to other factors is variable and subject to localized conditions in the reservoir.

1.4.10 Tar Sand Bitumen

Tar sands, also variously called *oil sands* or *bituminous sands*, are loose-to-consolidated sandstone or a porous carbonate rock, impregnated with bitumen, a high boiling asphaltic material with an extremely high viscosity that is immobile under reservoir conditions and vastly different from conventional crude oil (Meyer and Attanasi, 2003; Speight, 2008, Speight, 2013a,b, 2014a). It is therefore worth noting here the occurrence and potential supply of these materials. On an international note, the bitumen in tar sand deposits represents a potentially large supply of energy. However, many of the reserves are available only with some difficulty and optional refinery scenarios will be necessary for conversion of these materials to liquid products because of the substantial differences in character between conventional crude oil and tar sand bitumen (Table 1.5).

Because of the diversity of available information and the continuing attempts to delineate the various world tar sand deposits, it is virtually impossible to present accurate numbers that reflect the extent of the reserves in terms of the barrel unit. Indeed, investigations into the extent of many of the world's deposits are continuing at such a rate that the numbers vary from one year to the next. Accordingly, the data quoted here must be recognized as approximate with the potential of being quite different at the time of publication.

Table 1.5 Comparison of Tar Sand Bitumen (Athabasca) Properties and Crude Oil Properties

Property	Bitumen (Athabasca)	Crude oil
Specific gravity	1.03	0.85–0.90
Viscosity (cp)		
38C/100F	750,000	<200
100C/212F	11,300	
Pour point, °F	>50	ca. −20
Elemental analysis (%, w/w):		
Carbon	83.0	86.0
Hydrogen	10.6	13.5
Nitrogen	0.5	0.2
Oxygen	0.9	<0.5
Sulfur	4.9	<2.0
Ash	0.8	0.0
Nickel (ppm)	250	<10.0
Vanadium (ppm)	100	<10.0
Composition (%, w/w)		
Asphaltenes (pentane)	17.0	<10.0
Resins	34.0	<20.0
Aromatics	34.0	>30.0
Saturates	15.0	>30.0
Carbon residue (%, w/w)		
Conradson	14.0	<10.0

In spite of the high estimations of the reserves of bitumen, the two conditions of vital concern for the economic development of tar-sand deposits are the concentration of the resource, or the percent bitumen saturation, and its accessibility, usually measured by the overburden thickness. Recovery methods are based either on mining combined with some further processing or operation on the oil sands *in situ*. The mining methods are applicable to shallow deposits, characterized by an overburden ratio (i.e., overburden depth to thickness of tar-sand deposit). For example, indications are that for the Athabasca deposit, no more than 15% of the in-place deposit is available within current concepts of the economics and technology of open-pit mining; this 10% portion may be considered as the *proven reserves* of bitumen in the deposit.

1.4.11 Resources and Reserves

The majority of crude oil reserves identified to date are located in a relatively small number of very large fields, known as *giants*. In fact, approximately 300 of the largest oil fields contain almost 75% of the available crude oil. Although most of the world's nations produce at least minor amounts of crude oil, the primary concentrations are in Saudi Arabia, Russia, the United States (chiefly Texas, California, Louisiana, Alaska, Oklahoma, and Kansas), Canada, Iran, China, Norway, Mexico, Venezuela, Iraq, Great Britain, the United Arab Emirates (UAE), Nigeria, and Kuwait. The largest known reserves are in the Middle East.

To place the various reserves of crude oil in perspective (as of 2013) http://gulfbusiness.com/2013/04/top-10-countries-with-the-worlds-biggest-oil-reserves/#.U7q4afldWZ8), Venezuela holds the largest proven oil reserves in the world—the proven oil reserve as of January 2013 stood at 297.57 billion barrels (297.57 $\times$ 10^9 bbls) accounting for approximately 20% (v/v) of the world total of proven oil reserves. The country produced 2.8 million barrels of crude oil per day in 2012 with 149 active rigs. Oil reserves in Venezuela are mainly concentrated in the Maracaibo Basin, a sedimentary basin in the north-western part of the country possessing more than 40 billion barrels (40 $\times$ 10^9 bbls) of oil reserves, and the Orinoco Belt in central Venezuela, which is estimated to contain 235 billion barrels of extra heavy crude oil.

The Saudi Arabian offshore fields Safaniya, Manifa and Zuluf hold considerable crude oil and natural gas resources. The Ghawar oil field, with a production capacity of five million barrels a day and an estimated remaining reserve of 70 billion barrels (70 $\times$ 10^9 bbls), is considered to be the largest crude oil field in the world. The Safaniya oil field, with a production capacity of 1.5 million barrels a day and an estimated recoverable reserve of 36 billion barrels (36 $\times$ 10^9 bbls), is the world's largest offshore oil field. Manifa oil field, with 13 billion barrels (13 $\times$ 10^9 bbls) of recoverable reserves, is the third biggest offshore field in the world. The state-owned oil company Saudi Aramco is responsible for the oil exploration and production in the country. Saudi Arabia's first oil was produced from the Dammam oil field in 1938.

The proven crude oil reserves of Canada total 173.1 billion barrels (173.1 $\times$ 10^9 bbls), which are the third largest in the world. Canada produced 1.308 million barrels of crude oil per day in 2012 with 353 active rigs. The crude oil reserves mainly comprise the tar sand (oil sands) of Alberta, Western Canada Sedimentary Basin (WCSB), and the offshore oil deposits in the Atlantic. Oil sands comprise more than 90% of the country's oil reserve—the Athabasca deposit in Alberta is the major oil sands deposit of the country. The Western Canada sedimentary basin is the principal source of conventional crude oil production. Offshore oil production mainly occurs in the Jeanne d'Arc Basin, off the eastern coast of Newfoundland and Labrador. Hibernia, Terra Nova, White Rose, and the upcoming Hebron oil field are the major offshore oil fields in Canada. Oil production in Canada is privatized and includes the participation of both domestic and international oil companies. Suncor, Canadian Natural Resources Limited, and Imperial Oil are the major domestic oil companies. Foreign oil companies operating in Canada include Chevron, ConocoPhillips, Devon Energy, ExxonMobil, BP, Shell, Statoil, Total, as well as Chinese companies.

The first offshore well in Canada was drilled between 1943 and 1945 in approximately 25–30 ft water and approximately eight miles off Prince Edward Island—the Hillsborough No. 1 was drilled from an artificial island made of wood cribbing, rock, and concrete, and reached a depth of 14,500 ft before it was abandoned without encountering oil or natural gas. Geological and geophysical studies continued into the 1950s, and exploration really began in earnest in the 1960s—made possible by the introduction of jackup and semisubmersible drilling systems in the 1950s. The Grand Banks (off Newfoundland) were first drilled in 1966 followed by a decade of exploration from 1973 to 1983 in the iceberg-prone waters off Labrador where substantial natural gas reserves were discovered. The first well on Sable Island, Nova Scotia was drilled in 1967, but had to be abandoned when it encountered higher-than-expected downhole pressures in a natural gas reservoir that would have exceeded the safety limits of the equipment. Other minor discoveries were made during that first decade of exploration off the Nova Scotia coast—including the Cohasset oilfield in 1973 but the development of that oilfield did not occur until after the nearby Panuke crude oil discovery in the 1980s. In 1979, two findings marked the beginning of a string of crude oil and natural gas discoveries: (1) the Venture natural gas discovery near Sable Island was the first of six natural gas fields that now make up the Sable Offshore Energy Project, which began production in 1999, and (2) the Hibernia crude oil and natural gas discovery, which has been producing since 1997.

Iran possesses the fourth largest proven oil reserves in the world and the second largest among Middle East countries. The proven oil reserves as of December 2012 were estimated at 157 billion barrels (157.57×10^9 bbls). The country produced 3.74 million barrels of crude oil per day in 2012 with 123 active rigs. Out of 34 oil-producing fields in Iran, 22 are onshore fields accounting for more than 70% of the country's oil reserves, whereas more than half of onshore oil reserves are contained in five huge fields, which include Marun field that holds 22 billion barrels (22×10^9 bbls) of crude oil, Ahwaz field with 18 billion barrels (18×10^9 bbls) of crude oil, and the Aghajari field, which contains 17 billion barrels (17×10^9 bbls) of crude oil.

More than 80% of the Iranian onshore oil reserves are confined to the south-western Khuzestan Basin near the Iraqi border. The Ahwaz-Asmari onshore field is the largest producing oil field in Iran, followed by Marun and Gachsaran. North Azadegan and South Azadegan fields, which are due for completion by 2020, contain 26 billion barrels (26×10^9 bbls) of proven oil reserve. Major offshore oil fields in Iran include Bahregansar, Balal, Alvand, Hengam, and Reshadat. The state-owned National Iranian Oil Company owns and operates the Iranian oil exploration and production. The country's first oil well was drilled in 1908 in the Masjid-i-Solaiman field in the Khozestan province of Iran—it was also the first oil well in the Middle East.

Iraq, with 140.3 billion barrels (140.3×10^9 bbls) of proven oil reserves, has the fifth largest oil reserves in the world and the third largest in the Middle East. The country produced 2.942 million barrels of crude oil per day in 2012 with 92 active rigs. Iraq's oil reserves are mainly concentrated in Shiite areas of the south and Kurdish region in the north. Some reserves are also located in central Iraq. Five giant fields in the Southern Iraq account for about 60% of the proven oil reserves, while Northern Iraq accounts for 17%.

Most Iraqi crude oil production comes from the Kirkuk field in Northern Iraq, and the North Rumaila field and the South Rumaila field. Other major onshore fields in Iraq include West Qurna, Az Zubair, Halfaya, Garraf, Badra, and the Manjoon field, which is estimated to contain 14 billion barrels (14×10^9 bbls) of recoverable oil. West Qurna, with a proven reserve of 12.8 billion barrels (12.8×10^9 bbls) of crude oil, is believed to be the world's second largest undeveloped field. The country's Ministry of Oil oversees the oil production through four operating entities, namely the North Oil Company (NOC), the Midland Oil Company, the South Oil Company, and the Missan Oil Company. Crude oil production in Iraq commenced from the Baba dome of the Kirkuk oil field in 1927.

Kuwait holds the sixth largest proven oil reserves in the world and has been estimated to contain 101.5 billion barrels (101.5×10^9 bbls) of proven oil reserve as of December 2012. Most of the crude oil reserves of Kuwait are concentrated in a few mature oil fields. The Greater Burgan oil field, comprising the Burgan, Magwa, and Ahmadi reservoirs, is the largest oil field in Kuwait. It is also considered the world's second largest oil field, next only to Saudi Arabia's Ghawar field. The remaining large oil fields, including the country's second largest producing field Raudhatain and the Sabriya and al-Ratqa oil fields, are located in Northern Kuwait. The major oil fields in Southern Kuwait include Umm Gudair, Minagish, and Abduliyah. The state-owned Supreme Crude oil Council and Kuwait Crude oil Corporation (KPC) monitor oil exploration and production. The first commercial oil well was drilled in the Al Burqan oil field in 1938.

The UAE holds the world's seventh biggest oil reserve with 97.8 billion barrels (97.8 $\times$ 10^9 bbls) of proven oil reserves. It produced 2.652 million barrels of crude oil per day in 2012 with 13 active rigs. The first commercial oil in UAE was discovered in 1958. Approximately 94% (v/v) of the proven oil reserves are located in Abu Dhabi, one of the seven emirates of UAE. Dubai, with four billion barrels (4 $\times$ 10^9 bbls), holds the largest proven oil reserves among the rest six emirates. Sharjah possesses the third largest proven oil reserve in UAE.

Zakum oil deposit located offshore of Abu Dhabi is one of the largest oil reserves in the Middle East. The Upper Zakum oil field contains 21 billion barrels (21 $\times$ 10^9 bbls) of crude oil and is the second largest offshore oil field in the world. Other major offshore fields include the Lower Zakum, Ghasha-Butini, Nasr, Umm Lulu, and Umm Shaif. Bu Hasa, Murban Bab, and Sahil, Asab and Shah (SAS), all located in Abu Dhabi, are the major producing onshore oil fields. The major fields in other emirates include Fateh-Southwest Fateh-Falah fields in Dubai and the Mubarak field, in Sharjah. The Abu Dhabi National Oil Company (ADNOC) is the leading oil-producing company in UAE.

Russia, with 80 billion barrels (80 $\times$ 10^9 bbls) of proven oil reserves, has the eighth largest oil reserves in the world and proven oil reserves increased by 3.4% in 2012. Russia produced 10.043 million barrels of crude oil per day during the year with 320 active rigs. The oil reserves are mainly concentrated in the West Siberia region that accounts for more than 60% (v/v) of Russia's crude oil production. North Priobskoye, Samotlor, Mamontovskoye, and Salymskoye are some of the large oil fields in the region. The Urals-Volga is the second largest oil producing region in Russia and contains some giant oil fields such as Romashkinskoye. Other oil reserve containing regions in Russia are East Siberia, Yamal Peninsula, North Caucasus, Timan-Pechora and Barents Sea, and the Sakhalin Island off the coast of eastern Russia.

Libya ranks ninth, holding 48.47 billion barrels (48.47 $\times$ 10^9 bbls) of proven oil reserve. It has the largest endowment (around 38%) of the total oil reserve in the African continent. The country produced 1.450 million barrels of crude oil per day in 2012 with 72 active rigs. The majority (approximately 80%, v/v) of Libya's oil reserves are in the Sirte basin located in the country's eastern region. The other five large sedimentary basins containing oil reserves are Murzuk, Ghadames, Cyrenaica, Kufra, and Libya offshore. Major onshore oil fields in the country are Waha, Samah, Dahra, Gialo, Nafoura, Amal, Naga and Farigh in the central-east region, Sarir field in the eastern region, and El Sharara, El Feel (Elephant) oil fields in the western region. Bouri and Al-Jurf are the major offshore fields in the country.

The state-owned NOC carries out oil exploration and production operations in the country through its own operating entities or in participation with foreign oil companies. Marathon Oil, ConocoPhillips, Hess, Eni, Repsol, and OMV are the major foreign companies with stakes in Libyan oil assets.

Nigeria holds the tenth largest proven oil reserves in the world—the proven oil reserves as of December 2012 were posted at 37.14 billion barrels (18×10^9 bbls) of crude oil. Nigeria produced 1.954 million barrels of crude oil per day in 2012 with 44 active rigs. The crude oil reserves are mainly concentrated along Niger River Delta and offshore of the Bight of Benin, the Gulf of Guinea, and the Bight of Bonny. Nigerian oil exploration activities are currently focused in the deep and ultra-deep offshore areas, as well as in the Chad basin in the north-eastern part of the country. Gbaran Ubie is the major onshore field in Nigeria. The major offshore and deep water oil fields include Bonga, Agbami, and Ebok. The Usan field operated by Total is the latest major deep water oil field to come on stream. Production at Usan began in July 2012 with initial capacity of 100,000 barrels per day. Major upcoming oil fields in the country include Bonga North, Egina, Nsiko, and Nkarika. The Nigerian National Crude oil Company manages oil projects in the country through partnerships with foreign companies including Exxon-Mobil, Chevron, Total, Eni, Addax Crude oil, ConocoPhillips, Petrobras, and Statoil. Oil and gas currently account for about 70% of the country's exports revenue. The first oil in the country was discovered in 1956.

As impressive as these numbers may be, there is need for caution because of recent recalculation of reserves in several OPEC countries that were not explained at all well (Speight, 2011b). All discoveries are initially appraised for their size in terms of oil in place. Based on probabilistic estimates derived from selected (predominantly) geological and engineering parameters, there will subsequently be an initial declaration of oil and gas that can, with reasonable certainty, be recovered in the future under existing economic and operating conditions. These constitute the so-called proven reserves of the field at the time of *estimation* of the reserves. Typically all oil and gas fields are estimated to have additional volumes of *probable* and *possible* reserves, which are reputedly over 50% and under 50% probability, respectively, of being recoverable from the estimated total volumes of oil–in–place and gas–in–place in the reservoir.

The definitions that are used to describe crude oil *reserves* are often misunderstood because they were not adequately defined at the time of inception of such definitions. Therefore, as a means of alleviating this problem, it is pertinent at this point to consider the definitions used to describe the amount of crude oil that remains in subterranean reservoirs.

Crude oil is a *resource*; in particular, crude oil is a *fossil fuel resource*—a *resource* is the entire commodity that exists in the sediments and strata, whereas the *reserves* represent that fraction of a commodity that can be recovered economically (Nersesian, 2010; Khoshnaw, 2013). Furthermore, it is very rare that crude oil (the exception being tar sand deposits, from which most of the volatile material has disappeared over time) does not occur without an accompanying cover of gas (Fig. 1.3) leaving the nonvolatile constituents as nonvolatile oil constituents, resin constituents, and asphaltene constituents (Fig. 1.5) as a deposit of a high-viscous bitumen-like material (Table 1.5). It is therefore important, when describing reserves of crude oil, to also acknowledge the occurrence, properties, and character of the gaseous material, more commonly known as *natural gas*.

Much confusion often occurs in name of various accumulations as resources or reserves. For the purposes of the book, the resource is the fossil fuel such as crude oil and the amount of the crude oil in the Earth (usually measured in barrels or, under the SI system, as cubic meters; 1 cubic meter = 6.29 barrels). On the other hand, however, the use of the term *reserves* is descriptive of the resource that is available to be recovered, and is subject to much speculation. In fact, the description of the reserves is subject to a series of adjectives that are, in turn, subject to word variations! For example, reserves are commonly (without getting into other word variations) classed as *proved, unproved, probable, possible,* and *undiscovered* (Speight, 2011b). In addition, the term reserve estimate is often used because the real (actual) numbers only are available after recovery of the resource to exhaustion has been applied.

Proven reserves are those reserves of crude oil that are actually found by drilling operations and are recoverable by means of current available technology. The numbers have a high degree of accuracy and are frequently updated as the recovery operation proceeds. They may be updated by means of reservoir characteristics, such as production data, pressure transient analysis, and reservoir modeling.

Probable reserves are those reserves of crude oil that are nearly certain but slight doubt about the numbers exists. *Possible reserves* are those reserves of crude oil with an even greater degree of uncertainty about the numbers and recovery but about which there is some information. An additional term *potential reserves* is also used on occasion; these reserves are based upon geological information about the types of sediments where such resources are likely to occur and they are considered to represent an educated guess. Then, there are the so-called *undiscovered reserves*, which are little more than figments of the imagination! The terms *undiscovered reserves* or *undiscovered resources* should be used with caution, especially when applied as a means of estimating reserves of crude oil reserves. The data are very speculative and regarded by many energy scientists as having little value other than unbridled optimism.

The term *inferred reserves* is also commonly used in addition to, or in place of, *potential reserves*. The numbers for inferred reserves are regarded as of a higher degree of accuracy than potential reserves, and the term is applied to those reserves that are estimated using an improved understanding of reservoir frameworks. The term also usually includes those reserves that can be recovered by further development of recovery technologies.

The differences between the data obtained from these various estimates can be considerable because of the speculation that is inherent in deriving the numbers; it must also be remembered that any data about the reserves of crude oil (and, for that matter, about any other fuel resource or mineral resource) will always be open to questions about the degree of certainty. Thus, in reality, and in spite of the use of word smithing to describe reserves, *proven reserves* may be a very small part of the total hypothetical and/or speculative amounts of a resource.

At some time in the future, certain resources may become reserves (Speight, 2011b). Such a reclassification can arise as a result of improvements in recovery techniques that may either make the resource accessible or bring about a lowering of the recovery costs and render winning of the resource an economical proposition. In addition, other uses may also be found for a commodity, and the increased demand may result in an increase in price. Alternatively, a large deposit may become exhausted and unable to produce any more of the resource, thus forcing production to focus on a resource that is lower grade but has a higher recovery cost.

More recently, the Society for Crude oil Engineers has developed a resource classification system (Fig. 1.8) that moves away from systems in which all quantities of crude oil that are estimated to be initially-in-place are used

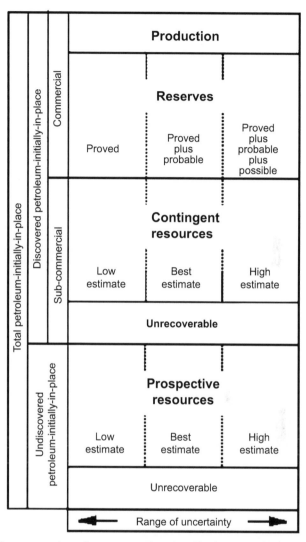

Figure 1.8 Representation of resource estimation. The horizontal axis represents the range of uncertainty in the estimated potentially recoverable volume for an accumulation, whereas the vertical axis represents the level of status/maturity of the accumulation. The vertical axis can be further subdivided to classify accumulations on the basis of the commercial decisions required to move an accumulation toward production.

(SPE, 2001, 2005). Some users consider only the estimated recoverable portion to constitute a resource. In these definitions, the quantities estimated to be initially-in-place are (1) total crude oil-initially-in-place, (2) discovered crude oil-initially-in-place, and (3) undiscovered crude oil-initially-in-place.

The recoverable portions of crude oil are defined separately as (1) reserves, (2) contingent resources, and (3) prospective resources. In any case and whatever the definition, reserves are a subset of resources and are those quantities of crude oil that are discovered (i.e., in known accumulations), recoverable, commercial, and remaining.

The *total crude oil-initially-in-place* is that quantity of crude oil that is estimated to exist originally in naturally occurring accumulations. The *total crude oil-initially-in-place* is, therefore, that quantity of crude oil that is estimated, on a given date, to be contained in known accumulations, plus those quantities already produced therefrom, plus those estimated quantities in accumulations yet to be discovered. The *total crude oil-initially-in-place* may be subdivided into *discovered crude oil-initially-in-place* and *undiscovered crude oil-initially-in-place*, with *discovered crude oil-initially-in-place* being limited to known accumulations.

It is recognized that the quantity of *crude oil-initially-in-place* may constitute *potentially recoverable resources* since the estimation of the proportion that may be recoverable can be subject to significant uncertainty and will change with variations in commercial circumstances, technological developments, and data availability. A portion of those quantities classified as *unrecoverable* may become recoverable resources in the future as commercial circumstances change, technological developments occur, or additional data are acquired.

Discovered crude oil-initially-in-place is that quantity of crude oil that is estimated, on a given date, to be contained in known accumulations, plus those quantities already produced therefrom. *Discovered crude oil-initially-in-place* may be subdivided into *commercial* and *sub-commercial* categories, with the estimated potentially recoverable portion being classified as *reserves* and *contingent resources*, respectively (as defined below).

Reserves are those quantities of crude oil that are anticipated to be commercially recovered from known accumulations from a given date forward.

Estimated recoverable quantities from known accumulations that do not fulfil the requirement of commerciality should be classified as *contingent resources* (as defined below). The definition of commerciality for an accumulation will vary according to local conditions and circumstances and is left to the discretion of the country or company concerned. However, reserves must still be categorized according to specific criteria and, therefore, proved reserves will be limited to those quantities that are commercial under current economic conditions, while probable and possible reserves may be based on future economic conditions. In general, quantities should not be classified as reserves unless there is an expectation that the accumulation will be developed and placed on production within a reasonable timeframe. In certain circumstances, reserves may be assigned even though development may not occur for some time. An example of this would be where fields are dedicated to a long-term supply contract and will only be developed as and when they are required to satisfy that contract.

Contingent resources are those quantities of crude oil that are estimated, on a given date, to be potentially recoverable from known accumulations, but which are not currently considered as commercially recoverable. Some ambiguity may exist between the definitions of contingent resources and unproved reserves. This is a reflection of variations in current industry practice, but if the degree of commitment is not such that the accumulation is expected to be developed and placed on production within a reasonable timeframe, the estimated recoverable volumes for the accumulation can be classified as contingent resources. *Contingent resources* may include, for example, accumulations for which there is currently no viable market, or where commercial recovery is dependent on the development of new technology, or where evaluation of the accumulation is still at an early stage.

Undiscovered crude oil-initially-in-place is that quantity of crude oil that is estimated, on a given date, to be contained in accumulations yet to be discovered. The estimated potentially recoverable portion of *undiscovered crude oil-initially-in-place* is classified as *prospective resources*, which are those quantities of crude oil that are estimated, on a given date, to be potentially recoverable from undiscovered accumulations.

Ultimate reserves consist of the initial ultimately recoverable reserves in discovered fields (see above) plus the recoverable oil from fields yet to be discovered. The usual estimates of the ultimate reserves are derived from the still generally accepted hypothesis for the origin of oil, viz. from organic materials that accumulated on the sea-bottom, were rapidly buried by the deposition of fine grained material such as mud and lime, and then converted to crude oil by subsequent processes involving heat and pressure. The resultant oil (and associated natural gas), it is believed, had its origin in sedimentary rocks through which it then migrated to form "reservoirs" in locations where conditions were favorably for such accumulations.

Estimated ultimate recovery (EUR) is the quantity of crude oil which is estimated, on a given date, to be potentially recoverable from an accumulation, plus those quantities already produced therefrom. EUR is not a resource category but it is a term that may be applied to an individual accumulation of any status/maturity (discovered or undiscovered).

Crude oil quantities classified as *reserves, contingent resources*, or *prospective resources* should not be aggregated with each other without due consideration of the significant differences in the criteria associated with their classification. In particular, there may be a significant risk that accumulations containing Contingent Resources or Prospective Resources will not achieve commercial production.

The *range of uncertainty* reflects a reasonable range of estimated potentially recoverable volumes for an individual accumulation. Any estimation of resource quantities for an accumulation is subject to both technical and commercial uncertainties, and should, in general, be quoted as a range. In the case of reserves, and where appropriate, this range of uncertainty can be reflected in estimates for *proved reserves* (1P), *proved plus probable reserves* (2P), and *proved plus probable plus possible reserves* (3P) scenarios. For other resource categories, the terms *low estimate*, *best estimate*, and *high estimate* are recommended.

The term *best estimate* is used as a generic expression for the estimate considered to be the closest to the quantity that will actually be recovered from the accumulation between the date of the estimate and the time of abandonment. If probabilistic methods are used, this term would generally be a measure of central tendency of the uncertainty distribution. The terms *low estimate* and *high estimate* should provide a reasonable assessment of the range of uncertainty in the *best estimate*.

For undiscovered accumulations (*prospective resources*) the range will, in general, be substantially greater than the ranges for discovered accumulations. In all cases, however, the actual range will be dependent on the amount and quality of data (both technical and commercial) that is available for that accumulation. As more data become available for a specific accumulation (e.g., additional wells, reservoir performance data), the range of uncertainty in the EUR for that accumulation should be reduced.

The *low estimate*, *best estimate*, and *high estimate* of potentially recoverable volumes should reflect some comparability with the reserves categories of *proved reserves*, *proved plus probable reserves*, and *proved plus probable plus possible reserves*, respectively. While there may be a significant risk that sub-commercial or undiscovered accumulations will not achieve commercial production, it is useful to consider the range of potentially recoverable volumes independently of such a risk.

Finally, as oil and gas are produced from the reservoir, the production process produces a continuing flow of information that enables frequent re-evaluation of the recoverable reserves of a field to be made. In some fields, the reserves are downgraded, as a result of production experience, but upgrading the reserve numbers is also possible. This process often leads to a situation in which the declarations of the proven reserves of oil in a field trend upward over time, in spite of the ongoing extraction of oil in the production process. The continuation of such reserve re-evaluation depends upon the rate of extraction on the one hand and, on the other, the re-evaluation of original *probable reserves* or *possible reserves* to the category of *proven reserves*. The re-evaluation of the reserve numbers upward is especially true if new information arising from the production process shows that the physical size of the reservoir available for exploitation and development is greater than first indicated leading to a re-evaluation of the volumes of oil and gas reserves in all categories. In addition, the changes to the reserve estimates—and the re-categorization of the reserves may also be modified by changes to the economics of the recovery process(es): on the one hand, lower prices and/ or higher costs will undermine the validity of earlier declarations of the reserves, whereas on the other hand, higher prices and/or lower costs will stimulate additional interest in the exploitation of the reservoir by increasing the recoverable reserves.

REFERENCES

Abraham, H., 1945. Asphalts and Allied Substances. Van Nostrand Publishers, New York.

Ancheyta, J., Speight, J.G., 2007. Hydroprocessing of Heavy Oils and Residua. CRC Press, Taylor & Francis Group, Boca Raton, Florida.

Barker, C., 1980. Problems of Crude oil Migration. In: Roberts, W.H., Cordell, R.J. (Eds.), Studies in Crude oil Geology. Bulletin No. 10. American Association of Crude oil Geologists, Tulsa, Oklahoma.

Barrow, M.P., McDonnell, L.A., Feng, X., Walker, J., Derrick, P.J., 2003. Determination of the nature of naphthenic acids present in crude oils using nanospray Fourier transform ion cyclotron resonance mass spectrometry: The continued battle against corrosion. Analytical Chemistry 75 (4), 860–866.

Barth, T., Høiland, S., Fotland, P., Askvik, K.M., Pedersen, B.S., Borgund, A.E., 2004. Acidic compounds in biodegraded crude oil. Organic Geochemistry 35 (11–12), 1513–1525.

Beauchamp, B., 2004. Natural gas hydrates: Myths, facts and issues. Comptes Rendus Geoscience 336 (9), 751–765.

Bennion, D.B., Mastmann, M., Moustakis, M.L., 2003. A case study of foamy oil recovery in the Patos-Marinza Reservoir, Driza Sand, Albania. Journal of Canadian Crude Oil Technology 42 (3), 21–28.

Bishnoi, P.R., Clarke, M.A., 2006. Encyclopedia of Chemical Processing. CRC Press, Taylor & Francis Group, Boca Raton, Florida, pp. 1849–1863.

Bjorøy, M., Hall, P.B., Loberg, R., McDermott, A., Mills, N., 1987. Hydrocarbons from Non-Marine Source Rocks. In: Mattavelli, L., Novelli, L. (Eds.), Advances in Organic Geochemistry 1987 (Proc. 13th Int'l. Mtg. on Org. Geochem. Venice), Part 1. Pergamon Press, Oxford, United Kingdom, pp. 221–224.

Carroll, J.J., 2009. Natural Gas Hydrates: A Guide for Engineers, 2nd Edition Gulf Professional Publishing, Burlington, Massachusetts.

Chadeesingh, R., 2011. The Fischer-Tropsch Process. In The Biofuels, Handbook, J.G., Speight, (Eds.), The Royal Society of Chemistry, London, United Kingdom. Part 3, Chapter 5, pp. 476-517.

Clemente, J.S., Yen, T.W., Fedorak, P.M., 2003a. Development of a high performance liquid chromatography method to monitor the biodegradation of naphthenic acids. Journal of Environmental Engineering and Science 2 (3), 177–186.

Clemente, J.S., Prasad, N.G., MacKinnon, M.D., Fedorak, P.M., 2003b. A statistical comparison of naphthenic acids characterized by gas chromatography–mass spectrometry. Chemosphere 50, 1265–1274.

Costantinides, G., Arich, G., 1967. Non-hydrocarbon compounds in crude oil. In: Nagy, B., Colombo, U. (Eds.), Fundamental Aspects of Crude oil Geochemistry. Elsevier Publishing Company, London, United Kingdom, pp. 109–175.

Dahlmann, U., Feustel, M., 2008. Corrosion and gas hydrate inhibitors having improved water solubility and increased biodegradability. US Patent 7341671. March 11.

EREC, 2010. Renewable Energy in Europe: Markets, Trends, and Technologies. European Renewable Energy Council (EREC). Earthscan, Washington, DC.

Fafet, A., Kergall, F., Da Silva, M., Behar, F., 2008. Characterization of acidic compounds in biodegraded oils. Organic Geochemistry 39 (8), 1235–1242.

Fink, J., 2007. Crude Oil Engineer's Guide to Oil Field Chemicals and Fluids. Elsevier B.V., Amsterdam, Netherlands, Chapter 13.

Forbes, R.J., 1958a. A History of Technology. Oxford University Press, Oxford, England.

Forbes, R.J., 1958b. Studies in Early Crude oil Chemistry. E. J. Brill, Leiden, Netherlands.

Forbes, R.J., 1959. More Studies in Early Crude oil Chemistry. E.J. Brill, Leiden, The Netherlands.

Forbes, R.J., 1964. Studies in Ancient Technology. E. J. Brill, Leiden, The Netherlands.

Gary, J.G., Handwerk, G.E., Kaiser, M.J., 2007. Crude Oil Refining: Technology and Economics, 5th Edition CRC Press, Taylor & Francis Group, Boca Raton, Florida.

Giampietro, M., Mayumi, K., 2009. The Biofuel Delusion: The Fallacy of Large-Scale Agro-Biofuel Production. Earthscan, Washington, DC.

Glasby, G.P., 2006. Abiogenic origin of hydrocarbons: an historical overview. Resource Geology 56 (1), 83–96.

Gold, T., 2013. The Deep, Hot Biosphere: The Myth of Fossil Fuels. Copernicus Publications; Springer Publishers, Göttingen, Germany.

Hamilton, M.S., 2013. Energy Policy Analysis: A Conceptual Framework. M.E. Sharpe Publishers, Armonk, New York.

Hell, C.C., Medinger, E., 1874. Ueber das Vorkommen und die Zusammensetzung von Säuren im Rohcrude oil. Berichte der deutschen chemischen Gesellschaft 7 (2), 1216–1223.

Herndon, J.M., 2006. Enhanced prognosis for abiotic natural gas and crude oil resources. Current Science 91 (5), 596–598.

Holm, N.G., Charlou, J.L., 2001. Initial indications of abiotic formation of hydrocarbons in the rainbow ultramafic hydrothermal system, Mid-Atlantic Ridge. Earth and Planetary Science Letters 191, 1–8.

Hsu, C.S., Robinson, P.R. (Eds.), 2006. Practical Advances in Crude oil Processing Volume 1 and Volume 2. Springer Science, New York.

Humphries, M., Pirog, R., Whitney, G., 2010. U.S. Offshore Oil and Gas Resources: Prospects and Processes. CRS Report No. R40645. Congressional Research Service. United States Congress, Washington, DC.

Kieft, T.L., McCuddy, S.M., Onstott, T.C., Davidson, M., Lin, L.H., Mislowack, B., Pratt, L., Boice, E., Sherwood Lollar, B., Lippmann-Pipke, J., Pfiffner, S.M., Phelps, T.J., Gihring, T., Moser, D., Van Heerden, A., 2005. Geochemically generated, energy-rich substrates and indigenous microorganisms in deep, ancient groundwater. Geomicrobiology Journal 22 (6), 325–335.

Kim, K., Stanford, L.A., Rodgers, R.P., Marshall, A.G., Walters, C.C., Qian, K., Wenger, L.M., Mankiewicz, P., 2005. Microbial alteration of the acidic and neutral polar NSO compounds revealed by Fourier transform ion cyclotron resonance mass spectrometry. Organic Geochemistry 36 (8), 1117–1134.

Khoshnaw, F.M. (Ed.), 2013. Crude oil and Mineral Resources. WIT Press, Billerica, Massachusetts.

Kumar, R., Mahadevan, J., 2012. Well-performance relationships in heavy-foamy-oil reservoirs. SPE Production & Operations 27 (1), 94–105.

Kvenvolden, K.A., 2006. Organic geochemistry – A retrospective of its first 70 years. Organic Geochemistry 37, 1–11.

Langeveld, H., Sanders, J., Meeusen, M. (Eds.), 2010. The Biobased Economy: Biofuels, Materials, and Chemicals in the Post-Oil Era. Earthscan, Washington, DC.

Larue, D.K., Yue, Y., 2003. How stratigraphy influences oil recovery: A comparative reservoir database study. The Leading Edge 22 (4), 332–339.

Lee, S., Shah, Y.T., 2013. Biofuels and Bioenergy: Processes and Technologies. CRC Press, Taylor & Francis Group, Boca Raton, Florida.

Lochte, H.L., 1952. Crude oil acids and bases. Industrial and Engineering Chemistry 44 (11), 2597–2601.

Luciani, G., 2013. Security of Oil Supplies: Issues and Remedies. Claeys and Casteels, Deventer, Netherlands.

Mackenzie, A.S., Wolff, G.A., Maxwell, J.R., 1981. Fatty acids in some biodegraded crude oils. Possible origins and significance. In: Bjorøy, M. (Ed.), Advances in Organic Geochemistry. John Wiley & Sons Inc., Chichester, United Kingdom, pp. 637–6439.

Magoon, L.B., Dow, W.G., 1994. The crude oil system. Memoir AAPG. (Ed.), The Crude oil System – Form Source to Trap, Volume 60, American Association of Crude oil Geologists, Tulsa, Oklahoma, pp. 3–24.

Maini, B.B., 1996. Foamy oil flow in heavy oil production. Journal of Canadian Petroleum Technology 35, 21–24.

Marcelle-De Silva, J., Thomas, A., De Landro, Clarke, W.L., Allum, M., 2012. Evidence of gas hydrates in Block 26 – Offshore Trinidad. Energies 5, 1309–1320.

Mastmann, M., Moustakis, M.L., Bennion, D.B., 2001. Predicting Foamy Oil Recovery. Paper No. SPE 69722. Society of Crude oil Engineers, Richardson, Texas.

Max, M.D. (Ed.), 2003. Natural Gas Hydrate in Oceanic and Permafrost Environments. Kluwer Academic Publishers, Dordrecht, Netherlands.

Meredith, W., Kelland, S.J., Jones, D.M., 2000. Influence of biodegradation on crude oil activity and carboxylic acid composition. Organic Geochemistry 31 (11), 1059–1073.

Meyer, R.F., Attanasi, E.D., 2003. Heavy Oil and Natural Bitumen—Strategic Crude oil Resources. Fact Sheet 70-03. United States Geological Survey, Washington, DC. August 2003. http://pubs.usgs.gov/fs/fs070-03/fs070-03.html.

Mokhatab, S., Poe, W.A., Speight, J.G., 2006. Handbook of Natural Gas Transmission and Processing. Elsevier B.V., Amsterdam, Netherlands.

Nersesian, R.L., 2010. Energy for the 21st Century: A Comprehensive Guide to Conventional and Alternative Sources. M.E. Sharpe Publishers, Armonk, New York.

Ney, W.O., Crouch, W.W., Rannefeld, C.E., Lochte, H.L., 1943. Crude oil acids. VI. Naphthenic acids from California crude oil. Journal of the American Chemical Society 65 (5), 770–777.

Nichols, L., 2013. Brazil plans major refinery expansions. Hydrocarbon Processing, September 1.

Parkash, S., 2003. Refining Processes Handbook. Elsevier-Gulf Professional Publishing Company, Burlington, Massachusetts.

Radovanović, L., Speight, J.G., 2011. Visbreaking: A Technology of the Future. Proceedings. First International Conference – Process Technology and Environmental Protection (PTEP 2011). University of Novi Sad, Technical Faculty "Mihajlo Pupin," Zrenjanin, Republic of Serbia. December 7. pp. 335–338.

Rodgers, R.P., Hughey, C.A., Hendrickson, C.L., Marshall, A.G., 2002. Advanced character-
    ization of crude oil crude and products by high field Fourier transform ion cyclotron
    resonance mass spectrometry. Preprints. Division of Fuel Chemistry, American Chemical
    Society 47 (2), 636–637.
Seifried, D., Witzel, W., 2010. Renewable Energy: The Facts. Earthscan, Washington, DC.
Sherwood Lollar, B., Westgate, T.D., Ward, J.A., Slater, G.F., Lacrampe-Couloume, G., 2002.
    Abiogenic formation of alkanes in the earth's crust as a minor source for global hydro-
    carbon reservoirs. Nature 416 (6880), 522–524.
Sloan, E.D., Koh, C.A., 2008. Clathrate Hydrates of Natural Gases, 3rd Edition CRC Press,
    Taylor & Francis Group, Boca Raton, Florida.
SPE, 2001. Guidelines for the Evaluation of Crude oil Reserves and Resources A Supplement
    to the SPE/WPC Crude oil Reserves Definitions and the SPE/WPC/AAPG Crude oil
    Resources Definitions. Society for Crude Oil Engineers, Richardson, Texas.
SPE, 2005. Comparison of Selected Reserves and Resource Classifications and Associated
    Definitions. Oil and Gas Reserves Committee (OGRC), Mapping Subcommittee Final
    Report. Society for Crude Oil Engineers, Richardson, Texas.
Speight, J.G., 2000. The Desulfurization of Heavy Oils and Residua, 2nd Edition Marcel
    Dekker Inc., New York.
Speight, J.G., Ozum, B., 2002. Crude oil Refining Processes. Marcel Dekker Inc., New York.
Speight, J.G., 2007. Natural Gas: A Basic Handbook. GPC Books, Gulf Publishing Company,
    Houston, Texas.
Speight, J.G., 2008. Handbook of Synthetic Fuels: Properties, Processes, and Performance.
    McGraw-Hill, New York.
Speight, J.G., 2009. Enhanced Recovery Methods for Heavy Oil and Tar Sands. Gulf Publishing
    Company, Houston, Texas.
Speight, J.G., 2011a. The Refinery of the Future. Elsevier-Gulf Professional Publishing,
    Oxford, United Kingdom.
Speight, J.G., 2011b. An Introduction to Crude oil Technology, Economics, and Politics.
    Scrivener Publishing, Salem, Massachusetts.
Speight, J.G., 2011c. Handbook of Industrial Hydrocarbon Processes. Gulf Professional
    Publishers, Elsevier, Oxford, United Kingdom, Chapter 4.
Speight, J.G., 2013a. Oil Sand Production Processes. Elsevier-Gulf Professional Publishing,
    Oxford, United Kingdom.
Speight, J.G., 2013b. Heavy Oil Production Processes. Elsevier-Gulf Professional Publishing,
    Publishing Company, Oxford, United Kingdom.
Speight, J.G., 2013c. Heavy and Extra Heavy Oil Upgrading Technologies. Elsevier-Gulf
    Professional Publishing, Oxford, United Kingdom.
Speight, J.G., 2013d. Shale Gas Production Processes. Elsevier-Gulf Professional, Oxford,
    United Kingdom.
Speight, J.G., 2013d. The Chemistry and Technology of Coal, 3rd Edition CRC Press, Taylor
    & Francis Group, Boca Raton Florida.
Speight, J.G., 2014a. The Chemistry and Technology of Crude Oil, 5th Edition CRC Press,
    Taylor & Francis Group, Boca Raton Florida.
Speight, J.G., 2014b. High Acid Crudes. Elsevier-Gulf Professional Publishing, Oxford, United
    Kingdom.
Speight, J.G., 2014c. Oil and Gas Corrosion Prevention. Elsevie-Gulf Professional Publishing,
    Oxford, United Kingdom.
Thorn, K.A., Aiken, G.R., 1998. Biodegradation of crude oil into nonvolatile organic acids in
    a contaminated aquifer near Bemidji, Minnesota. Organic Geochemistry 29 (4), 909–931.
Tissot, B.P., Welte, D.H., 1978. Crude oil Formation and Occurrence. Springer-Verlag,
    New York.

Tomczyk, N.A., Winans, R.E., Shinn, J.H., Robinson, R.C., 2001. On the nature and origin of acidic species in crude oil. 1. Detailed acidic type distribution in a California crude oil. Energy and Fuels 15 (6), 1498–1504.

Watson, J.S., Jones, D.M., Swannell, R.P.G., Van Duin, A.C.T., 2002. Formation of carboxylic acids during aerobic biodegradation of crude oil and evidence of microbial oxidation of hopanes. Organic Geochemistry 33 (10), 1153–1169.

Wilkes, H., Kühner, S., Bolm, C., Fischer, T., Classen, A., Widdel, F., Rabus, R., 2003. Formation of n-alkane- and cycloalkane-derived organic acids during anaerobic growth of a denitrifying bacterium with crude oil. Organic Geochemistry 34 (9), 1313–1323.

Zhao, B., Currie, R., Mian, H., 2012. Catalogue of Analytical Methods for Naphthenic Acids Related to Oil Sands Operations. OSRIN Report No. TR-21. Oil Sands Research and Information Network, University of Alberta, School of Energy and the Environment, Edmonton, Alberta, Canada.

CHAPTER TWO

Offshore Geology and Operations

Outline

2.1 INTRODUCTION

Petroleum (crude oil) is a flammable liquid that occurs naturally in reservoirs, most often found beneath the surface of the earth (Chapter 1) (Speight and Ozum, 2002; Parkash, 2003; Hsu and Robinson, 2006; Gary et al., 2007; Speight, 2011, 2014). Natural gas, a gaseous mixture of hydrocarbons and other constituents that are either obtained with petroleum or from natural gas reservoir, is also a flammable hydrocarbon mixture (Chapter 1) (Mokhatab et al., 2006; Speight, 2007). Over a period encompassing millions of years, a variety of geologic formation, plant, and animal remains collected on the floor of shallow seas and, as the seas receded, the organic material was covered by sedimentary layers such as silt, sand, and clay. With time and depth of burial under anaerobic conditions (absence of oxygen), the organic material decomposed, first into a petroleum precursor (called protopetroleum for convenience) with the concomitant production of gases and then to petroleum and natural gas (Speight, 2014). The petroleum and natural gas (or the partially formed petroleum and natural gas) usually migrated away from the source rock where it was formed to a reservoir rock—where any unfinished maturation chemistry was completed—in fact some of the maturation processes are still occurring at the present time since petroleum is known to change composition in the reservoir (Speight, 2014)—and awaited discovery.

The formation of traps in which the crude oil and natural gas are retained is due to the movement of strata by geological processes. The Earth is a dynamic geological and chemical system and as the tectonic plates moved, the rock is warped into folds or broken along fault lines, allowing the petroleum and natural gas to collect in a trap, which generally becomes the reservoir (Chapter 1), and the petroleum exists as a *pool*—not a pool in the usual sense of the word but a pool in the geological sense in which a liquid collects within the pores spaces of the rock in the sediment. In the Middle East (especially the Fertile Crescent in Mesopotamia—now modern Iraq), seepages and escaping natural gas burned continuously (giving rise to fire worship) and any residues that remained as result of evaporation (natural or intentional) of volatile constituents was used as a mastic for building mastic and for roadways (Forbes, 1958a,b, 1959, 1964; James and Thorpe, 1994). The volatile liquids were used, to a limited extent, for lighting (for the most past, the extreme flammability of naphtha was uncontrollable) but were primarily used for healing ailments varying from everything from headaches to deafness. Derivatives of petroleum were also used in war, in the form of *Greek Fire* and other variants of the flammable distillate (naphtha) (Cobb and Goldwhite, 1995; Speight, 2014). Generally, petroleum (or the nonvolatile constituents) was a land-based resource—other than lumps of bitumen (from which the volatile constituents had evaporated and which collected and floated on inland waters such as the Mesopotamian rivers (the Tigris and the Euphrates) and lakes as well as inland seas such as the Caspian Sea that were observed by travelers such as Marco Polo (Burgan, 2002; Bergreen, 2007)).

The majority (approximately 75%) of the surface of the Earth is covered by oceans and with the continuous use and depletion of land-based crude oil and natural gas resources, it seemed more than appropriate that further exploration and production of crude oil and natural gas have been extended into offshore regions, typically basins that border a continental land mass. Furthermore, to meet energy demand, operations are moving into ever deeper waters and crude oil natural gas is produced in many coastal areas where the depth of the water above the petroleum reservoir is on the order of 8,000 ft or more. As a results of such activities, current offshore production of crude oil accounts for approximately 60% (v/v) of the world crude oil production (with the associated natural gas) that comes from offshore operations in the offshore areas of more than half the nations that are regarded as *coastal nations*.

Although offshore crude oil natural gas production has a significant effect on local onshore economies as well as national economies (Morbey, 1996; Mason, 2009), there are also the ever-present environmental risks. In fact, recent environmental concerns have prevented or restricted offshore drilling in some areas, and the issue has been hotly debated at the local, state, and federal levels. On the one hand, there is general agreement that the environment must be protected, whereas on the other hand, there are groups who dictate that the environment should not be jeopardized or invaded by any form of energy production, particularly by offshore exploration and recovery operations. As nobble as this may seem, it is necessary that a balance be struck between energy production and protection of the environment. It is certain that without energy (from crude oil and natural gas), the possibility maintaining a recognizable standard of living is diminished and the possibility of freezing in the green darkness would not be a pleasant option.

In the current context, the geology of the seabed is an important factor in determining where marine crude oil and natural gas projects can be effectively located. Similar to onshore geology, there are mountain ridges and valleys (sometime referred to as basins) and during the geologic timescale, the movement of the tectonic plates has produced areas of the seabed better suited to formation containing crude oil and natural gas than others. Furthermore, of the two crusts, the continental crust is thicker and less dense than oceanic crust. When continental and oceanic plates collide, the heavier and thinner oceanic plate tends to get pushed underneath the continent. As new crust forms at the expanding oceanic ridges, old oceanic plates are consumed at deep ocean trenches found at the plate boundaries opposite the mid-ocean ridges. In these *subduction zones*, the edge of an ocean plate is driven below the continental plate and is eventually reabsorbed into the mantle as it continues downward. It is the summation of such effects and the collection of organic debris in formation that eventually leads to the formation of crude oil and natural gas.

Finally, there are two terms that need attention: (1) subsea and (2) deepwater. The former, *subsea*, is a general term frequently used to refer to equipment, technology, and methods employed to explore, drill, and develop crude oil natural gas fields that exist below the ocean floors, which may be in *shallow water* or in *deep water*. The latter term, *deepwater*, is a term often used to refer to subsea projects in water depths below 1,000 ft—some as deep as 10,000 ft—and may include floating drill vessels, semi-sub rigs, or semi-submersible platforms. The words *shallow* and *shelf* (usually referring to the *continental shelf*) (Chapter 1) are used for shallower depths and can include standing drilling rigs.

2.2 STRATA, SEDIMENTS, AND BASINS

Geologically, a *stratum* (plural: *strata*) is a layer of rock or soil with internally consistent characteristics that distinguishes it from contiguous layers and each stratum is generally one of several parallel layers that were deposited initially by natural events one above another. Strata typically occur as bands of different colored or differently structured material that can be observed when exposed in cliffs, river banks, and any man-made excavations such as quarries and roadways cut through hilly or mountainous areas. Strata, which may extend over hundreds or thousands of square miles, vary in thickness from fractions of an inch to hundreds of feet and each stratum represents a specific mode of deposition—such as river silt, sea sand, and coal swamp, and a variety of minerals. Strata are categorized (characterized) by the material in the strata and each distinct layer is usually assigned a *formation name* and distinctions in material in a formation may be described as *members* or *beds*. On a larger scale, the *formations* are collected into *groups* and the groups may be collected into *super-groups*.

A *sediment* is formed from any particulate matter that has been transported by fluid flow (river flow or sea flow) and which eventually is deposited as a layer of solid particles on the bed or bottom of a body of a body of water. *Sedimentation* is the deposition by settling of a suspended material. Sediments are also transported by wind (*aeolian sediments*) and glaciers (*glacial sediments*). For example, desert sand and loess (deposits of silt—sediment with particles 2–64 μm in diameter) are examples of aeolian transport and deposition. Glacial moraine deposits and till are ice transported sediments—the moraine is a deposit of such material left on the ground by a glacier—also a ridge, mound, or irregular mass of unstratified glacial drift, chiefly boulders, gravel, sand, and clay. Simple gravitational collapse also creates sediments such as talus (a deposit of material left on the ground by a glacier) and mountain-slide deposits as well as karst collapse features—karst is formed from the dissolution of soluble rocks such as limestone, dolomite, and gypsum and is characterized by underground drainage systems with sinkholes and caves.

Seas, oceans, lakes, and rivers accumulate sediment over time and the deposited material can be *terrestrial* (deposited on the land) or *marine* (deposited in the ocean). *Terrigenous* deposits originate on land, but may be deposited in either terrestrial environments, marine environments, or lacustrine (lake) environments. Deposited sediments are the precursors to source of sedimentary rocks, which can contain carbonaceous remains of the flora and fauna that lived and died in the body of water and that, upon death, became part of the sedimentary deposit.

Geologically, the *crust* is the outermost solid shell of the Earth and is chemically and mechanically different from underlying material having been generated by (predominantly) igneous processes, and these crusts are richer in incompatible elements than the underlying mantles. Igneous rock is a rock formed through the cooling and solidification of magma or lava and may form with or without crystallization—the other types of rocks are sedimentary and metamorphic, the latter type of rock (metamorphic rock) arises from the transformation of existing rock types, in a which the original rock (protolith, which may be sedimentary rock, igneous rock or older metamorphic rock) is subjected to heat (temperatures greater than 150–200°C, 300–390 °F) and pressure (approximately 22,000 psi), which causes chemical and physical change.

2.3 THE CRUST

The crust of the Earth is a relatively thin layer of rock that amounts to less than half 0.5% (w/w) of the total mass of the Earth. In some places, the crust is 50 miles thick, whereas in others, it may be less than 1 mile thick. Under the crust is the mantle that may be as much as 1700 miles thick. The crust is primarily made of granite and basalt, whereas the mantle beneath is composed of peridotite.

Peridotite is a generic name used for coarse-grained, dark-colored, ultramafic igneous rocks. Peridotite rocks usually contain olivine as their primary mineral, frequently with other mafic minerals such as pyroxenes and amphiboles. Their silica content is low compared with other igneous rocks and they contain very little quartz (SiO_2) and feldspar rocks composed of sodium, potassium, and calcium derivatives of the silicates ($NaAlSi_3O_8 \cdot KAl Si_3O_8 \cdot CaAl_2Si_2O_8$).

Basalt (basaltic rock) is a common extrusive igneous (volcanic) rock formed from the rapid cooling of basaltic lava exposed at or very near the surface of the Earth. By definition, basalt is an aphanitic igneous rock with less than 20% (v/v) quartz and less than 10% (v/v) feldspathoid by volume, and where at least 65% of the feldspar is in the form of plagioclase—in comparison, granite has more than 20% (v/v) quartz. Basaltic rocks underlie the seafloors and granitic rocks make up the continents. For general purposes and for the purposes of this text and for the crust is composed of two systems: (1) the continental crust and (2) the oceanic crust and each type exhibits differences from the other.

2.3.1 Continental Crust

The *continental crust* is thick and old—typically approximately 30 miles thick and approximately 2 billion (2×10^9) years old—and covers approximately 30–40% of the Earth. Whereas almost all of the oceanic crust is underwater, most of the continental crust is exposed to the air.

The continents slowly enlarge over geologic time as oceanic crust and seafloor sediments are pulled beneath them by subduction. The descending basalts have the water and incompatible elements squeezed out of them, and this material rises to trigger more melting in the so-called *subduction zone*. Briefly, subduction is the process that takes place at convergent plate boundaries where one tectonic plate moves under another tectonic plate and sinks into the mantle as the plates converge. Rates of subduction are typically on the order of inches (or less) per year, with the rate of convergence being on the order of 0.75–3 in. per year.

The continental crust is composed of granitic rocks, which have even more silicon and aluminum than the basaltic oceanic crust and are less dense than basalt. In terms of mineral composition, granite has more feldspar and less amphibole than basalt and almost no pyroxene or olivine, but has abundant quartz—geologically, the continental crust is described as felsic. Chemically (or physically), the continental crust is too buoyant to return to the mantle. The relatively thin skin of limestone rocks and other sedimentary rocks tend to stay on the continents, or in the ocean, rather than return to the mantle—the sand and clay that is washed off the continent into the sea through abrasion or attrition of land-based rocks by the action of the ocean eventually returns to the continents via the oceanic currents.

The *continental crust* is the layer of granitic, sedimentary, and metamorphic rocks, which form the continents and the areas of shallow seabed close to their shores (continental shelves). Consisting mostly of sialic rock (the assemblage of rocks, rich in silica and alumina, that comprise the continental portions of the upper layer of the crust), it is less dense than the material of the mantle of the Earth, which consists of mafic rock (rock that is relatively rich in magnesium and iron). The continental crust is also less dense than oceanic crust, although it is considerably thicker.

As a consequence of the density difference, when active margins of continental crust meet oceanic crust in subduction zones, the oceanic crust is typically subducted back into the mantle. Because of its relative low density, continental crust is only rarely subducted or recycled back into the mantle (for instance, where continental crustal blocks collide and over thicken, causing deep melting). For this reason, the oldest rocks on Earth are within the *cratons* (cores of the continents or the older and stable parts of the continental lithosphere), rather than in repeatedly recycled oceanic crust.

Continental margins are classified as *passive* or *active*. The Atlantic continental shelf of the United States is a passive margin where the continental and oceanic crusts are joined together far from the volcanic spreading forces of the Mid-Atlantic Ridge. The Atlantic margin is stable, geologically old, and notable for its broad and shallow slope and thick sediment deposits. The areas of the East Coast and the Gulf of Mexico are characterized by a gently sloping submerged continental shelf—the section closest to shore has a slope of approximately 0.5°—which changes at the shelf break to the steep and narrow continental slope (with an approximate gradient of 3.0°) that yields to the less sloped and wider continental rise. The shelf break is located at a fairly uniform depth of approximately 430 ft. The continental rise, which is formed by sediment that rolls off of the continental slope, ends at the relatively flat, deep seabed known as the *abyssal plain*.

Passive continental margins will typically have a large shelf—for example, the US East Coast New Jersey shelf extends approximately 92 miles from the shoreline. Off the coast of Maine, the shelf is approximately 460 miles from the shoreline. On the other hand, active margins (such as the Pacific Coast) are usually narrower—the Pacific Coast generally lacks a continental rise because the continental slope leads directly into a trench where the Pacific oceanic plate is subducted under the continental crust.

Finally, the high temperatures and pressures at depth, often combined with a long history of complex distortion, mean that much of the lower continental crust is metamorphic—the main exception to this being recent igneous intrusions. Igneous rock may also be *underplayed* to the underside of the crust, that is, adding to the crust by forming a layer immediately beneath it.

2.3.2 Oceanic Crust

The *oceanic crust* covers about 60% (v/v) of the surface of the Earth and is relatively thin and geologically young, being less than 12 miles thick and younger than approximately 180 million years—older rocks have disappeared under the oceanic crust by subduction. The oceanic crust is born at the mid-ocean ridges, where the tectonic plates are pulled apart (divergent boundary). As that happens, the pressure upon the underlying mantle is released and the peridotite there responds by starting to melt—basaltic lava, which rises and erupts, while the remaining peridotite becomes depleted. The mid-ocean ridges migrate over the Earth and extract the basaltic component from the peridotite of the mantle as they go. Thus, basaltic rocks contain more silicon and aluminum than the peridotite left behind, which has more iron and magnesium. Basaltic rock is also less dense than granitic rock. In terms of minerals, basalt has more feldspar and amphibole, less olivine and pyroxene, than peridotite. Geologically, the oceanic crust is mafic (composed of silicate minerals, magmas, and rocks that are relatively high in the heavier elements), whereas oceanic mantle is ultramafic (composed of igneous and meta-igneous rocks with a very low silica content—less than 45% and generally >18% magnesium oxide (MgO), high iron oxide (FeO), low potassium, and composed of usually greater than 90% mafic minerals (dark colored, high magnesium and iron content)).

Because of the thinness, the oceanic crust is a very small fraction of the Earth—approximately 0.1% (v/v)—the life cycle serves to separate the constituents of the upper mantle into a heavy residue and a lighter set of basaltic rocks. The oceanic crust extracts any incompatible elements, which do not fit into mantle minerals and move into the liquid melt. In turn, these minerals move into the continental crust thought the action of plate tectonics. Meanwhile, the oceanic crust reacts with seawater and carries some of it down into the mantle.

The *oceanic crust* is the part of lithosphere of the Earth that surfaces in the ocean basins and is composed primarily of mafic rocks, or sima (the lower layer of the crust that is composed primarily of rocks rich in magnesium silicate minerals). It is thinner than continental crust but is more dense. The oceanic crust is significantly simpler than continental crust and generally it can be divided in three layers.

Layer 1 consists of unconsolidated or semi-consolidated sediments, usually thin or even not present near the mid-ocean ridges but thickens farther away from the ridge. Near the continental margins, sediment is terrigenous, meaning derived from the land, unlike deep sea sediments that are made of tiny shells of marine organisms, usually calcareous and siliceous, or it can be made of volcanic ash and terrigenous sediments transported by turbidity currents. *Layer 2* could be divided into two parts: an uppermost volcanic layer of glassy to finely crystalline basalt usually in form of pillow basalt, and a lower layer composed of diabase dikes. *Layer 3* is formed by slow cooling of magma beneath the surface and consists of coarse-grained gabbros and cumulate ultramafic rocks.

By way of explanation, *gabbro* is dense green or dark colored and contains pyroxene, plagioclase, amphibole, and olivine—it is referred to as *olivine gabbro* when olivine is present in a large amount. The pyroxene is mostly clinopyroxene but small amounts of orthopyroxene may be present. If the amount of orthopyroxene is substantially greater than the amount of clinopyroxene, the rock is then a norite. Quartz gabbro is also known to occur and is probably derived from magma that was oversaturated with silica. Essexites represent gabbros whose parent magma was undersaturated with silica, resulting in the formation of the feldspathoid mineral nepheline. Silica saturation of a rock can be evaluated bynormative mineralogy. Gabbros contain minor amounts, typically a few percent, of iron–titanium oxides such as magnetite, ilmenite, and ulvospinel.

Gabbro is generally coarse grained, with crystals in the size range of 1 mm or greater. Finer grained equivalents of gabbro are called diabase, although the vernacular term *microgabbro* is often used when extra descriptiveness is desired. Gabbro may be extremely coarse grained to pegmatitic, and some pyroxene-plagioclase cumulates are essentially coarse-grained gabbro, although these may exhibit acicular crystal habits. Gabbro is usually equigranular in texture, although it may be porphyritic at times, especially when plagioclase oikocrysts have grown earlier than the ground-mass minerals.

Gabbro can be formed as a massive, uniform intrusion via *in situ* crystallization of pyroxene and plagioclase, or as part of a layered intrusion as a cumulate formed by settling of pyroxene and plagioclase. Cumulate gabbro is more properly termed pyroxene-plagioclase orthocumulate.

The oceanic crust generally does not last longer than 200 million years—it is continuously being created at oceanic ridges where hot magma rises into the crust and cools, pushing the crust apart at the ridge. The continuous formation of new oceanic crust pushes the older crust away from the mid-ocean ridge. As it moves away from the ridge, the crust becomes cooler and denser, while the sediment may build on top of it.

2.4 BASINS

The rock formations that hold petroleum reserves are found everywhere on earth. Ancient oceans and seas collected organic materials from plants and microscopic animals in the sediment at the bottom of the body of water. Later heating, bending, and folding these organic materials within the rock formations transformed this organic material into petroleum. Wherever the right set of circumstances occurs, there is a possibility that petroleum will be produced and trapped underground in fields or pools that can be accessed by drilling.

All petroleum resources are found in sedimentary basins. The easy way to access oil and gas is to find a location where petroleum products are leaking to the surface. These areas are called seeps, and petroleum products have been used from these areas for thousands of years. Not all sedimentary basins have petroleum-producing features, but geologists always are looking for new resources.

An oceanic basin may be any basin that is covered by sea water, but geologically oceanic basins are large geological basins that are below sea level. In addition, there are also other under-sea geomorphological features such as (1) the continental shelves, (2) mountain ranges, such as the mid-Atlantic Ridge, and (3) ocean trenches, some of which are very deep, and which are not considered to be part of the ocean basins.

Originally, the oceanic basins were considered to complement to the continents and erosion dominated the latter with the sedimentary material originating from continental erosion ending up in the ocean basins (Littlehales, 1930). Since that time, modification to these conclusions (with the more detailed investigations stimulated by deep-water offshore search for crude oil and natural gas as well as environmental issues related to oceanic pollution) has placed the oceanic basins in the realm of basaltic plains rather than sedimentary depositories, since most sedimentation occurs on the continental shelves and not in the geologically defined ocean basins (Floyd, 1991; Biju-Duval, 2002). On a hydrological basis, some geologic basins are both above and below sea level, such as the Maracaibo Basin (Venezuela), although geologically it is not considered a true oceanic basin (in the geological sense) because it is on the continental shelf and underlain by continental crust.

Because oceans lie lower than continents, the former serve as sedimentary basins that collect sediment eroded (clastic sediments) from the continents, as well as precipitation sediments. Ocean basins also serve as repositories for the skeletons of carbonate-secreting and silica-secreting organisms such as coral reefs, diatoms, radiolarians (protozoa), and foraminifera (forams—single-celled protozoa with shells). Geologically, an oceanic basin may be actively changing size or may be relatively, tectonically inactive, depending on whether there is a moving tectonic plate boundary associated with the basin. The elements of an active and growing-oceanic basin include an elevated mid-ocean ridge, flanking abyssal hills leading down to abyssal plains. The elements of an active oceanic basin often include an ocean trench associated with a subduction.

Briefly, the Atlantic Ocean and Arctic Ocean are examples of active, growing oceanic basins, whereas the Mediterranean Sea is shrinking and the Pacific Ocean is also an active, shrinking ocean basin, even though it has both spreading ridge and oceanic trenches. On the other hand, the Gulf of Mexico is an example of an inactive oceanic basin—the Gulf formed during the Jurassic Period (approximately 200 million years ago to 145 million years ago, from the end of the Triassic Period to the beginning of the Cretaceous Period) and has been doing nothing but collecting sediments that time (Huerta and Harry, 2012). The Aleutian Basin is another example of a relatively inactive oceanic basin (Verzhbitsky et al., 2007).

The origin of some of the largest oil and gas basins (such as those of the Persian Gulf, Venezuela, Algeria, Canada, and the North Slope of Alaska) may have been formed by migration of hydrocarbons from recent and old zones of under-thrusting of lithospheric plates (subduction zones). The hydrocarbons are generated in these zones via thermolysis of organic matter locked in sediments over-ridden by island arcs that are thrust over the passive continental margins.

Examples of such basins with dispersed hydrocarbons are the eastern shelves of North and South America, the West African shelf, and certain other basins located on passive continental margins. The mobilization of hydrocarbons dispersed in such rocks would require intensive tectonic action, such as that exerted by thrusting of island arcs on the passive margins of continents.

2.5 GEOCHEMISTRY AND MINERALOGY

Petrology is the branch of geology that deals with the origin, composition, structure, and alteration of rocks, which also include mineralogy—the study of the individual minerals in rocks and sediments. Organic petrology involves the study of organic matter in rocks including its origin and thermal maturity due to burial within the crust of the Earth. Often, these areas of study are included within the broader area of petroleum geochemistry (Shimoyama and Johns, 1971; Bailey et al., 1974; Tissot and Welte, 1978; Snowdon and Powell, 1982; Robinson, 1993; Hunt, 1996; Speight, 2014).

While the field of geochemistry involves study of the chemical composition of the Earth and other planets, the composition of rocks and soils, the cycles that involve the chemical components of the Earth, and the interaction of those cycles with land and water. Some issues within the field of geochemistry (as it relates to crude oil and natural gas) include: (1) the elements and chemicals present in various soils and rocks in different locations, (2) changes in rocks over time, and (3) the decomposition of flora and fauna after death and the resultant organic mass that is produced. With the inclusion of mineralogy, there is also the ability to assess the effects of mineral's on the decomposition process and product(s).

However, in terms of crude oil and natural gas petroleum geochemistry is multidisciplinary and involves (1) organic petrology, (2) geology, (3) petrophysics, (4) gas geochemistry, (5) microbiology, (6) reservoir engineering, and last but not least (7) mineralogy to derive molecular information from individual organic deposits that led to the formation of crude oil and natural gas. In fact, the multidisciplinary approach to organic petrology is key discipline to assessing petroleum exploration and involves evaluating the generation potential of source rocks by assessing source quality (organic matter typing) and thermal flux (maturity). The data are used as basic inputs for petroleum generation models to assess the amounts and timing of generation and expulsion. This section is concerned with the relatively little-studied mineralogy of various formations since minerals can play a role in the formation of the protopetroleum that is the precursor to crude oil and natural gas generation as well as aiding the identification of formation that could potentially contain crude oil and natural gas (Speight, 2014). Thus, the mineralogy of the oceans is an important aspect of offshore exploration and drilling for crude oil and natural gas. The minerals can affect the drilling operations and can also be instrumental in affecting the behavior of the reservoir. It is, therefore, necessary to be aware of the mineralogy of the various sub-oceanic stat before a drilling program is instituted.

Rivers carry material into the oceans each year and sea-floor springs and volcanic eruptions also add elements to the sea—winds also contribute solid materials to the oceans in appreciable quantities. Most of these sediments rapidly settle to the seafloor in near-shore areas, in some cases, forming potentially valuable mineral deposits. Because of the nature of the minerals and their mode of formation, it is convenient to consider the occurrence of ocean deposits in several environments: marine beaches, seawater, continental shelves, sub-seafloor consolidated rocks, and the marine sediments of the deep-sea floor.

Minerals that resist the chemical and mechanical processes of erosion in nature and that possess a density greater than that of the Earth's common minerals have a tendency to concentrate in gravity deposits (*placers*). During the Pleistocene glaciations, sea level was appreciably lowered as the ocean water was transferred to the continental glaciers. Because of the cyclical nature of the ice ages and the inter-glacial warm periods, a series of beaches were formed in near-shore areas both above and below present sea level. Also, when sea level was lowered in the past, the streams that currently flow into the sea coursed much further seaward, carrying placer minerals to be deposited in channels that are now submerged.

Mineral deposits in the sea not only have potential economic significance but also have the potential to influence the chemistry of crude oil maturation processes. Furthermore, minerals formed in the deep sea are frequently found in high concentrations because there is relatively little clastic material (geological detritus—pieces of rock broken off other rocks, often by physical forces such as weathering) generated in these areas to dilute the chemical precipitates.

The sediments in the ocean, which consist of three major components of detrital, biogenic, and antigenic (minerals that were generated where found or observed) origins, contain direct and/or indirect evidence of chemical and material inputs to the ocean and recycling within it (Kastner, 1999).

2.5.1 Detrital Minerals

Detrital sediments are transported by water, wind, or ice and aluminosilicate minerals, which are ultimately derived from the weathering and erosion of onshore rocks, comprise the bulk of detrital sediments. The distribution of these minerals in the ocean depends on (1) the effects of climate and weathering at the source, as well as (2) the geographic disposition of rivers, existence of glaciers, and prevailing winds. They are most abundant in continental margins but occur over the entire seafloor.

Detrital minerals and their spatial and temporal distribution contain extremely important information on climate and tectonics such as the changing patterns and intensities of winds, the nature and intensity of weathering and erosion, and detritus transport by rivers and ice. Mineral matter ejected from volcanoes is another type of detrital sediment carried by water and wind into the ocean. Volcanic sediments provide important information on periods of intense island arc and submarine volcanic activities, which can be compared with records of submarine tectonics and with climate and biological productivity records.

2.5.2 Biogenic Minerals

Dissolved weathering products constitute most of the salt in the sea. The major- and minor-element chemistry of biogenically produced minerals reflects the geologic processes and rates that control the chemistry of seawater, ocean circulation, and biologic evolution. The less reactive, long residence time elements such as chlorine (Cl), sodium (Na), potassium (K), magnesium (Mg), calcium (Ca), and strontium (Sr) do not vary throughout the ocean. Plants and organisms (such as those forming the precursors to crude oil and natural gas) preferentially extract some elements, primarily carbon, nitrogen, phosphorus, calcium, and silicon (Si), from seawater to form soft tissues and minerals. Some of these elements are consumed by other organisms or redissolve, thus they are internally recycled.

Organic matter-rich sediments (such as sediments in which crude oil and natural gas may occur) signify periods of higher productivity and/or higher preservation of organic carbon in low-oxygen waters, either in a more intense and expanded oxygen minimum zone or in low-oxygen bottom (benthic) waters. The stable carbon isotopic composition of organic matter in marine sediments helps to identify periods of high productivity or of high terrestrial organic matter input. Prime examples of periods with widely distributed organic carbon-rich sediments are (*a*) the mid-Cretaceous abnormally organic carbon-rich black sediments in all major ocean basins (Calvert and Pedersen, 1993).

2.5.3 Authigenic Minerals

Authigenic minerals form by *in situ* inorganic precipitation on the seafloor and within the sediment column—barite is the only mineral so far reported to also form in the water column. Some authigenic mineral reactions are bacterially mediated—the bacteria modify the immediate geochemical environment, inducing mineral formation. The concentrations and isotopic compositions of oceanic conservative components in these authigenic minerals that form at the seafloor are extremely informative for studies of the history of seawater chemistry.

The important marine clay minerals (which can influence crude l and natural gas formation) are smectite minerals (hydrous aluminum phyllosilicates, sometimes with variable amounts of iron, magnesium, alkali metals, alkaline earths, and other cations). The iron smectite, nontronite, is widespread in the floor of the Pacific Ocean floor where hydrothermal activity is prevalent and sedimentation rates are slow.

Authigenic calcite is widespread in the ocean and forms mostly from biogenic calcite recrystallization as well as being geochemically distinct in its minor and trace element concentrations from its precursor. This recrystallization process ultimately transforms calcareous ooze into chalk and limestone. Dolomite forms where pore-fluid seawater is modified bacterially or by physical–chemical processes. In continental margins where pore fluids become suboxic to anoxic, dolomite forms both as a primary precipitate and by replacing precursor calcium carbonate ($CaCO_3$) and calcium–magnesium carbonate ($CaCO_3 \cdot MgCO_3$). The chemical and isotopic compositions, primarily of carbon and oxygen, reveal the origins of the various dolomites. Manganese and iron mostly substitute for magnesium in the dolomite structure in suboxic to anoxic pore fluid environments where they are mobilized. In these environments, only in the presence of much detrital manganese and iron Fe, rhodochrosite and/or siderite may form during late diagenesis where the reactions have driven magnesium/calcium ratios in the pore fluids to levels below dolomite stability.

2.5.4 Submarine Hydrothermal Sulfide Deposits

Of particular interest to the petroleum geochemist is the submarine hydrothermal sulfide deposits forming at divergent plate boundaries that have been discovered only in the past three-to-four decades. These deposits primarily consist of the sulfides chalcopyrite, sphalerite and/or wurtzite, pyrrhotite, pyrite and/or marcasite, and the associated minerals anhydrite, barite, and opal-A. At the seafloor in oxidizing low-temperature environments, hydrothermal sulfides are unstable, they are oxidized to form Fe oxides and secondary sulfides such as covellite, digenite, and bornite, except if rapidly covered by new volcanic eruptions. As yet, observations mainly of surface seafloor sites have been conducted, and the magnitude and structures of these sulfide deposits have not yet been thoroughly characterized.

At fast- and slow-spreading centers, seawater penetrates several kilometers into oceanic basement (maximum depth is above the top of the magma chamber), where it is heated to 350–400°C (650–750°F), and interacts with oceanic basement and sediments. This fluid ascends by buoyancy to form hydrothermal vents and sulfide deposits. The discharging fluids support rich chemosynthetic, previously unknown biological communities.

2.5.5 Sedimentary Organic Matter

Sedimentary organic matter refers to the organic matter deposited with sediment that has amassed over geological time. Generally, sedimentary organic matter can be divided operationally into solvent-extractable material (sometime erroneously referred to as bitumen) bitumen and insoluble material, often referred to as kerogen. The range of components includes many biomarkers that retain structural remnants inherited from their source organisms, which carry evidence of their origins and history. Recognition of the specificity of biomarker structures initially helped confirm that crude oil was derived from organic matter produced by biological processes. Of the thousands of individual constituents of crude oil, many reflect precise biological sources of the original organic matter, which distinguish and differentiate their varied origins. The diagnostic suites of components may derive from individual families of organisms, but contributions at a species level can occasionally be recognized. Biomarker abundances and distributions help to elucidate sedimentary environments, providing evidence of depositional settings and conditions. The biomarkers also reflect the relative maturity of sediments and attesting to the progress of the successive, sequential transformations that convert biological precursors into geologically occurring products, such as crude oil and natural gas. Thus, specific biomarker characteristics permit assessment of the thermal history and the catalytic history of the formation of crude oil and natural gas in a variety of sedimentary basins (Welte et al., 2011).

2.6 DRILLING AND RECOVERY

The evolution of the crude oil natural gas industry and its movement to the offshore has been one of the highlights of the 20th. However, the history of the offshore crude oil natural gas is diverse because the industry is a complex collection of technologies that are used in areas related to exploration, drilling, production, and transportation (Kemp, 2012a,b). As the crude oil and natural gas industry moved (not always in step-wise manner) from land-based operations to water-based operations (swamp, lake, marsh, shallow sea, the Outer Continental Shelf, and then to deep water), the industry has evolved and changed considerably though a blending of the evolution of the necessary older and new technologies.

2.6.1 Historical Perspectives

The history of offshore crude oil natural gas can be traced to simple, if not accidental, humble beginnings—early documents contain fragmentary evidence of oil being obtained from the sea (unknown at the time but emanating from undersea seepages). For example, as early as the 1270s, Marco

Polo (Burgan, 2002; Bergreen, 2007), on his way to China, was reported to have observed crude oil taken from the Caspian Sea at Baku. However, the first so-called off-shore oil recovery operations were carried out during the 19th century at the Bibi-Eibat field—Bibi-Eibat (Bibiheybat) is a suburb of Baku (the capital and largest city of Azerbaijan, as well as the largest city on the Caspian Sea) and has been known (in modern times) as an oil field since 1803. In the late 19th century, oil recovery from subsea formations was documented in the United States—in 1896, at Summerland field on the California Coast. After a large number of wells were drilled, it soon became evident that wells in closest proximity to the Pacific Ocean were the best producers and several additional wells were actually drilled on the beach. As a result of continuing production, a wharf with a drilling rig erected on it was built and the first offshore drilling operation extended approximately 300 ft into the ocean. As expected, the well exhibited a steady production of crude oil and was followed shortly thereafter by construction of several more wharves—the longest of these wharves extended more than 1200 ft into the Pacific Ocean.

By 1910, at the same time as when Winston Churchill as First Lord of the Admiralty was advocating the conversion the coal-fired British battle-ships (and other ships of the fleet) to petroleum-based power, America was quickly turned to oil as its primary resource. The invention of the internal combustion engine boosted the consumption of gasoline. At this time, American companies and entrepreneurs were discovering faster and more efficient ways to retrieve crude oil from land-based and sea-based operations. Particularly noteworthy, the steel cable was used in place of rope for cable tool drilling and by 1919, the first diamond-tipped drill was used (Leffler et al., 2011).

In the late 1920s and into the 1930s, the lakes, marshes, and bayous of the southern United States began to rival the well-known Spindletop oilfield in Texas (discovered on a salt dome formation south of Beaumont in eastern Jefferson County on January 10, 1901) in the production of petroleum and natural gas—involving the Texas Company, the California Company, Humble, and Shell (Wooster and Sanders, 2010). Within a decade, the technology related to oil recovery advanced—the valves and controls that gauged the flow of oil at the wellhead, nicknamed the Christmas tree (Figure 2.1)—was developed in 1922, followed in 1925 by the creation of drilling control instrumentation. At the same time, the exploration for subterranean of oil-bearing formations moved ahead, and in 1926, modern seismologic techniques of exploration were added to the techniques used for exploration.

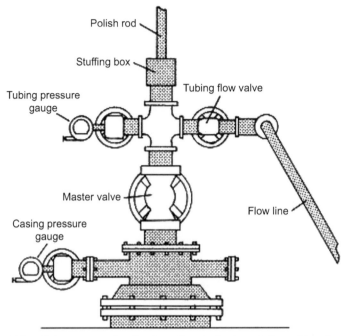

Figure 2.1 The Christmas tree: A collection of control valves at the wellhead.

In the more modern world, the search for crude oil natural gas continues and has not been contained by many boundaries. However, until the 1930s, the great uncertainty in terms of petroleum exploration and development was the point at which the land and the ocean met. Before that time, the petroleum industry had found innovative ways to develop oil fields beneath the waters of shallow lakes and protected bays but the characteristics of winds and waves of the open seas blocked the exploration and development of offshore oil and natural gas. For exploration and development that were activities close-to-inshore, elaborate systems of piers were created which extended, for example, from the land out into the Pacific Ocean (coastal California). Without ignoring the frustration weather system and the ocean currents, in the 1930s, several oil companies made the (courageous) decision to explore the possibilities for offshore exploration and production of crude oil and natural gas. Initially, most of the activity focused on offshore California, but in the late 1930s, several Texas-based oil companies took the first steps of exploring for crude oil and natural gas in the Gulf of Mexico.

The first real test of offshore construction in the Gulf of Mexico was located approximately 50 miles along the coast near Cameron, Louisiana.

In 1937 and 1938, the Superior Oil and Pure Oil companies jointly constructed a large wooden platform approximately 1 mile offshore in approximately 14 feet of water. This led to the discovery of the Creole offshore field (Superior Oil/Pure Oil) and on March 18, 1938, the Creole platform completed the first successful well in the Gulf on March 18, 1938. In April, 1938, the Creole oil field became the first producing property in the Gulf of Mexico.

The venture showed that margins (profits) could be made from offshore oil but also revealed the limitations of applying onshore technology in an offshore environment as well as the challenges posed by the ever-present hurricane during the hurricane season (hypothetically June 1 to December 1—give or take a week or so—in any given year in the Atlantic Ocean and the Caribbean Sea). As a result of the success of this venture, Superior Oil and Pure Oil then took further steps and constructed a much larger platform—320 ft by 180 ft—from which to perform exploratory drilling activities to produce any crude oil or natural gas. The primary activity in platform construction was to drive 300 treated yellow pine piles 14 ft into the sandy bottom using pile drivers mounted on barges. By driving the piles as far as possible into the sand, this *stick-building* approach gave the platform considerable strength through the clustering of a large number of wooden piles, and in terms of safety, the concept brought greater stability against wave forces. The companies also sought protection from hurricane winds by using design criteria developed for onshore buildings to construct a structure that could survive winds of up to 150 miles per hour (Austin et al., 2004).

The Creole oil field showed that offshore exploration could lead to the promise (or indications) of discovering large oil fields and gas fields under the waters of the Gulf of Mexico and offered economic incentives. Furthermore, the successes and failures of these ventures laid bare some of the challenges to be encountered by the oil and gas industry to the challenges of conducting operations offshore. And then along came World War II, which put a halt to most offshore efforts. Although out of the range of the long-range enemy aircraft of the time, the need to be alert to the real (and potential) presence of enemy warships and submarines was not conducive to healthy offshore drilling and exploration activities.

Once World War II was over and the world started to settle into various peace-time styles of living, offshore crude oil natural gas exploration and development became a strong focus once again. In the United States, peace-time affluence brought an increased public demand for crude oil and natural gas. Offshore exploration was increased and with the increased

activities came the ever-present challenges arising from underwater exploration, weather forecasting, tidal and current prediction, drilling location
determination, and (last but by no means least) offshore communications.
Nevertheless, the challenges and uncertainties notwithstanding, the Kerr-
McGee Corporation drilled the first well from a fixed platform offshore
and out-of-sight of land in 1947. The *barge and platform combination* as used
for this project was a major breakthrough in the design of the drilling-unit
for offshore use.

In fact, October 1947 is often considered to be the dawn of the modern
offshore-oil-and-gas era and by 1949, 11 fields were found in the Gulf of
Mexico with 44 exploratory wells and from there the people with know-
how, companies, and technologies spread. From negligible production in
1947, the total worldwide production of offshore crude oil natural gas grew
steadily to account. Indeed, offshore crude oil natural gas is recognized as a
series of noteworthy and recordable innovations of the 20th century. As the
industry progressed, offshore drilling and production rigs were developed to
meet the task of exploration and production and the drilling rigs used for
current offshore operations is far removed from that of the 1950s, when the
interest in offshore resources became a regular activity. By the early 1960s,
development of ship-shaped rigs had extended the offshore drilling capability to approximately 1,000 ft depth.

Thus, offshore oil exploration, as with most commercial venture, has
evolved even further and greater attention has been turned more and more
to the continental shelves of the world and the associated basins that are
filled with thick sedimentary sequences and, in many cases, crude oil and
natural gas. For example, exploration of the Gulf of Mexico was followed by
the exploration of the North Sea and then active exploration in areas such
as the Falkland Island (which are located on a projection of the Patagonian
continental shelf) and the South China Sea.

2.6.2 Modern Perspectives

The modern offshore crude oil and natural gas industry provides a substantial amount the crude oil and natural gas supply of the United States. Large
crude oil natural gas reservoirs are found under the waters of the Gulf of
Mexico (offshore from Louisiana and Texas), California, and Alaska. Other
notable world-wide offshore fields are located in the North Sea—which
is an extremely harsh environment—as well as in the Campos and Santos
basin off the coast of Brazil, several fields off West Africa (notably offshore
Nigeria and offshore Angola) as well as offshore fields in South East Asia and

the crude oil natural gas fields in the Barents Sea (located north of Norway and Russia).

As exploration and production of offshore oil and gas have advanced, the focus has evolved to deep-water sources (Chapter 6). It was once generally believed that oil could not be trapped in deep water, since reservoir rocks such as limestone and sandstone were not thought to occur in these deep waters but this line of thinking has been proven to be erroneous—sandstones formations do indeed occur in deep water. In fact, during the past 75 million years, the Gulf Coast has been progressively pushed further southward as sediments have piled up along the shore, moved here by the Mississippi River and other smaller streams. Furthermore, since the Gulf of Mexico has been a depositional basin for so long, there has been no shortage of mostly sandstone, siltstone, and shale that have been deposited. In fact, the deposition of potential reservoir rock continues to be deposited by the Mississippi River as sand and mud into the ocean is transported into the Gulf of Mexico. As the river shifts back and forth (*avulsing*), sand is deposited in some areas, whereas clay is deposited in other areas. These (and other) factors have created the sedimentary rocks that currently exist along the Gulf Coast, both above ground, and in the subsurface.

Since the 1950s, offshore structures have evolved to take on the size and appearance of small villages and the use of these structures for the exploration and production of offshore crude oil and natural gas reserves has increased to use in most continental shelf areas of the world. Offshore structures for the purpose of producing crude oil and natural gas reserves are located in waters ranging from shallow water to water approximately 10,000 ft deep. Subsea production facilities are being considered such that no surface structure is necessary, but current subsea wells are connected to surface producing offshore structure using flowlines, umbilicals, and manifolds. Undersea pipelines transport crude oil natural gas to the shore and are another type of offshore structure that are designed and installed to provide a means of transporting energy resources to land. Along with the expansion of offshore activities offshore, there has been a trend in the industry to a rationalization to lessen the risk of failure at all levels of the exploration process, and stimulating resource replacement and growth through the efficient exploration of basins and plays (Morbey, 1996).

At the same time, new opportunities have opened for the success of exploration activities success passive margin coastal basins, many have which had been considered to be beyond peak production capabilities. The recent ability for companies to explore beyond the known offshore in deeper waters and *passive margin exploration* has moved into deeper waters of the South Atlantic off West Africa, as well as the North Atlantic West of the Shetlands. These areas offer a high-risk, high-reward challenge, and drilling environments in need of innovative new E&P technology and creative initial exploration strategies (Morbey, 1996).

2.7 OFFSHORE DRILLING AND PRODUCTION

First and by way of clarification, throughout this book, the word rig is used to refer to a drilling rig (equipment for on-land oil drilling and for offshore oil drilling) as well as and oil platform (equipment for offshore oil drilling).

Offshore drilling processes and recovery operations are fundamentally the same as onshore drilling and recovery operations with the exception that special types of rigs are used and selection of the rig is very much dependent upon water depth and the location of the drilling and recovery operations (Chapter 3). Rigs (jack-up rigs) with legs attached to the ocean floor are used in shallow water with depths to 200 ft. In depths up to 4,000 ft, drilling takes place on semisubmersible rigs that float on air-filled legs and are anchored to the bottom. Drillships (drill-ships) with very precise navigational instruments are used in deep water with depths to 8,000 ft. Once a promising area has been identified, a huge fixed platform is constructed that can support 40 or more offshore wells, along with living quarters for the drilling crew.

However, in spite of the apparent ease with which the previous paragraph conveys, offshore drilling and production operations for crude oil natural gas is much more challenging than for land-based installations and much of the innovation in the offshore petroleum sector revolves around overcoming these challenges. This includes (1) manned facilities that need to be kept above sea-level, and (2) undersea operations, facilities both of which present serious challenges.

An offshore structure has no fixed access to dry land and may be required to stay in position in all weather conditions. Offshore structures may be fixed to the seabed or may be floating. Floating structures may be moored to the seabed, dynamically positioned by thrusters or may be allowed to drift freely. The engineering of structures that are mainly used for the transportation of goods and people, or for construction, such as marine and commercial ships, multiservice vessels and heavy-lift crane vessels used to support field development operations as well as barges and tugs are not discussed in detail in this book. While the majority of offshore structures support the exploration and production of crude oil natural gas, other major structures, for example, for harnessing power from the sea, offshore bases, offshore airports, are also coming into existence.

Manned facilities above sea-level can be accomplished with enormous platforms that have the *feet* placed on the bottom of the sea. An example is the Troll A platform operated by Statoil Hydro in the North Sea, which is standing on a depth of 300 m. Other platforms may be floating, only anchored to the bottom of the sea. While this cuts construction costs, extra security measures are needed, as well as mechanisms for mitigating the effects of the movement due to waves. In both cases, the ocean adds several hundred meters to the fluid column in the drill-string. This increases bottom-hole pressure as well as increasing the energy needed to lift sand and cuttings for oil-sand separation on the platform. The current trend is for more of the production to be done subsea, such as facilities to separate sand from oil and reinject sand before it is pumped up to the platform, or even pumping it onshore, with no installations visible above the sea. Subsea installations further the goal of exploiting resources at progressively deeper waters, which have previously been inaccessible. It also circumvents many of the challenges related to sea ice, such as in the Barents Sea.

An offshore oil platform is a small society with all of the necessary support functions to keep personnel comfortable at sea. In the North Sea, for example, personnel are transported in by helicopter, in for a 2-week shift. Supplies and waste are transported by ship and need to be delicately choreographed because floor area on the platform is limited. Much effort goes into moving as much of the personnel as possible onshore, where management and technical experts are in touch with the platform by video conferencing. The increased use of subsea facilities of course goes hand in hand with the goal of moving people onshore.

However, offshore oil production involves environmental risks (Chapter 11)—in particular oil spills from (1) oil tankers transporting oil from the platform to onshore facilities, and (2) pipelines doing the same and leaks and accidents on the platform. There is also the impact of produced water (water from well drilling or production) that contains varying amounts of oil and other chemicals used in or resulting from oil production. The platform is typically allowed a quota of produced water that can be returned to the ocean. The platforms themselves also present environmental issues when decommissioned.

The most common type of platform is the steel-jacket platform, which is suitable for a wide range of oceanic conditions, which vary from the relatively calm waters off the west coast of Africa to the extremely unpredictable and wild North Sea. Steel-jacket platforms are constructed from welded steel pipes and can Steel jackets can weigh up to 20,000 tons, and support similar weights—they are assembled onshore and floated out to sea when ready for use are at depths to approximately 500 ft and may support structures that are 100 ft above the surface of the ocean. Once in place, they are pinned to the seafloor with steel pipes (piles).

Concrete and steel gravity platforms may be used in similar water depths as steel jacketed platforms—the principal difference between steel-jacketed units and gravity platforms is that the gravity unit relies on the weight (of the unit) to secure it to the seabed, which eliminates the need for driving piles supports into hard sea beds. In addition, concrete units are constructed with hollow ballast tanks and hollow concrete legs, which allow the platform support structures to be floated into position without the use of a barge, and then sunk in place at the designated location. The caissons of these platforms may be used as temporary storage for crude oil, which (on some platforms) may be on the order of 1 million barrels.

On the other hand, floating platforms (such as tension leg platforms, SPAR platforms, and semi-submersible units) are typically used in deep waters where fixed platforms would be vulnerable or very expensive. These types of platforms may also be referred to a deep-water floating platforms or deep-water floaters.

REFERENCES

Austin, D., Carriker, R., McGuire, T., Pratt, J., Priest, T., Pulsipher, A.G., 2004. History of the Offshore Crude oil Natural Gas Industry in Southern Louisiana Interim Report Volume I: Papers on the Evolving Offshore Industry OCS Study MMS 2004-049. Contract 1435-01-02-CA-85169. Minerals Management Service, Gulf of Mexico OCS Region, United States Department of the Interior, New Orleans, Louisiana.

Bailey, N.J.L., Evans, C.R., Milner, C.W.D., 1974. Applying petroleum geochemistry to search for oil: Examples from Western Canada Basin. Bulletin American Association of Petroleum Geologists 58, 2284–2294.

Bergreen, L., 2007. Marco Polo: From Venice to Xanadu. Quercus Books, London, united Kingdom.

Biju-Duval, B., 2002. Sedimentary Geology: Sedimentary Basins, Depositional Environments, Petroleum Formation. Editions Technip, Paris, France.

Burgan, M., 2002. Marco Polo: Marco Polo and the Silk Road to China. Mankato: Compass Point Books, North Mankato, Minnesota.

Calvert, S.E., Pedersen, T.F., 1993. Geochemistry of recent oxic and anoxic marine sediments: Implications for the geological record. Marine Geology 113 (1–2), 67–88.

Cobb, C., Goldwhite, H., 1995. Creations of Fire: Chemistry's Lively History from Alchemy to the Atomic Age. Plenum Press, New York.

Floyd, P.A., 1991. Oceanic Basalts. Blackie Publishers, Glasgow, United Kingdom.

Forbes, R.J., 1958a. A History of Technology. Oxford University Press, Oxford, United Kingdom.

Forbes, R.J., 1958b. Studies in Early Petroleum Chemistry. E. J. Brill, Leiden, Netherlands.

Forbes, R.J., 1959. More Studies in Early Petroleum Chemistry. E. J. Brill, Leiden, Netherlands.

Forbes, R.J., 1964. Studies in Ancient Technology. E. J. Brill, Leiden, Netherlands.

Gary, J.G., Handwerk, G.E., Kaiser, M.J., 2007. Petroleum Refining: Technology and Economics, 5th Edition CRC Press, Taylor & Francis Group, Boca Raton, Florida.

Hsu, C.S., Robinson, P.R. (Eds.), 2006. Practical Advances in Petroleum Processing Volume 1 and Volume 2. Springer Science, New York.

Huerta, A.D., Harry, D.L., 2012. Wilson cycles, tectonic inheritance, and rifting of the North American Gulf of Mexico continental margin. Geosphere 8 (2), 374–385.

Hunt, J.M., 1996. Petroleum Geochemistry and Geology, 2nd Edition W.H. Freeman and Co., New York.

James, P., Thorpe, N., 1994. Ancient Inventions, Reprint Edition Ballantine Books, New York.

Kastner, M., 1999. Oceanic minerals: Their origin, nature of their environment, and significance. Proceedings of the National Academy of Science of the United States of America 96 (7), 3380–3387.

Kemp, A., 2012a. The Official History of North Sea Crude Oil and Natural Gas. Volume I: The Growing Dominance of the State. Routledge Publishers, Taylor & Francis Group, New York.

Kemp, A., 2012b. The Official History of North Sea Crude Oil Natural Gas. Volume II: Moderating the State's Role. Routledge Publishers, Taylor & Francis Group, New York.

Mason, J.R., 2009. The Economic Contribution of Increased Offshore Oil Exploration and Production to Regional and National Economies. Hermann Moyse Jr., Louisiana Bankers Association Endowed Chair of Banking, Louisiana State University, E.J. Ourso College of Business for the American Energy Alliance, Washington, DC.

Leffler, W.L., Pattarozzi, R., Sterling, G., 2011. Deepwater Petroleum Exploration and Production: A Nontechnical Guide, 2nd Edition PennWell Corporation, Tulsa, Oklahoma, pp. 1–8.

Littlehales, G.W., 1930. The Configuration of the Oceanic Basins. Graficas Reunidas, Madrid, Spain.

Mokhatab, S., Poe, W.A., Speight, J.G., 2006. Handbook of Natural Gas Transmission and Processing. Elsevier B.V, Amsterdam, Netherlands.

Morbey, S.J., 1996. Offshore Brazil: Analysis of a successful strategy for reserve and production growth. In: Doré, A.G., Sinding-Larsen, R. (Eds.), Quantification and Prediction of Petroleum Resources, Norwegian Petroleum Society (NPF) Special Publication No. 6. Elsevier B.V., Amsterdam, Netherlands, pp. 123–133.

Parkash, S., 2003. Refining Processes Handbook. Gulf Professional Publishing Company, Elsevier Publishing Company, Burlington, Massachusetts.

Robinson, A.G., 1993. Inorganic Geochemistry: Applications to Petroleum Geology. Wiley-Blackwell, John Wiley & Sons Inc, Hoboken, New Jersey.

Shimoyama, A., Johns, W.D., 1971. Catalytic conversion of fatty acids to petroleum-like paraffins and their maturation. Nature (London, United Kingdom) 232, 140–144.

Snowdon, L.R., Powell, T.G., 1982. Immature oil and condensates modification of hydrocarbons generation model for terrestrial organic matter. Bulletin American Association of Petroleum Geologists 66, 775–788.

Speight, J.G., Ozum, B., 2002. Petroleum Refining Processes. Marcel Dekker Inc., New York.

Speight, J.G., 2007. Natural Gas: A Basic Handbook. GPC Books, Gulf Publishing Company, Houston, Texas.

Speight, J.G., 2011. The Refinery of the Future. Gulf Professional Publishing, Elsevier Publishing Company, Oxford, United Kingdom.

Speight, J.G., 2014. The Chemistry and Technology of Petroleum, 5th Edition CRC Press, Taylor & Francis Group, Boca Raton, Florida.

Tissot, B.P., Welte, D.H., 1978. Petroleum Formation and Occurrence. Springer-Verlag, New York.

Verzhbitsky, E.V., Kononov, M.V., Kotelkin, V.D., 2007. Plate tectonics of the northern part of the Pacific Ocean. Oceanology (Okeanologiya) 47 (5), 705–717.

Welte, D.H., Horsfield, B., Baker, D.R., 2011. Petroleum and Basin Evolution: Insights from Petroleum Geochemistry, Geology and Basin Modeling. Springer Publishing Company, New York, Reprint of 1st (1997) Edition.

Wooster, R., Sanders, C.M., 2010. Spindletop Oilfield. Texas State Historical Association, Denton, Texas.

CHAPTER THREE

Offshore Platforms

Outline

3.1 INTRODUCTION

The offshore exploration and production of crude oil and natural gas has advanced far beyond the 100-by-300-ft platform secured on a foundation of timber piles that served as the base of an early offshore well that was drilled in the Gulf of Mexico in 1938 (Chapter 2) (Pratt et al., 1997; Chakrabarti et al., 2005; Gudmestad et al., 2010; Hunter, 2011). However, the construction of offshore structures is an expensive proposition and mobile exploratory drilling rigs are used to drill wells to determine the presence or absence (dry hole) of crude oil and natural gas at the offshore site. If crude oil and/or natural gas are/is present in sufficient quantity, then the well is plugged until a permanent production platform is installed.

Handbook of Offshore Oil and Gas Operations. http://dx.doi.org/10.1016/B978-1-85617-558-6.00003-9

Modern offshore crude oil and natural gas exploration—the search for likely environments where crude oil and natural gas may exist in the rock formations that are beneath the surface of the waterways of the world. In addition, offshore operations include transporting crude oil and natural gas from their point of production offshore to refineries and plants on land. Very little refining of the crude oil and natural gas is carried out on the production platform—the level of processing is restricted to preparation of the crude oil for transportation. The natural gas may also be transported but is often reinjected into the reservoir to maintain the reservoir energy and stimulate the production of more crude oil.

Exploration, drilling, and production of crude oil and natural gas involve a wide range of technologies, which are similar in many cases to those used land-based exploration and production. However, offshore activities need to include additional technologies that relate to a marine environment. Unlike crude oil and natural gas operations on land, offshore operations involve meteorology, naval architecture, mooring, and anchoring techniques, as well as buoyancy, stability, and trim of the vessel or platform.

3.2 DRILLING OPERATIONS AND DRILLING PLATFORMS

A whole range of different structures are used offshore, depending on size and water depth and which require significant construction expertise and outlay (El-Reedy, 2012). In the last few years, we have seen pure sea bottom installations with multiphase piping to shore and no offshore top-side structure at all. Replacing outlying wellhead towers, deviation drilling is used to reach different parts of the reservoir from a few wellhead cluster locations.

In general terms, offshore drilling operations generally match the onshore operations but with changes that are adequate to meet the conditions of the offshore locale. Thus, offshore wells are drilled by lowering a drill string (consisting of a drill bit, drill collar, and drill pipe) through a conduit (riser) that extends from the drill rig to the sea floor (Sparks, 2007). The drill string consists of steel drill pipe sections—each section is typically 30 ft long and weighs approximately 600 lb. As the well deepens, additional sections of drill pipe are inserted into the string, which is lowered through the riser to the sea floor, where it passes through a system of safety valves (the blowout preventer), which contains the pressure within the well and to prevents a blowout as well as the potential for the ever-wasteful and environmentally disastrous-and hazardous *gusher*.

At the water surface (onboard the platform or drilling unit), a rotary table turns the drill string; and drilling fluid (drilling mud) is pumped into the drill pipe from a mud tank. As the drill bit teeth penetrate the sea floor sediment and rock formations, the mid flows through small holes in the drill bit and collects the cuttings of rock cut (drill cutting) released from the rock by the drill bit and carries the cuttings to the surface.

The drilling mud serves a purpose of well control as well as cleanliness—the weight of the mud exerts a pressure greater than the pressure in the rock formations, and therefore, maintains control of the well. As the bit penetrates further into the rock formations, strings of steel pipe casing are inserted into the well and then cemented in place in order to seal off the walls of the well and prevent collapse of the well. On occasion (or if necessary), instruments may be sent down a wireline into the well to determine the existence of crude oil or gas. If crude oil and natural gas are present, steel production casing is set in place. This production casing is used as the conduit for bringing crude oil and natural gas safely to the surface.

Briefly and by way of explanation, in the crude oil and natural gas industry, the term *wireline* refers to a cable technology used by operators of crude oil and natural gas wells to lower equipment or measuring devices into the well for the purposes of well intervention, reservoir evaluation, and pipe recovery. In addition, *well intervention* (well work) is any operation carried out on a crude oil or natural gas well during or at the end of its productive life, which alters the state of the well and/or well geometry, provides well diagnostics, or manages the production of the well.

The drilling mud program is a *return-journey program* in which the mud flows to the surface through the annulus between the well casing and the drill string below the sea bottom (the mud line) and the riser and the drill string above the mud line. A filter-strainer at the surface removes the cuttings from the drilling mud and the mud is then recirculated through the mud tank and pumped once more into the drill string. Discharge of these fluids and cuttings into the oceans of the world is governed by various environmental regulations—in the United States, the discharge of the cutting is administered and regulated by the US Environmental Protection Agency (Chapter 11, Chapter 12).

There are two types of offshore drilling rigs: (1) those movable from place to place, allowing for drilling in multiple locations, and (2) those that are permanently placed. Moveable rigs are often used for exploratory purposes because they are much cheaper to use than permanent platforms. Once large deposits of petroleum and natural gas have been found, a permanent platform is built to allow their extraction.

Some of the desirable characteristics applicable to exploratory drilling units, such as limited structure motions and good station-keeping characteristics in relatively severe environment are equally applicable to crude oil and natural gas production units. Some drilling platforms are capable of being used for drilling and producing crude oil and natural gas. However, whatever the purpose, the modern offshore should be a large structure that is capable of carrying the machinery necessary to drill for crude oil and natural gas through wells in the ocean bed and housing the workers (ABS, 2014). Depending on the circumstances, the drilling platform may be attached to the ocean floor, consist of an artificial island, or be floating.

The fluids from the well contain a mixture of oil, gas, sand, and brine water. This mixture is processed on the platform for (at least) water and solids so that the fluid meets pipeline specification for fluid transportation; the well fluid may also be transported to shore by a tanker. In terms of offshore on-platform refining, the way lies open for the future installation of simple visbreaking equipment to prepare heavier oils to meet the specifications for transportation by pipeline (Bai and Bai, 2005).

Mobile offshore drilling units (MODUs) must accommodate highly variable deck loads (VDLs) due to the different drilling requirements and are usually designed for relatively high transit speeds to minimize the costs of moving to another site.

The most common forms of drilling structures (alphabetically) are (1) drilling barges and drillships, (2) jackup platforms, (3) semisubmersible platforms, and (4) submersible units. Drillships are ship shaped and self-propelled and sufficiently large and stable to accommodate the drilling equipment on board. Drillships also have the advantage of rapid transit between stations and can take up and leave stations quickly, especially if they are dynamically positioned instead of being moored in place. However, the large motions and thruster (or anchor) capacity limit the weather conditions in which drilling operations can occur. The semisubmersible gravity unit is suitable for use in shallow water and these units have buoyant legs and pontoons and are set on seafloor by ballasting, thus allowing the structure to be deballasted, and moved to another location.

3.2.1 Drilling Barges and Drillships

A *drilling barge* is a large, floating platform, which must be towed by tugboat from location to location and are used mostly for onshore, shallow water drilling, which typically occurs in lakes, swamps, rivers, and canals. Because of the design and suitability for use in relatively calm shallow water ways, drilling barges are not built to withstand the water movement experienced in large open water (oceanic or rough sea) situations.

A *drillship* is a maritime vessel that has been equipped with drilling apparatus and is often built on a modified tanker hull and outfitted with a dynamic positioning system to maintain its position over the well. This type of vessel is capable of operating in deep water and typically carries larger payloads than semisubmersible drilling vessels. A drillship must stay relatively stationary on location in the water for extended periods of time—the positioning may be accomplished with multiple anchors, dynamic propulsion (thrusters), or a combination of these. In addition, the ship-shaped design gives the unit greater mobility and can move quickly under its own propulsion from drill site to drill site in contrast to jackup platforms and semisubmersible platforms, which require towing. Drillships typically carry larger payloads than semisubmersible drilling vessels and the vessel must stay relatively stationary on location in the water for extended periods of time.

A drillship is most often used for exploratory drilling of new oil or gas wells in deep water but can also be used for scientific drilling. A drillship is also employed for deep-water drilling and has been used to drill wells at water depths in excess of 5,000 ft. In Arctic regions, wells have been drilled from artificial islands constructed of material dredged from the ocean floor, earth-filled steel structures, ice-strengthened conventional drillships, and even from ice platforms. In addition, the drillship can be used as a platform to carry out well maintenance or completion work such as casing and tubing installation, subsea tree installations, and well capping. Some of the modern drillships have larger derricks that allow dual activity operations, for example, simultaneous drilling and casing handling.

All drillships have what is called a *moon pool*, which is an opening on the base of the hull and depending on the mission the vessel is on, drilling equipment, small submersible crafts, and divers may pass through the moon pool. Since the drillship is also a vessel, it can easily relocate to any desired location but, due to the mobility, drillships are not as stable compared with semisubmersible platforms. To maintain its position, drillships may utilize their anchors or use the ship's computer-controlled system on board to run off their dynamic positioning.

3.2.2 Jackup Rig

A jackup rig (sometime referred to as a jackup barge) are similar to a drilling barge, with one difference—when the jackup rig is towed to the drilling site, three or four legs (that can be raised or lowered independently of each other) are lowered until they rest on the sea bottom. This allows the working platform to rest above the surface of the water, as opposed to a floating barge. However, jackup rigs are suitable for shallower waters, as extending these legs down too deeply would be impractical. These rigs are typically safer to operate than drilling barges, as their working platform is elevated above the water level.

The jackup is used in relatively low depths, are designed to move from place to place, and then anchored by deploying the jack-like legs. A jackup unit is self-elevating—the legs are stationed on ocean floor and the drilling equipment is jacked up above the water's surface. The jackup platform provides a very stable drilling environment, in comparison with other types of offshore drilling rigs, and can drill in waters up to 350 ft deep. However, when drilling in deeper water is required, semisubmersibles and drillships become a more logical choice for exploration and development operations.

The first jackup platform was constructed in 1954 and, since that time, jackup platforms have become the most popular type of MODU for offshore exploration and development purposes. The jackup design can utilize three legs or four legs and there are two main types of legs (1) open-truss legs that resemble electrical towers, which are fabricated from tubular steel sections that are crisscrossed, making them strong and light weight, and (2) columnar legs, which are fabricated from steel tubes. Columnar legs are less expensive than open-truss legs to fabricate but may be less stable and cannot adapt to stresses in the water as well as open-truss legs and, hence, columnar-leg jackup platforms are not usually recommended for use in water with a depth in excess of 250 ft.

Upon arrival at the drilling location, the legs of the unit (preloaded to securely drive them into the sea bottom,) are jacked down on to the seafloor. Since the legs have been preloaded and will not penetrate the seafloor further, continued jacking down of the legs once the sea floor has been reached has the effect of raising the jacking mechanism, which is attached to the barge and drilling package. In this manner, the entire barge and drilling structure are slowly raised above the water to a predetermined height above the water, so that wave, tidal, and current loading acts only on the relatively small legs and not the bulky barge and drilling package.

Jackup platforms are a more logical choice for drilling environments that have soft floors and mat-supported jackup platforms are used to distribute the weight of the rig across the sea floor. The A-shaped mat supports are connected to the bottom of each leg of the jackup, ensuring that the rig does not penetrate the bottom of the ocean. Alternatively, spud cans (cylindrically shaped steel shoes with pointed ends) are used on independent-legged jackup platforms and are attached to the bottom of each leg, and the spike in the can is driven into the ocean floor, adding stability to the rig during operations.

In addition, jackup platforms have two types of elevating devices—when the jackup is on location, the legs are lowered to the ocean floor and the rig hull and drilling equipment is elevated well above the water's surface and away from any potential waves. The first type of elevating device uses hydraulic cylinders equipped with moving and stationary pins—the cylinders extend and retract to climb up and down the legs of the jackup. The other type of elevating device employs a rack and two pinion gears that are turned to move the legs up and down. Whichever type of leg that is used in the design, the legs of the jackup rise up through holes in the hull of the drilling rig—a deck is used to support the drilling derrick and other equipment.

There drilling equipment is mounted on the hull—the most popular design for drilling equipment is a cantilevered jackup in which the drilling derrick is mounted on an arm that extends outward from the drilling deck. In this way, drilling can be performed through existing platforms, as well as without them. Because of the range of motion that the cantilever provides, most modern jackup platforms have cantilevered jackup platforms. The other type of jackup platform is the slot-type jackup (also known as a *keyway jackup*). Drilling slot jackup platforms are built with an opening in the drilling deck, and the derrick is positioned over it. While exploration wells can be drilled with drilling slot jackups, the rig can also be jacked up over another smaller facility, drilling through its hull.

The primary advantage of the jackup platform design is that it offers a steady and relatively motion-free platform in the drilling position and mobilizes relatively quickly and easily. Although they originally were designed to operate in very shallow water, some newer units, such as the *ultra-harsh environment* Maersk MSC C170-150 MC can be operated in water 550 ft deep and the unit has 673.4-ft leg length, a hull dimension of $291 \times 336 \times 39$ ft^3, and a VDL up to 10,000 tons (www.GustoMSC.com).

In addition, when the support legs are not deployed, a jackup platform can float, which makes this type of MODU quite readily transportable from one drilling location to another. While units some are capable of self-propulsion and do not need an outside source for movement, most jackup platforms are transported via tug boats or submersible barges.

3.2.3 Semisubmersible Rig

The semisubmersible rig is a floating platform that is supported primarily on large pontoon-like structures submerged below the sea surface—the operating decks are elevated 100 or more feet above the pontoons on large steel columns. Semisubmersible units can operate in a wide range of water depths, including deep water and are usually anchored with six to 12 anchor chains, which are computer controlled to maintain the placement of the unit. A typical design has four columns connected at the bottom by pontoon with a nominally rectangular cross-section. A truss structure connects the column tops and supports topsides modules. The platform can be used in depths from 600 to 6,000 ft.

Semisubmersible platforms offer a number of benefits, including large payload capacity, limited sensitivity to water depth, quayside integration, and the ability to relocate the platform after field abandonment. However, drillships are capable of holding more equipment; but semisubmersibles are typically chosen for their stability. The design concept of partially submerging the rig lessens both the rolling and the pitching on semisubmersible platforms in heavy seas. In transit to the site (usually by tugboats or barges or even under their own power), semisubmersible platforms are not lowered into the water and it is only during drilling operations that semisubmersible platforms partially submerged.

The semisubmersible platform is designated as being column stabilized and the columns primarily provide floatation stability. Variable in the design include items such as column dimensions, column spacing, pontoon size, the ratio of pontoon width to pontoon height, and draft of the hull. The columns are sized to provide adequate water plane area to support all anticipated loading conditions and are spaced to support topside modules. The columns are supported by two parallel pontoons or a ring pontoon, which are of a size that will provide adequate buoyancy to support all weights and vertical loads, and proportioned to maximize heave damping.

There are two main types of semisubmersible platforms that are based on the manner in which the unit is submerged in the water: (1): bottle-type semisubmersible platforms and (2) column-stabilized semisubmersible platforms. Bottle-type semisubmersible platforms consist of bottle-shaped hulls below the drilling deck that can be submerged by filling the hulls with water. The bottle-type unit offers the stability required for drilling operations—rolling and pitching from waves and wind is greatly diminished. In addition to occasional weather threats, such as storms, hurricanes (so-called in the Atlantic Ocean), cyclones or typhoons (so-called in the Pacific Ocean), some drilling locations are always harsh with constant rough waters. Being able to drill in deeper and rougher waters, semisubmersible platforms opened up a new avenue for exploration and development operations.

A more popular design for semisubmersible platforms is the column-stabilized model in which two horizontal hulls are connected by means of cylindrical or rectangular columns to the drilling deck above the water. Smaller diagonal columns are used to support the structure and submerging this type of semisubmersible platforms is achieved by controlled (partially) filling the horizontal hulls with water until the rig is submerged to the desired depth. Mooring lines anchor the rig above the well, and dynamic positioning uses different motors or propulsion units on the vessel to counteract against the motions of the water and can be used in addition to mooring lines to keep the rig in place.

Semisubmersible platforms and other mobile offshore facilities are moored in systematic ways, but there are many different designs according to the location of the platform. Mooring is similar to multiple anchors, and a number of spread mooring patterns are used to keep the floating rig in place, including symmetric six-line, symmetric eight-line, symmetric nine-line, and symmetric 12-line patterns. These mooring spreads are chosen depending on the shape of the vessel being moored and the sea conditions in which it will be moored.

Since the wellbore is extremely precise, it is very important that the semisubmersible platform is kept in position, despite the buffeting from wind and waves. Furthermore, working in ultra-deep water, the drilling riser pipe may span thousands of feet from the semisubmersible platform to the stationary subsea well equipment located on the ocean floor. However, the drilling equipment does require some flexibility to overcome slight movements caused by the wind and waves, but the drilling risers must not be bent beyond the specification or rupture and release of crude oil and natural gas will occur.

Semisubmersible rigs are the most common type of offshore drilling rigs, combining the advantages of submersible rigs with the ability to drill in deep water. A semisubmersible rig works on the same principle as a submersible rig: through the *inflating* and *deflating* of its lower hull. The main difference with a semisubmersible rig, however, is that when the air is let out of the lower hull, the rig does not submerge to the sea floor. Instead, the rig is partially submerged, but still floats above the drill site. When drilling, the lower hull, filled with water, provides stability to the rig. Semisubmersible rigs are held in place by huge anchors, each weighing upwards of 10 tons. These anchors, combined with the submerged portion of the rig, ensure that the platform is stable and safe enough to be used in turbulent offshore waters. Semisubmersible rigs can be used to drill in much deeper water than the rigs mentioned above.

The Thunder Horse project (a joint of BP and ExxonMobil) use a semi-submersible platform on location over the Mississippi Canyon Thunder Horse oil field (Block 778/822), in deepwater Gulf of Mexico, 150 miles southeast of New Orleans and is moored in water having a depth of 6,040 ft. The Cheviot production platform is another recent example of semisubmersible drilling, production, and storage facility being fabricated for the Cheviot offshore field (UK North Sea). The platform can be flexibly used for dry-tree completions as well as steel catenary risers (SCRs).

3.2.4 Submersible Rig

Submersible rigs, also suitable for shallow water, are like jackup rigs as they come in contact with the ocean or lake floor. These rigs consist of platforms with two hulls positioned on top of one another. The upper hull contains the living quarters for the crew, as well as the actual drilling platform. The lower hull works much like the outer hull in a submarine—when the platform is being moved from one place to another, the lower hull is filled with air—making the entire rig buoyant. When the rig is positioned over the drill site, the air is let out of the lower hull, and the rig submerses to the sea or lake floor. This type of rig has the advantage of mobility in the water but the use is usually limited to shallow water areas.

3.3 OFFSHORE PRODUCTION PLATFORMS

An oil production platform is a large structure used to house workers and machinery needed to drill and then produce oil and natural gas wells in the ocean. In contrast to a drilling platform, a production platform is required to stay on station during the lifetime of the platform or the project—typically 20–30 years.

Progress in offshore technology is exemplified by advances in the technology and size of production platforms, which provide a base for production operations. For many years, the standard method for offshore development was to utilize a fixed structure based on the sea bottom, such as an artificial island or man-made platform but use of such an approach in deep-water environments is hindered by technical difficulties and economic disadvantages that increase dramatically (almost logarithmically) with depth of the water. Crude oil and natural gas fields reside in shallow water and also in deep water in various locales around the world—perhaps the best known examples of the many currently operating examples are the exploration and recovery activities in the Gulf of Mexico, the North Sea, and the Niger Delta.

Generally, oil platforms are located on the continental shelf, though as technology has improved, drilling and production in deeper waters have become feasible and economically viable. A typical platform may have around 30 wellheads located on the platform and directional drilling allows reservoirs to be accessed at both different depths and at remote positions up to 5 miles from the platform. Many platforms also have remote wellheads attached by pipeline connections (sometime called umbilical pipelines), which may be single wells or a collection of multiple wells.

Depending on the circumstances, the platform may be attached to the ocean floor, consist of an artificial island, or be floating (Chakrabarti et al., 2005). Generally, oil platforms are located on the continental shelf, though as technology improves, drilling and production in deeper waters become both feasible and profitable. A typical platform may have around 30 wellheads located on the platform and directional drilling allows reservoirs to be accessed at both different depths and at remote positions up to 5 miles (8 km) from the platform. Many platforms also have remote wellheads attached by umbilical connections—depending upon the need or design, these may be single wells or a manifold center for multiple wells.

Floating production platforms must be moored to maintain their position over the crude oil and/or natural gas reservoir and, more specifically, over the riser or the well that carries the crude oil and natural gas to the platform. Several types of mooring systems are employed, including (1) multi-leg catenary spread mooring, (2) multi-leg semi-taut and taut mooring, (3) multiple tension leg tendon mooring, (4) dynamic positioning system, (5) single anchor leg mooring (SALM), (6) catenary anchor leg mooring, and (7) external and internal turret mooring. The materials used in such systems include chain, wire rope, fiber rope, buoys, tendons, and the necessary connectors: shackles, thimbles, and swivels. The floating platform (such as the compliant tower (CT) platform, the semisubmersible platform, and the SPAR platform) typically has winches, chain jacks, and chain stoppers that are used to adjust the tension and scope of the mooring line. An anchor unit (such as a drag embedment, driven pile, or suction pile) is connected to the mooring line to hold the floating platform in position (API, 1994, 2005, 2011). As an initial part of the procedure, the loading for the floating platform must be determined and the data are used to evaluate the movement of the platform and the required size of the mooring components to moor the vessel within acceptable limits of movement due to ocean waves and variable winds.

An oil production platform must be self-sufficient in energy and water needs, housing electrical generation, water purifiers (desalinators), and all of the equipment necessary to process crude oil and natural gas for delivery onshore by pipeline or to a floating storage unit (FSU) and/or tanker loading facility. Elements in the crude oil and natural gas production process include items such as the wellhead, production manifold, production separator, glycol process equipment to dry gas, gas compressors, water injection pumps, and equipment for oil/gas export metering, and mail oil line pumps. In addition, all production facilities must be designed to exert minimal environmental impact. The larger platforms are assisted by smaller emergency support vessels (ESVs) that necessary in the event of an accident and, for day-to-day operations, platform support vessels (PSVs) which keep the platforms provisioned and supplied, and anchor handling tug supply (AHTS) vessels can also supply the platforms as well as tow them to location and serve as standby rescue and firefighting vessels.

Larger lake-based and sea-based oil platforms and oil rigs are among the largest moveable structures in the world and there are distinct types of platforms and rigs (Figure 3.1): (1) CT, (2) Condeep platform, (3) fixed platform (FP) (4) floating production system (FPS), (5) SPAR platform, (6) subsea system, (7) tension-leg platform (TLP), and (8) other installations.

3.3.1 Compliant Tower

A CT consists of a narrow, flexible tower and a piled foundation that can support a conventional deck for drilling and production operations. The

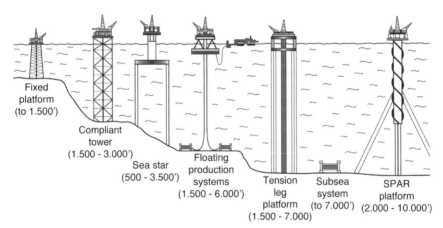

Figure 3.1 Offshore production systems. *Source: Energy Information Administration, US Department of Energy, Washington, DC.*

use of flex elements such as flex legs or axial tubes, resonance is reduced and wave forces are deamplified. This type of rig structure can be configured to adapt to existing fabrication and installation equipment. Compared with floating systems, such as TLP and SPAR platforms, the production risers are conventional and are subjected to less structural demands and flexing. However, because of cost, it becomes uneconomical to build CTs in depths greater than 3000 ft. In such a case, a FPS is more appropriate, even with the increased cost of risers and mooring. Thus, CTs are designed to sustain significant lateral deflections and forces, and are typically used in water depths ranging from 1,500 to 3,000 ft. The CT design is more flexible (hence the name) than conventional land structures to cope better with sea forces—it can deflect (sway) in excess of 2% of height. Despite its flexibility, the CT system is sufficiently sturdy to withstand hurricane conditions.

CTs are similar to FPs as they have a steel tubular jacket that is used to support the surface facilities and, like a FP, the compliant tower is secured to the seafloor with piles. The jacket of a CT has smaller dimensions than those of a FP and may consist of two or more sections—there may also be buoyant sections in the upper jacket with mooring lines from jacket to seafloor (guyed-tower designs) or a combination of the two. The water depth at the intended location dictates platform height (Figure 3.2).

Figure 3.2 Concrete gravity platform.

Once the lower jacket is secured to the seafloor, it acts as a base (CT) for the upper jacket and surface facilities. Large barge-mounted cranes position and secure the jacket and install the surface facility modules. These differences allow the use of CTs in water depths ranging up to 3,000 ft, which is generally considered to be beyond the economic limit for fixed jacket-type platforms. The portion of the tower that contains the production and crew quarter modules is the surface facility. Individually, size is dictated by the dimensions needed to handle production operations, and crew accommodations (ABS, 2014). The surface facilities are smaller by design on CTs than on FPs because of the decreased jacket dimensions that support them.

The Chevron Petronius platform (a deep water CT) is a crude oil and natural gas platform in the Gulf of Mexico, which stands 2,000 ft above the ocean floor and the structure is partially supported by buoyancy. The multideck topsides are approximately 210-by-140 ft and hold 21 well slots, and the entire structure weighs around 43,000 tons. The platform is situated to exploit the Petronius field, discovered in 1995 in Viosca Knoll (block VK 786 and the seabed is 1754 ft below the platform. Approximately 60,000 barrels of oil and 100 million cubic feet of natural gas are produced daily by the Petronius platform.

Another compliant structure is the guyed tower, which is a slender truss-steel structure supported on the sea floor by a spud-can foundation and held upright by multiple wire or chain guy lines. The guy lines connect to anchor piles and are equipped with heavy clump weights between the anchor and tower. The guy wires also restrain the platform motion during typical operating weather conditions without lifting the clump weights off the bottom. During more extreme weather conditions, the guy wires are designed to lift the clump weights off the bottom, and the clump weights create a larger restoring force to resist the larger wave forces.

The Lena guyed tower was installed in 1983 in the Gulf of Mexico with a steel weight of 24,000 tons and deck weight of 19,640 tons. The tower was placed in 1000 ft of water and utilized 20 galvanized spiral wound steel wire ropes as guy lines with 179-ton clump weights placed in a symmetrical pattern around the platform.

For the development of small reservoirs below water at a water depth of approximately 650 ft, an articulated column a single anchor leg storage and tanker system can be used in relatively calm weather areas. The SALM system uses a yoke structure, buoyancy tank, and tensioned riser. Crude oil and

natural gas are moved up the articulated tower and transferred to the tethered tanker for processing and storage. A shuttle tanker is brought alongside
to receive the processed crude oil and transport it to shore by means of a
shuttle tanker or an export oil pipeline.

3.3.2 Condeep Platform

The Condeep platform (which is no longer manufactured) is the name
used for a series of platforms that were developed in Norway for crude oil
and natural gas in the North Sea and the Norwegian continental shelf. The
name is a derivation of the term *concrete deep water structure* and consists of a
base of concrete oil storage tanks from which one, three, or four concrete
shafts rise approximately 100 ft above sea level. The platform deck itself is
not a part of the construction. The Condeep platform is used for a series of
production platforms introduced for crude oil and natural gas production
in the North Sea.

This gravity-based structure (GBS) for platform was unique as it was
built from reinforced concrete instead of steel, which was the norm up to
that point. This platform type was designed for the heavy weather conditions and the great water depths often found in the North Sea. The platform
has the advantage that it allows for storage of crude oil and natural gas at sea
in its own construction. It further allows for equipment installation in the
hollow legs well protected from the sea. In contrary, one of the challenges
with the steel platforms is that they only allows for limited weight on the
deck—with a Condeep, the weight allowance for production equipment
and living quarters is seldom a problem.

The advantage of such a structure is the ability to act as an oil storage
unit. In addition, construction and testing can be completed before floating
the structure and towing it to an offshore location. This type of structure
was tolerant to overloading and degradation resulting from exposure to sea
water than steel platforms. However, disadvantages included greater costs
compared with a similar steel structure—copious amounts of steel is often required for reinforcing the concrete members than is required for an
equivalent steel jacketed structure.

In the Condeep design Beryl A concrete gravity structure (North Sea;
installed in 1975), the concrete caisson at the base is used for oil storage.
Three water-piercing support towers (310 ft high) were used for drilling
and oil production conductors. Although concrete platforms are more expensive, they did offer the advantages of lower maintenance and higher
deck payload (Figure 3.3).

Figure 3.3 Tow boat. *Source: Figure reproduced from El-Reedy, 2012, fig. 5.35.*

3.3.3 Fixed Platform

A FP consists of a jacket (a tall vertical section made of tubular steel members supported by piles driven into the seabed) with a deck placed on top. The deck provides space for crew quarters, drilling rigs, and production facilities. The FP is economically feasible for installation in water depths up to about 1,650 ft. An example of a FP is the Bullwinkle platform (Shell Oil Company) in Green Canyon block 65 installed in mid-1988 with an increased capacity to handle production from the Troika prospect in Green Canyon Block 244, which began production in late 1997. The platform is built on concrete and/or steel legs anchored directly onto the seabed, supporting a deck with space for drilling rigs, production facilities and crew quarters. Such platforms are, by virtue of their immobility, designed for very long term use (for instance the Hibernia platform off the coast of Canada). Typically, a FP is economically feasible for installation in water at up to approximately 1,700 ft.

The Hibernia platform, which stands approximately 730 ft high, is located on the Jeanne D'Arc basin, in the Atlantic Ocean off the coast of Newfoundland. The GBS, which sits on the ocean floor, is 364 ft high and has storage capacity for 1.3 million barrels of crude oil in its 278.8 ft high caisson (Dorel Iosif). The platform acts as a small concrete island with serrated outer edges designed to withstand the impact of an iceberg. The GBS contains production storage tanks and the remainder of the void space is filled with ballast with the entire structure weighing in at 1.2 million tons.

Briefly, gravity-based platforms take advantage of the large size and heavy mass to support large facilities in water depths of up to 1,000 ft. The platforms can also be designed to resist severe arctic conditions, such as multi-year ice and even icebergs in shallow waters and in depths of up to around 200 ft. GBSs can be made of steel or concrete, and provide support for heavy drilling rigs and production equipment. They function similarly to gravel islands and jacket structures, but can be used in deeper water than gravel islands and can resist ice much better than jacket structures. They effectively act as steel or concrete islands.

In early 2000, BP Exploration (Alaska) Inc. successfully enlarged Seal Island, a man-made gravel island offshore Alaskan Beaufort Sea, and installed the Northstar pipeline system, the first trenched subsea oil pipeline in the Arctic. Development drilling began in December 2000 with planned production to start in fourth quarter 2001 when crude oil will be transported to the Trans-Alaskan Pipeline System (TAPS) Pump Station No. 1 via the 10-in. OD offshore pipeline and an elevated overland pipeline section. A second 10-in. diameter pipeline was used to transport natural gas from existing Prudhoe Bay facilities to Seal Island for fuel and will be used for future reservoir management purposes (Lanan et al., 2001).

The Northstar Project location is characterized by long, ice-covered winter seasons followed by short summer open water seasons, when the natural environment is most sensitive (USACE, 1999). As a result of the initial planning for the pipeline construction, it was concluded that winter ice-based installation methods were best suited to these conditions, which set the direction for pipeline permitting, design, and construction. Overland pipelines on the Alaskan North Slope have typically been constructed during the winter from temporary ice roads that were constructed by spreading water on the tundra to protect the fragile vegetation. Thus, thee 11-mile overland section of the Northstar oil pipeline is supported 5 ft aboveground on conventional vertical support members to protect the tundra from melting and allow caribou passage. The light crude oil (42° API gravity) is first cooled on Seal Island to an average annual temperature of 10°C (50°F) before being pumped through the uninsulated subsea pipeline. The overland pipe has 2-in. thick polyurethane foam insulation with a galvanized metal jacket to minimize further heat loss in winter air temperatures that reach −45°C (−50°F) (Palmer et al., 1990; Nogueira et al., 2000).

Designing a structure to resist arctic ice requires a thorough understanding of ice forces that is gained through a combination of actual ice force measurements, ice model tests, and engineering analyses. The configuration of an appropriate platform will depend on the severity of the ice at the location. In mild ice conditions, the platform may have multiple columns but for severe ice conditions, the platform may be composed of a single, large column. In all cases, protection of the wells from the ice by containing them within the structure is essential and materials suitable for the Arctic conditions must be used to make sure that the structure can maintain its strength in the extreme cold. Broad foundations contain extra ballast weight for stability and cut into the seafloor to resist ice loading.

Various types of structure are used, steel jacket, concrete caisson, floating steel, and even floating concrete. Steel jackets are vertical sections made of tubular steel members, and are usually piled into the seabed. Concrete caisson structures, pioneered by the Condeep concept, often have in-built oil storage in tanks below the sea surface and these tanks were often used as a flotation capability, allowing them to be built close to shore (Norwegian fjords and Scottish firths—a firth is equivalent to a fjord and vice-versa—are popular because they are sheltered and deep enough) and then floated to their final position where they are sunk to the seabed. The concrete caisson structures often have in-built oil storage in tanks below the sea surface and these tanks were often used as a flotation capability, allowing them to be built close to shore (in locations such as the Norwegian fjords and Scottish firths are popular because they are sheltered and sufficiently deep) and then floated to their final position where they are sunk to the seabed.

In shallow waters, the most common type of production platforms is the fixed piled structure, commonly known as jackets. These are tubular structures fixed to the seafloor by means of driven or drilled and grouted piles and the economic water depth limit for FPs varies by environment. For example, in the North Sea, the deepest fixed jacket platform, the BP Magnus platform, is in 610 ft of water. The deepest concrete structure in the North Sea, the Shell Troll Gravity Base Structure, extends the fixed structure limit there to nearly 1000 ft water depth. In the Gulf of Mexico, the Shell Bullwinkle sits at a water depth of approximately 1350 ft of water. When the water depth exceeds these limits, CTs or floating production platforms are more functional and efficient. An example of a CT is the ChevronTexaco Petronius platform in 1750 ft water depth.

A fixed jacketed structure consists of a steel framed tubular structure that is attached to the sea bottom by piles, which are driven into the sea floor through pile guides (sleeves) on the outer members of the jacket. The topside structure consists of drilling equipment, production equipment, crew quarters, eating facilities, gas flare stacks, revolving cranes, survival craft, and a helicopter landing pad. Drill pipes and production pipes are brought up to topside through conductor guides within the jacket framing, and the crude oil and natural gas travel from the reservoir through the production riser to topside for processing prior to transport to an onshore storage facility or to a refinery. The detailed design of the frame varies widely and depends on the requirements of strength, fatigue, and launch procedure. The design life of the structure is typically 10–25 years. This is followed by the requirement to remove and dispose of the platform once the reservoir is depleted (Chapter 11).

Launching of these structures is usually accomplished with barges. Some structures are floated off the barge and righted using barge cranes, and others are designed with flotation to be self-righting. Since these structures are made of steel, the effects of corrosion must be considered because of the exposure to the ocean environment (Chapter 9, Chapter 10). Anodic and cathodic protection systems are employed and maintained to protect against corrosion of the structure.

3.3.4 Floating Production System

A FPS is typically a large ship equipped with processing facilities and moored to a location for a long period. The *floating production, storage, and offloading systems* (FPSOs) have been used in offshore production since the 1970s—typically in the North Sea, offshore Brazil, Asia Pacific, the Mediterranean Sea, and offshore West Africa (Angus, 2009). Furthermore, because these systems can be moved, they are a more economical solution for more marginal fields, as the vessel can be moved to another development and redeployed once the original field has been depleted. Also, the FPSOs are an optimal choice for development when there are no existing pipelines or infrastructure to transfer production to shore. Adding to the economic advantages of these systems, existing tankers are frequently converted into FPSOs (Pelletier, 2000a,b).

Floating production storage and offloading vessels are offshore production facilities that house both processing equipment and storage for produced hydrocarbons. The basic design of most FPSs involves a ship-shaped vessel, with processing equipment, or topsides, aboard the vessel's deck and hydrocarbon storage below in the double hull. After processing, the system stores oil or gas before offloading periodically to shuttle tankers or transmitting processed petroleum via pipelines.

A FPS consists of a semisubmersible that has drilling and production equipment. It has wire rope and chain connections to an anchor, or it can be dynamically positioned using rotating thrusters. Wellheads are on the ocean floor and connected to the surface deck with production risers designed to accommodate platform motion. The FPS can be used in water depths from 600 to 6,000 ft.

The FPS is moored in place by various mooring systems—a central mooring system allows the vessel to rotate freely to best respond to weather conditions, or weathervane, whereas spread-mooring systems anchor the vessel from various locations on the seafloor. The system is usually tied to multiple subsea production wells and gathers the oil and/or natural gas through a series of in-field pipelines. Once tapped by subsea wells, the crude oil and natural gas are transmitted through flow lines to risers, which transport the crude oil and natural gas from the seafloor to the vessel's turret and then to the FPS on the surface of the sea.

The processing equipment aboard the floating production and storage system is similar to that on a production platform and can consist of water separation, gas treatment, oil processing, water injection, and gas compression, among others (Figure 3.4). Crude oil is then transferred to the vessel's double-hull for storage. Crude oil that is stored onboard is frequently transferred to shuttle tankers or ocean barges going ashore, via a loading hose. Regular loading oil from the stern of the FPS to the bow of the shuttle tanker (tandem loading) is practiced while gas is may be transferred to shore via pipeline or reinjected into the field to boost production.

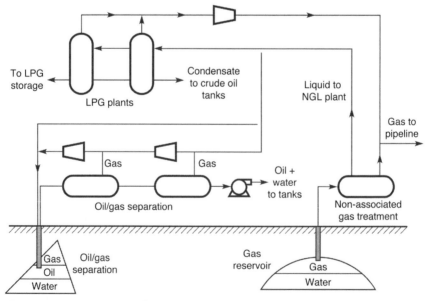

Figure 3.4 Representation of a crude oil–natural gas separation system.

In addition, to the FPSOs, similar floating systems include *floating storage and offloading systems* (FSOs), FPSs, and FSUs. The first *floating drilling production storage and offloading vessel* was developed in 2009 for the Azurite field (Murphy Oil) offshore Republic of Congo. This unit incorporates deepwater drilling equipment that will help to develop the field and can be removed and reused after all the Azurite production wells have been drilled. Furthermore, the first *floating liquid natural gas* (FLNG) or vessel has also been developed.

A variation of the FPSO is the Sevan Marine design that uses a circular hull which shows the same profile to wind, waves, and current regardless of direction. This design shares many of the characteristics of the ship-shaped FPSO such as high storage capacity and deck load, but does not rotate and therefore does not need a rotating turret.

3.3.5 SPAR Platform

The *Spar Platform* consists of a large-diameter single vertical cylinder supporting a deck. It has a typical FP topside (surface deck with drilling and production equipment), three types of risers (production, drilling, and export), and a hull moored using a taut catenary system of 6–20 lines anchored into the sea floor. Spars are available in water depths up to 3,000 ft, although existing technology can extend refers to the analogy of a spar on a ship. In September 1996, Oryx Energy installed the first Spar production platform in the Gulf in 1,930 ft of water in Viosca knoll Block 826. This is a 770-ft long, 70-foot broad diameter cylindrical structure anchored vertically to the sea floor.

The cylinder does not extend all the way to the seafloor, but instead is tethered to the bottom by a series of cables and lines. The large cylinder serves to stabilize the platform in the water, and allows for movement to absorb the force of potential hurricanes. The first Spar platform in the Gulf of Mexico was installed in September of 199—the cylinder measured 770 ft long and was 70 ft in diameter, and the platform operated in 1,930 ft of water. Similar to an iceberg, the majority of a spar facility is located beneath the water's surface, providing the facility increased stability.

SPAR platforms (such as the Brent SPAR platform) were previously used as floating storage offloading systems. The Devil's Tower SPAR platform (Dominion Oil) is located in 5,610 ft (1,710 m) of water, in the Gulf of Mexico. The first Truss SPAR platforms were the Boomvang and Nansen SPAR platforms (Kerr-McGee). The first cell SPAR platform spar is the Red Hawk (Kerr-McGee). The Perdido platform (Shell Oil Company) is situated in the Gulf of Mexico (Perdido fold belt) on water at a depth of 8000.

A SPAR platform is a platform that is moored to the seabed like the TLPs but, whereas the TLP has vertical tension tethers, the SPAR has more conventional mooring lines. SPAR platforms have been designed in three configurations: (1) the conventional one-piece cylindrical hull, (2) the truss spar in which the midsection is composed of truss elements connecting the upper buoyant hull—called a hard tank—with the bottom soft tank containing permanent ballast, and (3) the cell spar that is built from multiple vertical cylinders.

In the truss spar version, the midsection is composed of truss elements connecting the upper buoyant hull (called a hard tank) with the bottom soft tank containing permanent ballast. On the other hand, the cell spar version is built from multiple vertical cylinders. The Spar may be more economical to build for small and medium sized rigs than the TLP, and has more inherent stability than a TLP since it has a large counterweight at the bottom and does not depend on the mooring to hold it upright. It also has the ability, by use of chain-jacks attached to the mooring lines, to move horizontally over the oil field. The first production spar was Kerr-McGee's Neptune, which is a floating production facility anchored in 1,930 ft (588 m) in the Gulf of Mexico, however spars (such as Brent Spar) were previously used as FSOs. Dominion Oil's Devil's Tower is located in 5,610 ft (1,710 m) of water, in the Gulf of Mexico, and is the world's deepest spar. The first Truss spars were Kerr-McGee's Boomvang and Nansen. The first (and only) cell spar is Kerr-McGee's Red Hawk.

Truss Spar platforms are different from both semisubmersible platforms and TLPs with regards the mechanism of motion control. One of the distinctions of the Truss Spar is that its center of gravity is always lower than the center of buoyancy which guarantees a positive GM. This makes the Truss Spar unconditionally stable. The Truss Spar derives no stability from its mooring system, so it does not list or capsize even when completely disconnected from its mooring.

The deep draft is a favorable attribute for minimal heave motions, its deep draft and large inertia filter wave frequency motions in all but the larger storms. The natural period in heave and pitch are above the range of wave energy periods. The long response periods for Truss Spars mitigate the mooring and riser dynamic responses, which are common to ship shaped FPSOs and Semisubmersible platforms. The deep draft, along with protected center well, significantly reduce the current and wave loading on the riser system These loads normally control the tension and fatigue requirements of the production risers on TLP or Semis.

One of the principal advantages of the Truss Spar over other floating platforms lies in its reduced heave and pitch motions. Low motions in these degrees of freedom permit the use of dry trees. Dry trees offer direct vertical access to the wells from the deck, which allows the Truss Spar to be configured for full drilling, workover operations, production operations, or any combination of these activities.

The SPAR platform may be more economical to build for small and medium sized rigs than the TLPs, and has more inherent stability than a TLP since it has a large counterweight at the bottom and does not depend on the mooring to hold it upright. The platform also has the ability, by use of chain-jacks attached to the mooring lines, to move horizontally over the oil field.

3.3.6 Subsea Production Systems

Subsea production systems are wells located on the sea floor, as opposed to at the surface. Just as in a FPS, the petroleum is extracted at the seafloor, and then "tied-back" to an already existing production platform. The well is drilled by a moveable rig, and instead of building a production platform for that well, the extracted natural gas and oil are transported by riser or even by undersea pipeline to a nearby production platform. This allows one strategically placed production platform to service many wells over a reasonably large area. Subsea systems are typically in use at depths of 7,000 ft or more, and do not have the ability to drill, only to extract and transport.

Subsea systems comprise of the well system (includes the downhole completion system and the subsea tree), the production system (includes protective structures, manifolds, templates, intervention systems, and subsea processing systems), and the pipeline system (includes tie-ins, umbilicals, risers, injection pipelines, and production pipelines).

A *subsea production system* (FPS)—which is typically wells located on the sea floor in shallow or deep water—to extract petroleum can be associated with an already existing production platform or an onshore facility. The oil well is drilled by a movable rig and the extracted oil or natural gas is transported by pipeline under the sea and then to rise to a processing facility. The system ranges from a single subsea well producing to a nearby platform to multiple wells producing through a manifold and pipeline system to a distant production facility. These systems are being applied in water depths of at least 7,000 ft or more.

Subsea developments have been made possible by technologies such as subsea trees, risers, and umbilical lines. The production equipment is located on the seafloor rather than on a fixed or floating platform, subsea processing provides a less-expensive solution for myriad offshore environments. Originally conceived as a way to overcome the challenges of extremely deepwater situations, subsea processing has become a viable solution for fields located in harsh conditions where processing equipment on the water's surface might be at risk. Additionally, subsea processing is an emergent application to increase production from mature or marginal fields.

Subsea production systems can range in complexity from a single satellite well with a flow line linked to a FP or an onshore installation, to several wells on a template or clustered around a manifold. Subsea production systems can be used to develop reservoirs, or parts of reservoirs, which require drilling of the wells from more than one location (Chapter 6). Deep water conditions, or even ultra-deep water conditions, can also inherently dictate development of a field by means of a subsea production system, since traditional surface facilities such as on a steel-piled jacket, might be either technically unfeasible or uneconomical due to the water depth, because the development of subsea crude oil and natural gas fields requires specialized equipment, which must be sufficiently reliable to safe guard the environment, and make the exploitation of the subsea hydrocarbons economically feasible. The deployment of such equipment requires specialized and expensive vessels, which need to be equipped with diving equipment for relatively shallow equipment work (i.e., a few hundred feet water depth maximum), and robotic equipment for deeper water depths. Any requirement to repair or intervene with installed subsea equipment is thus normally very expensive. This type of expense can result in economic failure of the subsea development.

Subsea processing can encompass a number of different processes to help reduce the cost and complexity of developing an offshore field. The main types of subsea processing include subsea water removal and reinjection or disposal, single-phase and multi-phase boosting of well fluids, sand and solid separation, gas/liquid separation and boosting, and gas treatment and compression. Subsea separation reduces the amount of production transferred from the seafloor to the surface of the water, debottlenecking the processing capacity of the development. Also, by separating unwanted components from the production on the seafloor, flow lines and risers are not lifting these ingredients to the facility on the surface but are used to direct unwanted ingredients them back to the seafloor for reinjection. Reinjection of produced gas, water, and waste increases pressure within the reservoir that has been depleted by production. Also, reinjection helps to decrease unwanted waste, such as flaring, by using the separated components to boost recovery.

On deepwater or ultra-deepwater fields, subsea boosting is needed to move the crude oil and natural gas from the seafloor to the facilities on the water's surface (Leffler et al., 2011). Subsea boosting negates backpressure that is applied to the wells, providing the pressure needed from the reservoir to transfer production to the sea surface. Even in mid-water developments, subsea boosting, or artificial lift, can create additional pressure and further increase recovery from wells, even when more traditional enhanced oil recovery methods are being employed.

Most subsea processing will increase the recovery from the field and, by enhancing the efficiency of flow lines and risers, subsea processing contributes to flow management and assurance and also enables development of challenging subsea fields. Furthermore, subsea processing converts marginal fields into economically viable developments. As a result, many offshore fields have included subsea processing into the development protocols. Whether the fields are mature and the subsea processing equipment has been installed to increase diminishing production, or the fields incorporated subsea processing from the initial development to overcome deepwater or environmental challenges, the subsea processing concept has enabled the fields to achieve higher rates of production.

For example, with the successful start-up of a full-field subsea separation, boosting and injection system on the Tordis field (StatoilHydro) in the North Sea in 2007, the mature Tordis oil field (StatoilHydro) increased recovery by an extra 35 million barrels of oil and extended the life of the field by 15–17 years. In addition, the Parque das Conchas (BC-10) project (offshore Brazil, Shell Oil Company) used gas/liquid separation and boosting and developed via 13 subsea wells, six subsea separators and boosters. Pazflor project (offshore West Africa, Total SA) is utilizing a gas/liquid separation system. Another example of a subsea system development is this to about 10,000 ft. A SPAR is located in the Mensa field located (Mississippi Canyon Blocks 686, 687, 730, and 731) (Shell Oil Company) which started producing in July 1997 in 5,376 ft of water, shattering the then depth-record for production. Consisting of a subsea completion system, the field is tied back through a 12-in. flow line to the shallow water platform West Delta 143. The 68-mile tieback has the world record for the longest tieback distance to a platform.

As offshore oil and gas regions mature, the focus of exploration and development shifts toward smaller hydrocarbon accumulations, which are in many cases better suited to subsea development solutions than to the more costly stand-alone platform developments.

Finally, there has been an increasing focus on deepwater prospects that have previously been out of reach. The use of floating production facilities and subsea technology has made exploitation of such fields possible, but a number of technical challenges remain.

3.3.7 Tension-Leg Platform

A TLP refers to the platform that is held in place by vertical, tensioned tendons connected to the sea floor by pile-secured templates. TLPs consist of floating rigs tethered to the seabed in a manner that eliminates most vertical movement of the structure. The platform is buoyant and held in place by a mooring system. TLPs are similar to conventional FPs except that the platform is maintained on location through the use of moorings held in tension by the buoyancy of the hull. The mooring system is a set of tension legs or tendons attached to the platform and connected to a template or foundation on the seafloor. The template is held in place by piles driven into the seafloor. This method dampens the vertical motions of the platform, but allows for horizontal movements. The topside facilities (processing facilities, pipelines, and surface trees) of the TLP and most of the daily operations are the same as for a conventional platform. TLPs are used in water depths up to about 6,000 ft. The typical TLP is a four-column design which looks similar to a semisubmersible.

The basic design of a TLP includes four air-filled columns forming a square, which are supported and connected by pontoons, similar to the design of a semisubmersible production platform. The buoyant hull supports the topside of the platform and an intricate mooring system keeps the platform in place. The buoyancy of the hull of the platform offsets the weight of the platform, requiring clusters of tight tendons, or tension legs, to secure the structure to the foundation on the seabed. The foundation is then kept stationary by piles driven into the seabed.

Since their inception in the mid-1980s, TLP designs have changed according to development requirements—more recent designs also comprise the E-TLP, which includes a ring pontoon connecting the four air-filled columns; the Moses TLP, which centralizes the four-column hull; and the SeaStar TLP, which includes only one central column for a hull—the SeaStar TLP is a widely used floating production facility since TLPs are ideal for a broad range of water depths.

The main idea behind the design of the TLP is to assure that the vertical forces acting on the platform are in balance, that is, fixed and variable platform loads plus tendon tension are equal to its displacement. Positive displacement is obtained by locking the platforms draft below the fixed and variable payload displacement draft. This will result in an upward force applied to the tendons, thereby keeping them in constant tension. As a consequence the vertical platform motion (heave) is almost eliminated, except for motions resulting from tendon elasticity and vertical motion as result of environmental introduced lateral platform motions. The tendons do allow a lateral motion of the platform as a result of wind, wave, and current. This motion is similar to an inverted pendulum except for the fact that the displacement variation by pulling the hull down is giving a restoring force to the lateral movement. The tendon tension is set within predefined values, or window of operation. If the variable load of the platform exceeds these values by adding risers or drilling loads, the tendon pretension is adjusted by reballasting of the platform. Consequently, the hull is compartmented into void, machinery, and ballast spaces.

The tension leg mooring system allows for horizontal movement with wave disturbances, but does not permit vertical, or bobbing, movement, which makes TLPs a suitable choice for stability, such as in the hurricane-prone Gulf of Mexico.

The platform deck is located atop the hull of the TLP—the topside of a TLP is the same as a typical production platform, consisting of a deck that houses the drilling and production equipment, as well as the power module and the living quarters. Dry tree wells are common on TLPs because of the lessened vertical movement on the platforms. Most wells producing to TLPs are developed through rigid risers, which lift the crude oil and natural gas from the seafloor to dry trees located on the deck of the TLP—SCRs are also used to tie-in the subsea flow lines and export pipelines.

An example of a TLP is Ursa platform (Shell Oil Company), anticipated to begin production in 1999. This platform is installed in 4,000 ft of water, will have the depth record for a drilling and production platform, and will be the largest structure in the Gulf of Mexico. Other versions of the TLP include Proprietary versions include the SeaStar and MOSES, which are mini TLPs and are relatively low cost, used in water depths between 600 and 3,500 ft (200 and 1,100 m). The *SeaStar platform* is a relatively low cost developed for production of smaller deep-water reserves that would be uneconomic to produce using more conventional deep-water production systems. It can also be used as a utility, satellite, or early production platform for larger deepwater discoveries. SeaStar platforms can be used in water depths ranging from 600 to 3,500 ft. British Borneo is planning to install the world's first SeaStar in the Gulf of Mexico in the Ewing Bank area at a water depth of 1,700 ft. British Borneo refers to this prospect as Morpeth.

Mini TLPs can also be used as utility, satellite or early production platforms for larger deepwater discoveries that would be uneconomic to produce using more conventional deep-water production systems. It can also be used as a utility, satellite, or early production platform for larger deepwater discoveries. SeaStar platforms can be used in water depths ranging from 600 to 3,500 ft.

3.3.8 Other Installations

3.3.8.1 Unmanned Installations

Normally unmanned installations, are small platforms, consisting of little more than a well bay, helipad and emergency shelter. They are designed for operate remotely under normal operations, only to be visited occasionally for routine maintenance or well work. These installations (sometimes called *toadstools*), are small platforms, consisting of little more than a well bay, helipad, and emergency shelter. They are designed to operate remotely under normal operations, only to be visited occasionally for routine maintenance or well work.

The type of platform used depends on water depth (and the risk of icebergs) as well as the economics, characteristics and production life of the resource. In shallow water, less than 85 ft in depth, the platform could rest on pilings planted on the sea floor. Jackup platforms have legs that can be raised or lowered for use in moderate depths, up to 500 ft but generally 100–250 ft depth in waters off the East Coast. In deeper waters, semisubmersible platforms or drillships are used for exploration, and floating facilities are used for production.

Another type of installation is the *artificial island*, which is exemplified by Northstar Island—a 5-acre artificial island in the Beaufort Sea 12 miles northwest of Prudhoe Bay (Alaska) and 6 miles north of the Alaska coast. The island was created to develop the Northstar Oil Pool (discovered on January 30, 1984 by Royal Dutch Shell), which is located approximately 12,500 ft below the seabed.

The island was built because a standard oil-drilling platform, such as those used in the Gulf of Mexico and the North Sea, was not feasible due to the annual formation of pack ice close to the northern Alaska coast. A stable, year-round artificial island was the only way to provide the permanent structures needed for a production oil well. In addition, protection from the erosive force of masses of ice is provided by covering the shores of the island with concrete mats that extend 4 above and 18 ft below the typical waterline.

Artificial islands provide and alternate option when considering a strategy for offshore drilling for a number of reasons—the island provide space, environmental impacts, safety, cost and number of well heads. However, these advantages are limited by a number of factors, such as: water depth, sand quality and reservoir size which have, and will no doubt continue to, limit their future use.

3.3.8.2 Storage

During the production of offshore oil, it may be desirable to store the crude temporarily at the offshore site before its transportation to the shore for processing. Storage capacity is dictated by the size of shuttle tankers and frequency of their trips. Historically, storage capacities have typically been between 15 and 25 days at peak production. These values are appropriate for the FPSOs employed on remote, marginal fields. During the 1990s, FPSOs became more popular for large fields in more developed areas. Storage requirements and shuttle tanker specifications could be optimized.

North Sea shuttle tankers, for example, are usually purpose-built vessels and the storage requirements can be optimized for a particular project. West African FPSOs are generally sized to store and load very large crude carriers (VLCCs) for long voyages. Southeast Asian FPSOs typically offload to tankers of opportunity. Storage capacities for recent FPSOs projects range from as low as 3 days production (BP Foinhaven in the North Atlantic) to as much as eleven days production (CNOC Liuhua in the South China Sea).

The ship-shaped production platforms (i.e., FPSOs) possess large enclosed volume, and are ideally suited for the combination of production and storage. FSO vessels, that is, without processing, FSOs may also be used in conjunction with floating or fixed production platforms. The Shell Expro Brent Spar was used for this purpose for 20 years in the North Sea. Oil storage tanks are usually maintained at atmospheric pressure with an inert gas blanket. According to international regulations, water ballast tanks are required to be segregated from cargo tanks. Cargo and ballast management is an important aspect of the operations of FPSOs.

Oil storage may also be accommodated at ambient pressure in tanks, which are open to the sea at the tank bottom. Oil—unless it is some type of heavy oil, extra heavy oil, or tar sand bitumen—is lighter than water and can be pumped into the compartment displacing seawater out the bottom. The tank wall pressure is only a function of the height of the oil column and the difference in density between the water and the oil. The same storage principle was also used on the floating structure Brent Spar, and on most of the large concrete gravity-base structures built in Norway in the 1980s and 1990s (Condeep platforms). The Gullfaks C platform (North Sea) is capable of storing up to 2 million barrels of oil.

3.4 PLATFORM MAINTENANCE

Fixed and compliant platforms support conductor pipes, which are essentially extensions of the well casing from the seafloor, and are supported along their length and are not free to move with the dynamics of the waves. The wellhead is at the deck of the platform and well operations are similar to land-based operations. One of the most important requirements for FPSs is their interface with risers. Production may originate from wellheads on the sea floor (wet trees), or from wellheads located on the structure (dry trees). The selection is driven by reservoir characteristics and has a significant impact on the selection of the structure (Table 3.1). Dry tree risers are nearly vertical steel pipes, which must be designed to contain well pressure in all operational conditions, which places limits on the motions of the production platform. Subsea production risers are typically composite flexible risers, which are more tolerant to vessel motions. These risers have been used with all types of floating platforms. SCRs have been employed from TLPs, SPAR platforms, and semisubmersible platforms.

3.4.1 Risers

Production risers (which may be single or dual pipes and which range in diameter from 6 to 30 in.) provide the means for the crude oil and natural gas to flow from the ocean floor wellhead the production platform on the ocean surface. These annulus in the dual pipes serves as a means for other flowlines, electrical instrument wires, chemical injection lines, and wireline service devices to enter the well Typically, the riser pipes are constructed in sections up to 50 ft in length with the requisite connectors at each end and are designed to withstand hurricanes (or, in the Pacific Ocean, typhoons). The risers are secured at the surface platform with hydraulic tensioners for SPAR platforms and TLPs or are attached to riser porches for semisubmersible platform and FPSOs.

Table 3.1 Factors that Influence Selection of Platforms for Field Development Projects

Fixed steel platform:
- Established technology.
- Limited water depth.
- Platform wells.
- Rigid (supported) risers.
- Comprehensive marine operations required.
- Offshore installation of topsides required.
- No oil storage.

Jackup platform:
- Established technology.
- Ideal for shallow water.
- Platform wells.
- Rigid tensioned risers.
- Simple marine operations and self-installing platform.
- Vulnerable to the seabed conditions.
- Movable for reuse at another location.

Fixed concrete platform:
- Established technology.
- Long-life.
- Limited water depth.
- Platform wells.
- Rigid (protected) risers.
- Deep water construction site and transportation route may be required.
- Comprehensive marine operations required.
- Oil storage possible.

Tension leg platform:
- Established technology.
- Suitable for deep water, less suitable for ultra-deep water.
- Platform wells.
- Top-tensioned (exposed) rigid risers.
- Comprehensive marine operations required.
- No oil storage.

Spar platform:
- Proven technology.
- Limited deck area.
- Conventional mooring.
- Suitable for deep water.
- Complex dynamic behavior, particularly in case of large currents.
- Platform wells.
- Top-tensioned (exposed) rigid risers.
- Risers protected in wave zone.
- Possibility for oil storage.

(Continued)

Table 3.1 Factors that Influence Selection of Platforms for Field Development Projects *(cont.)*

Floating platform (semisubmersible):
- Proven technology.
- Conventional mooring.
- Suitable for deep water.
- Well-understood dynamic behavior.
- Subsea wells.
- Flexible risers.
- Simple marine operations.
- No oil storage.

Ship-shaped units (floating production, storage, and offloading unit—FPSO; floating production, storage, drilling and offloading unit—FPSDO; floating storage unit—FSU):
- Proven technology.
- Large deck area.
- Conventional mooring.
- Suitable for deep water.
- Well-understood dynamic behavior and large motions.
- Subsea wells.
- Flexible risers and swivel.
- Simple marine operations.
- Oil storage usually included.

The most commonly used risers are top tension risers (Tutors), SCRs, rigid risers, flexible risers, and hybrid risers. TLPs and SPAR platforms typically use top tension risers because these floating platforms have relatively low heave and use dry trees to control the well. SCRs and flexible risers are commonly used for semisubmersible floating platforms and floating production, storage, and offloading platforms. SCRs are used on floating production platforms that experience more heave motion and use subsea trees (wet trees) that are located on the seafloor. Flexible risers are used when wells are distributed over a large distance on the seafloor.

The crude oil and natural gas from the wells are processed onboard the FPSOs with the production equipment on the main deck. The processed oil is stored in tanks and eventually offloaded on to a shuttle tanker which will deliver oil to the onshore destination.. The typical storage capacity of a FPSOs is 1–2 (or more) million barrels of oil. The gas produced is usually reinjected into the reservoir via a riser, and it is sometimes used to provide power for the vessel.

Risers are subjected to ocean currents and can move significantly as a result of induced vibrations caused by ocean currents moving past the riser.

Where such movement is anticipated, flexible risers connect vertically to the wellhead and use buoyancy-supported arches or other buoyancy modules to support the riser. Flexible risers form an *S* shape (*lazy S*, *steep S*, *lazy wave*, or *steep wave* risers) are also used to connect the free-standing fixed risers to the production platform. Helical strakes can be installed to reduce riser motion and to prevent risers from contact with each other.

3.4.2 Maintenance and Supply

A typical oil production platform is self-sufficient in energy and water needs, housing electrical generation, water desalinators and all of the equipment necessary to process crude oil and natural gas such that it can be either delivered directly onshore by pipeline or to a FSU and/or tanker loading facility. Elements in the oil/gas production process include wellhead, production manifold, production separator, glycol process to dry gas, gas compressors, water injection pumps, oil/gas export metering, and main oil line pumps. All production facilities are designed to have minimal environmental impact.

Larger platforms are assisted by smaller ESVs like the British Iolair that are summoned when something has gone wrong, for example, when a search and rescue operation is required. During normal operations, PSVs (platform supply vessels) keep the platforms provisioned and supplied, and AHTS vessels can also supply them, as well as tow them to location and serve as standby rescue and firefighting vessels.

A tender support vessel (TSV) is used to support a drilling operation on a platform, and usually has the form of a barge or a semisubmersible. It is mainly used in offshore areas where the water depths are limited, wave conditions are mild, and many small platforms can be built cheaply. The TSV drops anchor alongside a small wellhead platform, and have all the drilling equipment onboard in order to drill a well from the platform while alongside. When the well(s) have been completed, the TSV is moved to the next platform.

3.4.3 Personnel

The size and composition of the crew of an offshore installation will vary greatly from platform to platform (Table 3.2). Because of the cost intensive nature of operating an offshore platform, it is important to maximize productivity by ensuring work continues 24 h a day. This means that there will essentially two complete crews onboard at a time, one for day shift and the other for night shift. Crews will also change out at regular intervals, nominally 2 weeks.

Table 3.2 Typical Production Platform Personnel

Title	Role
Offshore installation manager (OIM)	Ultimate authority during his shift Makes the essential decisions Responsible for platform operation
Operations team leader (OTL)	Senior member of the on-shift team
Offshore operations engineer (OOE)	Senior technical authority on the platform
Operations manager (PTSL)	Coordinator for managing crew changes
Catering crew	Cooking, laundry, cleaning accommodations
Crane operators	Operation of cranes Moving cargo around the platform Moving cargo to and from boats
Scaffolders	Rig scaffolding when required
Coxswains	Maintaining the lifeboats Manning lifeboats if necessary
Control room operators	Monitor platform operations
Production technicians	Production operations
Drill crew personnel	On board if drilling operations are active
Tool pusher	
Roughnecks	
Roustabouts	
Company man	
Mud engineer	
Derrick hand	
Geologist	
Well services crew	On board for well work
Well services supervisor	
Wireline operators	
Coiled tubing operators	
Pump operator	

REFERENCES

ABS, 2014. Guide for Crew Habitability on Mobile Offshore Drilling Units (MODUs). American Bureau of Shipping (ABS), Houston, Texas.

Angus, M., 2009. FPSO Handbook. M&I Books, London, United Kingdom.

API, 1994. API. 2005. Recommended Practice for Design, Analysis, and Maintenance of Moorings for Floating Production Systems. Report No. API RP 2FP1. American Petroleum Institute, Washington, DC.

API, 2005. Recommended Practice for Station Keeping, 3rd Edition. Report No. API RP 2SK American Petroleum Institute, Washington, DC.

API, 2011. Planning, Designing, and Constructing Floating Production Systems, 2nd Edition. Report No. API RP 2FPS American Petroleum Institute, Washington, DC.

Austin, D., Carriker, R., McGuire, T., Pratt, J., Priest, T., and Pulsipher, A.G. 2004. History of the Offshore Crude oil and natural gas Industry in Southern Louisiana Interim Report Volume I: Papers on the Evolving Offshore Industry OCS Study MMS 2004-049. Contract 1435-01-02-CA-85169. Minerals Management Service, Gulf of Mexico OCS Region, United States Department of the Interior, New Orleans, Louisiana.

Bergreen, L., 2007. Marco Polo: From Venice to Xanadu. Quercus Books, London, united Kingdom.

Bai, Y., Bai, Q., 2005. Subsea Pipelines and Risers, 2nd Edition Elsevier B.V., Amsterdam, Netherlands.

Burgan, M., 2002. Marco Polo: Marco Polo and the Silk Road to China. Mankato: Compass Point Books, North Mankato, Minnesota.

Chakrabarti, S., Halkyard, K., Capanoglu, C., 2005. Historical development of offshore structures. In: Chakrabarti, S. (Ed.), Handbook of Offshore Engineering. Amsterdam, Netherlands, Chapter 1.

Cobb, C., Goldwhite, H., 1995. Creations of Fire: Chemistry's Lively History from Alchemy to the Atomic Age. Plenum Press, New York.

El-Reedy, M.A., 2012. Offshore structures: Design, Constructions, and Maintenance. Gulf Professional Publishers, Elsevier, Boston, Massachusetts.

Forbes, R.J., 1958a. A History of Technology. Oxford University Press, Oxford, United Kingdom.

Forbes, R.J., 1958b. Studies in Early Petroleum Chemistry. E. J. Brill, Leiden, Netherlands.

Forbes, R.J., 1959. More Studies in Early Petroleum Chemistry. E.J. Brill, Leiden, Netherlands.

Forbes, R.J., 1964. Studies in Ancient Technology. E. J. Brill, Leiden, Netherlands.

Gudmestad, O.T., Zolotukhin, Jarlsby, E.T., 2010. Petroleum Resources with Emphasis on Offshore Fields. WIT Press, Billerica, Massachusetts.

Hunter, N., 2011. Offshore Oil Drilling. Capstone Press, North Mankato, Minnesota.

James, P., Thorpe, N., 1994. Ancient Inventions, Reprint Edition Ballantine Books, New York.

Kemp, A., 2012a. The Official History of North Sea Crude Oil and Natural Gas. Volume I: The Growing Dominance of the State. Routledge Publishers, Taylor & Francis Group, New York.

Kemp, A., 2012b. The Official History of North Sea Crude Oil and Natural Gas. Volume II: Moderating the State's Role. Routledge Publishers, Taylor & Francis Group, New York.

Lanan, G.A., Ennis, J.O., Egger, P.S., Yockey, K.E., 2001. Northstar Offshore Arctic Pipeline Design and Construction. Proceedings. Offshore Technology Conference, Houston, Texas. April 30–May 3. SPE Publication, SPE, Richardson, Texas.

Mason, J.R. 2009. The Economic Contribution of Increased Offshore Oil Exploration and Production to Regional and National Economies. Hermann Moyse Jr./Louisiana Bankers Association Endowed Chair of Banking, Louisiana State University, E.J. Ourso College of Business for the American Energy Alliance, Washington, DC.

Leffler, W.L., Pattarozzi, R., Sterling, G., 2011. Deepwater Petroleum Exploration and Production: A Nontechnical Guide, 2nd Edition PennWell Corporation, Tulsa, Oklahoma, Page 1-8.

Morbey, S.J., 1996. Offshore Brazil: Analysis of a successful strategy for reserve and production growth. In: Doré, A.G., Sinding-Larsen, R. (Eds.), Quantification and Prediction of Petroleum Resources. Norwegian Petroleum Society (NPF) Special Publication No. 6. Elsevier B.V, Amsterdam, Netherlands, pp. Page 123–133.

Nogueira, A.C., Lanan, G.A., Even, T.M., Fowler, J.R., Hormberg, B.A., 2000. Northstar Development Pipelines Limit State Design and Experimental Verification. Proceedings. International Pipeline Conference, Calgary, Alberta, Canada. October 1–5. ASME international Petroleum Technology Institute, Houston, Texas.

Palmer, A.C., Ellinas, C.P., Richards, D.M., Guijt, J., 1990. Design of Submarine Pipelines Against Upheaval Buckling. Proceedings. 1990 Offshore Technology Conference, Houston Texas. SPE Publication, SPE, Richardson, Texas.

Pelletier, J.L., 2000a. Offshore Oil Platforms: Rigs and Offshore Oil Support Vessels. Owners, Operators and Managers (Mariner's Directory & Guide, Volume 3). Marine Techniques Publishing, Augusta, Maine.

Pelletier, J.L., 2000b. Offshore Oil Platforms: Rigs and Offshore Oil Support Vessels. Owners, Operators and Managers (Mariner's Directory & Guide, Volume 6). Marine Techniques Publishing, Augusta, Maine.

Pratt, J.A., Priest, T., Castaneda, C.J., 1997. Offshore Pioneers: Brown & Root and the History of Offshore Oil and Gas. Elsevier-Gulf Professional Publishing, Amsterdam, Netherlands.

Sparks, C.P., 2007. Fundamentals of Marine Riser Mechanics: Basic Principles and Simplified Analysis. PennWell Corporation, Tulsa, Oklahoma.

Speight, J.G., 2014. The Chemistry and Technology of Petroleum, 5th Edition CRC Press, Taylor & Francis Group, Boca Raton, Florida.

USACE, 1999. Final Environmental Impact Statement: Beaufort Sea Oil and Gas Development/Northstar Project, Alaska District, Anchorage. US Army Corps. of Engineers, Washington, DC.

Wooster, R., Sanders, C.M., 2010. Spindletop Oilfield. Texas State Historical Association, Denton, Texas.

CHAPTER FOUR

Exploration

Outline

4.1 INTRODUCTION

Exploration for crude oil and natural gas originated in the latter part of the 19th century when geologists began to map land features to search for sediments that had a high potential to contain crude oil and/or natural gas (Unsworth, 1994; Constable et al., 1998; Eidesmo et al., 2002). In fact, the challenge of addressing the marine environment as a source of crude oil and natural gas long before serious thoughts were given to exploring the Gulf of Mexico. Exploring such environments tended to be a gradual and incremental process, involving the adaptation of land-based equipment and technologies to particular locations.

Initially, the search for crude oil and natural took the form of drilling activities based on topographical features and also, on occasion, the evidence of seepages to the surface. Then, of particular interest to geologists were the surface outcrops that provided evidence of alternating layers of porous and impermeable rock. The porous rock (typically a sandstone, SiO_2, limestone, $CaCO_3$, or dolomite, $CaCO_3 \cdot MgCO_3$) provides the reservoir for the petroleum and natural gas, whereas the impermeable rock (typically impermeable clay or impermeable shale) acts as a basement cover or top cover of the porous sediment trap and prevents migration of the petroleum and natural gas from the reservoir.

Handbook of Offshore Oil and Gas Operations. http://dx.doi.org/10.1016/B978-1-85617-558-6.00004-0

By the early part of the 20th century, many of the areas where surface structural characteristics offered the promise of oil had been investigated and the era of subsurface exploration for oil began in the early 1920s (Forbes, 1958) and the discipline of petroleum geochemistry and the associated disciplines added more detail to the study of rocks with the potential of containing petroleum (Chapter 2). New geological and geophysical techniques were developed for areas where the strata were not sufficiently exposed to permit surface mapping of the subsurface characteristics but the efforts using exploration technology (as designated at that time by the current status of the technology) continued with notable successes. As the 20th century evolved and matured, the development of geophysics (starting in the 1950s and the 1960s) provided newer and more accurate methods for the exploration and identification of sub-surface formations that had a high probability of containing crude oil and natural al gas—but the actual proof of the actual presence of these two fossil fuels comes from subsequent drilling operations.

Thus, the goal of exploration activities to identify formations that contain crude oil and natural gas operations is to obtain an image of the reservoir prior to drilling. By using exploration techniques, geological data are collected that provide information related to subsurface conditions for evaluation of the *potential* of a formation to contain crude oil and natural gas. From there, the helps the exploration teams to make the decision as to whether or not there is a hydrocarbon reserve and whether or not more detailed exploration activities should take be undertaken to prove the initial deductions of the presence of a substantial reservoir. As a result, an image of the reservoir can be obtained prior to drilling. By using exploration techniques, geological information data are collected on subsurface conditions to evaluate the potential for the presence of oil and natural gas. The information thus assembled helps a decision to be made about the potential and size of a crude oil or natural gas and where more intensive exploration activities should be initiated.

Briefly, in a typical seismic exploration operation, a line of data receivers (*geophones* for terrestrial operations, *hydrophones* for aquatic operations) are laid out. Explosives or mechanical vibrators are commonly used on land and air guns are used in aquatic or offshore environments. Seismic surveys are used by the offshore oil and gas industry to help determine the location of oil and gas deposits beneath the seafloor.

Interest in the concept of crude oil and natural gas in reservoirs that lay beneath the oceans (sub-sea formations) continued to receive attention with discoveries leading to further impetus in the search for the presence of offshore crude oil and natural gas (Chapter 1, Chapter 2). Since the oceans cover approximately three quarters of the surface of the Earth (Chapter 2)—the proportion of land to water is 2:3 in the northern hemisphere and 1:4.7 in the southern hemisphere; there was (in essence) a great deal of real estate (more correctly, water estate) where crude oil and natural gas could exists. Furthermore, the ocean basins (Chapter 1) were recognized as areas of high probability where suitable reservoir rocks could be found—approximately 76% of the ocean basins have a depth of 1.8–3.7 miles below sea level and the greatest depth sounded is on the order of 37,000–38,000 ft below sea level—specifically in the Mindanao Trench, east of the Philippines. The deepest sounding in the Atlantic Ocean is approximately 31,365 ft in the Puerto Rico Trench. In some ocean basins, the sea floor is relatively smooth, and stretches of the abyssal plain (Chapter 1) in the northwestern Atlantic have been found to be relatively flat within distances up to 62 miles. As the continents are approached, the abyssal plain rises through the continental slope to the continental shelf. The width of the continental shelf varies considerably but the typical width is on the order of 40 miles with a depth on the order of 425 ft (130 m). As a point of interest and for comparison, Mount Everest has been variously reported to be 29,029 or is 29,118 ft high.

Just as on land, the type of exploration technique employed for oceanic exploration to detect the presence of crude oil and natural gas depends upon the nature of the site—in fact, the techniques applied to a specific site are dictated by the nature of the site and are *site specific*. Thus, there is no one technique that will supply all of the necessary information to precisely define the formations and encourage immediate and successful drilling. For example, in areas where little is known about the subsurface formations, preliminary reconnaissance techniques are necessary to identify potential reservoir systems that then warrant further detailed (and more expensive) geophysical investigation.

Once an area has been selected for further investigation, more detailed methods (such as the *seismic reflection* method) are employed—drilling is the final stage of the exploration program and is in fact the only method by which a petroleum reservoir (worthy of development) can be identified. However, in keeping with the concept of site specificity, on some cases drilling may be the only workable option for commencement of the exploration program. The risk (often, the high risk) involved in the drilling operation depends upon previous knowledge of the site subsurface—remembering that site specify is the operative catch phrase. In fact, the is a strong need to relate the character of the exploratory wells at a given site to the characteristics of the reservoir, even though there may be similarity of the new site to other sites where successful exploration for crude oil and natural gas has occurred.

The principal methods used for site definition and characterization are *magnetism (magnetometer)*, *gravity (gravimeter)*, and *sound waves (seismograph)*. These techniques are based on the physical properties of the subsurface sediments and structures that can be utilized for measurement and estimation of the occurrence/production potential of the formation and include those that are responsive to the methods of applied geophysics. Furthermore, the methods can be subdivided into those that focus on *gravitational properties*, *magnetic properties*, *seismic properties*, *electrical properties*, *electromagnetic properties*, *properties*, and *radioactive properties*. These geophysical methods can be subdivided into two groups: (1) those methods without depth control and (2) those methods with depth control.

In the first group of methods, which are not subject to depth control, the measurements incorporate spontaneous effects from both local and distant sources—for example, gravity-based techniques and the data produced therefrom measurements are affected by the variation in the radius of the earth with latitude. Other effects, such as (1) the elevation of the site relative to sea level, (2) the thickness of the earth's crust, and (3) the configuration and density of the underlying rocks, as well as (4) any abnormal mass variation that might be associated with a mineral deposit also play a role in data interpretation.

In the second group of measurements—those methods that are subject to the depth-induced influence of the formation—seismic energy or electrical energy is introduced into the ground and variations in transmissibility with distance from the surface are observed and interpreted in terms of geological formations. Thus, depths to formations having significant differences in seismic transmissibility can be computed on a quantitative basis and the physical nature of the geological horizons deduced. The accuracy, ease of interpretation, and applicability of all methods falling into this group are not the same, and there are natural and economic considerations to be take into account to determine whether or not the data produced by first group of techniques are preferable for exploration, despite any perceived limitations of the techniques.

However, it must be recognized that geophysical exploration techniques cannot be applied indiscriminately. Knowledge of the geological parameters associated with the mineral or subsurface conditions being investigated is essential both in choosing the method to be applied and in interpreting the results obtained. It should also be noted that such terms as *geophysical borehole logging* can imply (and often does mean) the use of one or more of the geophysical exploration techniques. This procedure involves drilling a well and using instruments to log or make measurements at various levels in the hole by such means as *gravity (density)*, *electrical resistivity*, or *radioactivity*. In addition, formation samples (cores) are usually taken from such investigations for physical and chemical tests.

4.2 METHODS

Thus, the primary search for crude oil and natural gas frequently begins with observation of surface terrain and the location of geological faults in sedimentary formations are an important aspect of exploration insofar a fault can indicate where potential structural traps may lie beneath the surface in reservoir rocks (Figure 4.1). The observation of anticlines, another type of structural trap, also indicates potential traps at depth. Thence, geophysical techniques applied to the father definition and identification of potential reservoirs utilize specialized equipment to measure properties such as: electrical currents, gravitational and magnetic anomalies, heat flow, geochemical relationships, and density variations from deep within the Earth. Each technique records a different set of characteristics that can be used to locate the potential reservoirs in the sub-surface of the Earth.

Briefly, and using the geological anticline (Figure 4.2) as an example, the anticline is a structural phenomenon that is convex up and has the oldest sediments at the core—if age relationships between various strata are unknown, the term *antiform* is often used in place of the name anticline. On a geologic map, an anticline is usually recognized by a sequence of rock layers that are progressively older toward the center of the fold because the uplifted core of the fold is preferentially eroded to a deeper stratigraphic level relative to the topographically lower flanks and the strata dip away from the

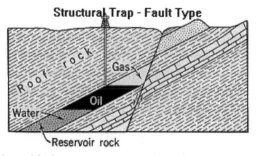

Figure 4.1 A geological fault causing a structural trap for crude oil and natural gas.

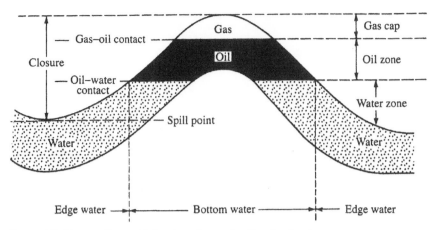

Figure 4.2 The syncline–anticline trap for crude oil and natural gas.

center (the *crest*) of the fold. If an anticline plunges (i.e., is inclined to the surface of the Earth), the surface strata will form a V-shape that points in the direction of incline. An anticline is typically flanked by synclines although faulting die to geological forces can complicate the relationship between the aniline and syncline. Folds often form during crustal deformation as the result of shortening that accompanies orogenic activity (the variety of processes that occur during mountain-building) such as folding and faulting of the crust of the Earth. In many cases, anticlines are formed by movement on nonplanar faults during both shortening and extension, such as a ramp anticline (an anticline formed in a thrust sheet as a result of movement up a ramp) and a *rollover anticline* (formed when the hanging wall slumps into the low pressure extensional zone associated with the faulting shape and displacement of normal faults).

In the current context of offshore surveys in the search for formations with a high probability of containing crude oil and natural gas, geophones (which are used for land-based exploration) are replaced by hydrophones for water-based exploration—a hydrophone converts acoustic energy into electrical energy and is used in passive underwater systems to listen only. On land, sound waves used to be created by detonation of explosives (such as dynamite) but such methods not only detrimental to land-based activities but are extremely detrimental to sea life. The most modern technique that is acceptable is to generate sound waves uses pulses of compressed air. This creates large air bubbles that burst beneath the water surface and create sound waves, which travel down to the sea floor, penetrate the rocks beneath, and return to the surface where they are intercepted by the hydro-phones. Data processing and illustration of the water-based data are much the same as processing the land-based data.

4.2.1 Gravity Methods

The *gravimeter* (*gravitometer*) is a device for measuring variations in the gravi-tational field of the Earth, useful in prospecting for oil and minerals. In the early days of gravity prospecting, both the *torsion* balance and the pendulum apparatus were extensively employed, but these have been supplanted by spring balance systems (*gravimeters*). The latter can be read in a matter of minutes, in contrast to the several hours required in obtaining readings with the earlier instruments. Thus, the *gravimeter* detects differences in gravity and gives an indication of the location and density of underground rock forma-tions. Differences from the normal can be caused by geological influences and such differences provide an indication of the presence subsurface struc-tural formations. In one form, the gravimeter consists of a weight suspended from a spring; variations in gravity cause variations in the extension of the spring. A number of different mechanical and optical schemes have been developed to measure this deflection, which in general is very small.

Gravimeters have been developed that can detect variations in the Earth's gravitational field as small as one part in 10,000,000. Furthermore, the use of watertight housings with automatic leveling and electronic read-ing devices allow gravimeter surveys to be carried out in aqueous environ-ments—the instrument can also be employed in shallow water. Other gra-vimeters have been developed for use in submarines and on gyro-stabilized platforms on surface ships as well as in aircraft.

Gravity methods are based upon the measurement of physical quantities related to the gravitational field, which in turn are affected by differences in the mode of deposition and, hence, the density of the underlying geologi-cal strata. In the context of undersea exploration for formation containing crude oil and natural gas, exploration is based on the mapping of geological structures to determine situations that might localize the promising forma-tion. In such cases, the significant density values are on the order of: (1) salt, 2.1–2.2, (2) igneous rocks, 2.5–3.0, and (3) sedimentary rocks, 1.6–2.8. The density of the sedimentary rocks increases with depth owing to geological age, which is accompanied by sediment consolidation and, as a result, struc-tural deformation associated with faults and folding can be detected. How-ever, other factors such as compaction of sediments over edges or knolls on the underlying crystalline rock surface can also lead to localized increases in density, as does the development of calcareous cap rock over the tops (heads) of various types of intrusive salt columns.

There is a variety of *gravimeters*, but those in common use consist essentially of a weighted boom that pivots about a hinge point. The boom is linked to a spring system so that the unit is essentially unstable and hence very sensitive to slight variations in gravitational attraction. Deflections of the boom from a central (zero) position are measured by observing the change in the tension in the spring system required to bring the boom back to that position. Readings are taken from a graduated dial on the head of the instrument that is attached to the spring system through a screw. There must be an accurate calibration of the screw, reading dial, and spring response for the readings to have gravitational significance.

4.2.2 Magnetic Methods

Magnetic methods are based upon measuring the magnetic effects produced by varying concentrations of ferromagnetic minerals, such as magnetite, in geological formations. Instruments used for magnetic prospecting vary from the simple mining compass used in the seventeenth century to sensitive airborne magnetic units permitting intensity variations to be measured with an accuracy greater than 1/1000 part of the magnetic field of the Earth. The detection instrument (the magnetometer) is used in exploration for crude oil and natural gas to show locations of geologic features that give a picture of the stratigraphy. In addition, magnetometers are used in directional drilling for crude oil or natural gas to detect the azimuth of the drilling tools near the drill. They are most often paired with accelerometers in drilling tools so that both the inclination and azimuth of the drill can be found.

The *magnetometer* is a specially designed magnetic compass and detects minute differences in the magnetic properties of rock formations, thus helping to find structures that might be favorable to presence of crude oil and natural gas—such as the layers of sedimentary rock that may lie on top of the much denser igneous, or basement, rock. The data provide indications of formations that might conceal anticlines or other structures favorable to the collection of crude oil and natural gas. Of even more value is the determination of the approximate total thickness of the sedimentary rock, which can save unwarranted expenditures later or application of further geophysical prospecting or even well drilling when the sediment may not contain sufficient oil to warrant further investigation.

One of the most widely used magnetic instruments is the *vertical magnetometer*. It consists of a pair of blade magnets balanced horizontally on a quartz knife edge. The balance is oriented at right angles to the magnetic meridian. The deflection from the horizontal is observed, giving the variation in magnetic vertical intensity with gravity. The *torsion fiber magnetometer* is also a vertical component instrument but typically has a greater operating range than the vertical magnetometer. The torsion fiber magnetometer also has an advantage in that it is easier and quicker to read. For data production, the instrument values are referred to a base, corrected for temperature and diurnal variation, and corrected for the normal geographic variation of the earth's magnetic field.

The *nuclear precession magnetometer* (proton precession magnetometer) is another continuous recording magnetic instrument that measures the total magnetic field of the Earth by observing the free precession (progressive movement) frequency of the protons in a sample of water. By adding free radicals to the measurement fluid, the nuclear Overhauser effect can be exploited to significantly improve upon the proton precession magnetometer. Rather than aligning the protons using a solenoid, a low power radio-frequency field is used to align (polarize) the electron spin of the free radicals, which then couples to the protons via the Overhauser effect. The *Overhauser effect magnetometer* or *Overhauser magnetometer* uses the same fundamental effect as the *proton precession magnetometer* to take measurements.

The interpretation of magnetic measurements is subject to the same fundamental issues as noted for the gravitometer method: (1) the contrast in physical properties of the formations, (2) the depth of origin and integrated contributions from many sources, (3) the changes in strength and direction of the earth's field with location, and (4) canceling effect related to proximity of opposite induced poses at the boundaries of finite geological bodies.

4.2.3 Seismic Methods

The earliest seismic surveys, during the 1920s, were analog recorded and produced two-dimensional (2-D) analyses. Digital recording was introduced in the 1960s, and then, as computer technology burgeoned, so did geophysical signal processing. During the past 30 years, computer-intensive techniques have evolved. The data are achieved by placing sources and receivers along a straight line so that it can be assumed that all the reflection points fall in a 2-D plane formed between the line of traverse and the vertical. This is known as 2-D seismic. 3-D seismic is a method of acquiring surface seismic data by placing sources and receivers in an aerial pattern. One example of a simple 3-D layout is to place the receivers along a line and shoot into these receivers along a path perpendicular to this line.

The search for petroleum and natural gas relies heavily on the use of seismic technology, which is based on reading data initiated from energy sources, such as explosions, air guns (offshore use), vibrator trucks, or well sources. These sources produce waves that pass through the subsurface and are recorded at strategically placed geophones or hydrophones. The *seismograph* measures the shock waves from explosions initiated by triggering small controlled charges of explosives in the bottom of shallow holes in the ground. The formation depth is determined by the time elapsed between the explosion and detection of the reflected wave at the surface. In the offshore, seismic responses are usually read from streamers towed behind modern seismic vessels, recorded, and processed later by computers that analyze the data.

In a typical seismic exploration operation, a line of data receivers (geophones) are laid out for terrestrial programs or hydrophones for aquatic operations. Explosives or mechanical vibrators are commonly used on land and air-guns are used in aquatic or offshore environments. Seismic surveys are used by the offshore crude oil and natural gas industry to help determine the location of crude oil and natural gas deposits beneath the seafloor. In fact, prospecting for crude oil and natural gas using exploration seismology involves studying body waves such as compressional and shear waves propagating through the interior of the Earth. Generally, seismic methods are preferred (to methods that use magnetometers) as the primary survey method for oil exploration although magnetic methods can give additional information about the underlying geology and in some environments evidence of leakage from traps.

Seismic methods are based on determinations of the time interval that elapses between the initiation of a sound wave from detonation of a dynamite charge or other artificial shock and the arrival of the vibration impulses at a series of seismic detectors (*geophones*). The arrivals are amplified and recorded along with time marks (0.01 s intervals) to give the seismogram. The method depends upon (1) the velocity within each of the layers penetrated at depth is greater than that in the layers above; (2) the layers are bounded by plane surfaces; and (3) the material within each layer is essentially homogeneous. The major differences between earthquake seismology and petroleum exploration seismology are scales and knowledge of the location of seismic disturbances. Earthquake seismology studies naturally generated seismic waves, which have periods in minutes and resolution in kilometers. In exploration seismology, artificial sources are used that have periods of tenths of a second and tens of meters of resolution. Production seismology requires higher-frequency seismic waves and better resolution, often resolution in the order of a few meters.

The seismic method as applied to exploration of crude oil and natural gas involves (1) field acquisition, (2) data processing, and (3) geologic interpretation. *Seismic field acquisition* requires placement of strings of acoustic receivers (hydrophones) in the water in the case of marine exploration and geophones on the surface, in the case of land exploration. The end result of seismic *data processing* is the production of a subsurface profile similar to a geologic cross section, which can be illustrated in the form of a time scale or shown as a function of depth, either of the profiles can be are used for (structural and stratigraphic) geologic interpretation. Structural interpretation of seismic data involves mapping of the different subsurface strata by using seismic data as well as information from boreholes and outcrops. Stratigraphic interpretation investigates the various attributes within a common stratum and is used to interpret changes to infer varying reservoir conditions such as lithology, porosity, and fluid content.

The depths and media reached by seismic waves depend on the distance between the shot point and the receiving point(s). The first *impulses* or *breaks* in a seismograph are caused by waves that have traveled quickly between the shot point and the receiving point(s). At short distances, this is usually also the shortest path, but beyond a certain distance it is quicker for a refracted pulse to travel via a longer path involving underlying layers with a higher velocity. From a plot of the time of travel as a function of surface distance, data are obtained for determining both the velocity of the sound wave through the material and number of layers present. From the distances at which changes in velocity are indicated, the depth of each layer can be assessed. In general the deeper, older formations as a result of higher compression have a higher density and also a higher seismic velocity than the overlying material. Observed differences in velocity not only define the direction of slope of the rock surfaces but also provide information for computing the degree of slope present.

Commercial 3-D seismology began in the early 1980s on a limited basis. Recent innovations that were essential to the development of 3-D seismology are satellite positioning, new processing algorithms, and the interpretative workstation. The 3-D seismic survey provides a more accurate and detailed image of the subsurface and offers a significantly higher quality signal than the 2-D data that are (have been) commonly acquired. The 3-D technology also improves both spatial and temporal resolutions. New data processing techniques are pre-stacked 3-D depth migration, interpretation of multiple 3-D surveys in different times (4-D seismic), and reservoir characterization of horizons.

Seismic geophysical has significantly aided the search for crude oil and natural gas on the subsea continental shelves and other areas covered by water. A marine seismic project moves continually, with detectors towed behind the boat at a constant speed and at a constant depth. Explosive charges are detonated at a position and time determined by the speed of the boat, so that a continuous survey of the reflecting horizons can be obtained. The technique for mapping the 3-D structure beneath the seafloor (*seismic reflection profiling*) relies on the subsurface penetration and reflection of sound waves produced by a loud acoustic source—an explosion of dynamite historically and an air-gun in modern times—whose differential travel times can be detected using hydrophones. These acoustic reflections reveal different layers of sediments, rocks, and faults, among other subsurface features, that can be combined using computer techniques to provide an image of the subsurface seafloor structures.

To obtain the seismic data, an exploration vessel tows one (or two) air guns (usually 3–400 ft behind the vessel) along with several kilometers of *streamers* (cables to which are attached the necessary number of hydrophones). While the vessel is underway (at, for example, approximately 5 knots), the air guns are triggered at regular intervals (every 10 to 12 s), which produces a pulse of sound that is reflected from the seabed before being echoed back to the hydrophones.

The air guns used in this technology produce sounds on the order of 235–240 decibels (dB). The environmental (noise) threat posed by seismic reflection profiling depends on the number and frequency of surveys and the seasonal behaviors of cetaceans that may be affected in a given region. Such surveys need to be approached with caution and to avoid regions where affected species congregate.

4.2.4 Electrical Methods

The use of *electrical logging* is based on the fact that the resistivity of a rock layer is a function of its fluid content—sand containing crude oil has a higher resistivity than the sand itself. The method consists of passing a current between an electrode at the surface and one that is lowered into the hole, the latter being uncased and filled with drilling mud, but there are environmental regulations that apply to the use of drilling muds (Chapter 11) (Wills, 2000). Any change in the resistivity conditions around the moving electrode affect the flow of current and voltage distribution around it. Voltage fluctuations can be measured by a pair of separate electrodes used in conjunction with the moving electrode. In fact, cuttings recovered from drilling operations should also be analyzed for the presence of crude oil. A well-drilling program may include *measuring while drilling* (MWD) and a suite of wireline logs, sidewall cores, and a testing program can be designed to measure the physical parameters of the rocks so that the potential for the occurrence of crude oil and natural gas can be assessed with a reasonable degree of certainty.

Electrical prospecting is the ability to determine the presence of insulating rather than conducting media. Use is made of resistivity determinations at various electrode spacing together with inductance and resistivity slope functions. The use of electrical methods for prospecting for crude oil and natural gas is one are the oldest methods that have been used as well as for reservoir description and monitoring (Heiland, 1932; Karcher and McDermott, 1935; Rust, 1938; Keller 1968; Haren, 1984; Morrison et al., 1995; Evans et al., 1999). However, electrical methods are more frequently used in searching for metals and minerals than in exploring for crude oil (crude oil at shallow depths is the exception), mainly because most of the methods have proved to be less effective only for deep exploration.

Crude oil is a dielectric medium (Speight, 2014a,b) but electrical methods of prospecting for crude oil and natural gas focus on the properties of rocks. For example, the resistivity, or inverse conductivity, of rock governs the amount of current that passes through the rock when a specified potential difference is applied. Another example is the electrochemical activity of the rock with respect to electrolytes in the ground, which is the basis of the self-potential method. The third property is the dielectric constant of the rock, which provides information on the capacity of a rock material to store an electric charge, which must be taken into consideration when high-frequency alternating currents are introduced into the earth, as in inductive prospecting techniques.

Electrical prospecting methods depend upon differences in *electrical conductivity* between the geological bodies under study and the surrounding rocks. In general, metallic minerals, particularly the sulfides, range in resistivity from 1.0 to several Ω cm, whereas consolidated sediments of low water content average about 10^4 cm, igneous rocks range from 10^4 to 10^6 Ω cm, and saturated unconsolidated sediments from 10^2 to 10^4 Ω cm. The resistivity of the last depends largely on the amount and electrolytic nature (salinity) of the included water. On the other hand, the self-potential method makes use of the fact that most metallic sulfide minerals are easily oxidized by downward-percolating groundwater. As a result of this surface oxidation, the elements of a simple chemical battery are established and an electrical current flows down through the ore body and back to the surface through the surrounding water-saturated ground, which acts as the electrolyte. It is possible to locate these localized electrical fields and, hence, ore bodies by mapping points of equal electrical potential at the surface using nonpolarizing electrodes and a sensitive ammeter, or a milli-ammeter. Alternatively, measuring the potential differences between successive profile stakes forming a grid over an area using a potentiometer can also be employed.

A special application of electrical methods is in the study of subsurface stratigraphy by measuring the potential differences between the surface and an electrode lowered in a borehole and by also measuring variations in electrical resistivity with depth (electrical logging). This method produces a measure of porosity and permeability since the data are affected markedly by the ability of the drilling fluid to penetrate the formation. The resistivity measurements define the position of formation boundaries and the lithological character of the sediments.

In the resistivity method, three resistivity logs are usually taken: (1) one having a shallow penetration to define the location of the formation boundaries and two others having (2) intermediate and (3) deep penetration. These last two logs are used to determine the extent to which the drilling fluid has penetrated into the formations and the true resistivity of the formation present. The various measurements taken in conjunction provide a valuable tool not only for studying conditions in a given well but also for carrying out correlation studies between wells and thus defining geological structure and horizontal changes in lithology.

4.2.5 Electromagnetic Methods

Just as in seismic exploration, electromagnetic geophysics can contribute to effective hydrocarbon exploration in two distinct ways. Most often, electromagnetic methods are used to image structures that could host potential reservoirs and/or source rocks. In certain cases, they may also give evidence for direct indication of the presence of hydrocarbons. Electromagnetic methods are based upon the concept that an *alternating magnetic field* causes an *electrical current* to flow in conducting material. Measurements are carried out by connecting a source of alternating current to a coil of wire, which acts as a source for a magnetic field similar to that which will be produced by a short magnet located on the axis of the coil. A receiving system consisting of a second coil connected to a voltmeter is mounted so that there is free rotation about a horizontal axis.

The receiving coil should be mounted so that rotation is on an axis perpendicular to that of the induced magnetic field. In this case, the induced voltage (in the absence of a conductor) will vary from zero (when the coil plane is parallel to the plane of the applied field) to a maximum (when the coil plane is perpendicular to the plane of the applied field). However, if a conductor is present, the induced current in the conductor sets up a secondary magnetic field that distorts the primary field and gives a value that is not horizontal except directly over the conductor. By using an inclinometer to record the angle of the moving search coil when in the null position, the location of a conductor can be determined as the crossover (inflection) point on a profile across the body. Another variation of this method is to have both the receiver and transmitting coils in the horizontal plane. In this arrangement, the voltage developed over non-conducting ground is a function of the construction of the coils, which are usually moved across the ground with a constant separation. The presence of a conductor is indicated by changes in the voltage values from the normal values for this configuration.

Marine electromagnetic methods (controlled-source electromagnetic methods and magnetotellurics) have as a way to reduce risk and improve exploration efficiency (Unsworth, 2005). Estimates of seafloor electrical conductivity obtained using electromagnetic surveys can be used to generate geological and structural models and even, given appropriate conditions, detect the existence of hydrocarbon reservoirs. While electromagnetic methods may never achieve the resolution of seismic methods, they are intrinsically more powerful than potential field methods (gravity and magnetics) and provide a more direct estimation of fluid content than using acoustic properties alone. In fact, marine electromagnetic methods have the potential to play an important role in any integrated exploration strategy.

4.2.6 Other Methods

The natural radioactive properties of many constituents of rock have made it possible to develop and use nuclear radiation detectors (*radioactive logging*) in the borehole or even in holes that have already had casing inserted. In addition to prospecting for radioactive minerals, the radioactive method is extensively applied in borehole studies of subsurface stratigraphy as might be deemed necessary when prospecting for crude oil and natural gas.

Two commonly used methods are γ-ray logging and neutron logging. In the first case (γ-ray logging), the natural radiation from the rock is used, whereas in the second case (neutron logging), a neutron source is employed to excite the release of radiation from the rock. The neutron source is usually a mixture of elements (of which of beryllium and radium have been commonly used) and the method is a means for determining the relative porosity or rock formations. An additional benefit of γ-ray logging is that the method assists in defining tight formations such as shale.

In the disintegration of radioactive minerals three spontaneous emissions take place, the election of an electron (β-ray), a helium nucleus (α-ray), and short-wavelength electromagnetic radiation (γ-rays). Therefore, the instruments used in radioactive exploration are the Geiger counter and the scintillometer. Different sedimentary rocks are naturally characterized by different concentrations of radioactive materials. Tight formations such as shale—currently popular because of the occurrence of crude oil and natural gas in such formations (Speight, 2013, 2014a)—and volcanic ash give the highest γ-ray count and the limestone the lowest γ-ray count.

Another valuable exploration method is *geophysical borehole logging*, which involves drilling a well and use of instruments to log or make measurements at various levels in the hole by such means as electrical resistivity, radioactivity, acoustics, or density. In addition, formation samples (cores) are taken for physical and chemical tests. On the other hand, the *acoustic logging* method is quite similar to surface seismic work—instead of explosives an electrically operated acoustic pulse generator is used. In one instrument, the generator is separated from the receiver by an acoustic insulator. The design permits automatic selection and recording of the travel times of the onsets of pulses that travel through the rock wall of the hole as the instrument moves down or up. Signals are recorded continuously at the surface, being transmitted through a cable on which the instrument is suspended. The velocity log provided by the instrument helps to define beds and evaluate formation porosity.

Density can now be logged with a new technique that uses radioactivity (*density logging*). The instrument consists of a radioactive cobalt source of γ-rays and a Geiger counter as a detector, which is shielded from the source. The rock formation is bombarded with the γ-rays, some of which are scattered back from the formation and enter the detector. The degree to which the original radiation is adsorbed is a function of the density of the rock.

Test well sampling is another important method used in the search for oil (*core sampling*). Well data obtained from the examination of formation samples taken from various depths in the borehole are of considerable value in deciding further exploratory work. These samples can be cores, which have been taken from the hole by a special coring device, or drill cuttings screened from the circulating drilling mud. The major purpose of sample examination is to identify the various strata in the borehole and compare their positions with the standard stratigraphic sequence of all the sedimentary rocks occurring in the specific basin in which the hole has been drilled.

4.2.7 Data Acquisition

Thus, exploration for crude oil and natural gas is a multidisciplinary approach insofar as data acquisition requires expertise in geology, geophysics, physics and chemistry, navigation and mapping, statistics, computer science and economics, engineering, and mathematics. Thus, the process of data collection, analysis, and field evaluation is an iterative process—when a field is discovered, more data are gathered using remote-sensing techniques and subsurface sampling. The oil and gas industry has long favored remote sensing platforms for exploration and production. A mobile, remotely controlled sensor platform has been developed to provide persistent coverage over an offshore area. Powered by wave motion and sunlight, it can be fitted with numerous sensors that continuously monitor ocean parameters and collect and transmit real-time data to support offshore exploration and production activities (Carragher et al; 2003).

Useful interpretation of geophysical data can be obtained without recourse to well data, and this is normally required in the beginning stage of exploring a new basin. The reflection amplitudes from seismic surveys and their geometric relationships can help to assess structural leads, identify stratigraphic leads, and provide clues on general stratigraphy and gross lithology (Larue and Yue, 2003). However, at some stage of the work, a well will be required in order to prove the nature (and contents) of the formations of interest. Downhole surveys such as check shots and vertical seismic profiling will give geological meaning to the information produced by seismic profiling and it is very necessary to understand that geological formations are not homogenous and the lithology can vary significant over the extent of a formation.

The objective of geophysical interpretation is to identify some attribute of the available data that might imply an accumulation of hydrocarbons that *might* support the initiation of economic recovery operations but a recommendation to produce a field must be an integrated interpretation of all available data. Thus, the various downhole tools are used to measure an assortment of basic rock properties: (1) *logs*, which give information about porosity, permeability, density, velocity, resistivity, neutron and gamma activity, bed dip, mineral content, grain size, and sorting, (2) *drill cuttings*, which give information about the presence of hydrocarbon shows, fossils, and minerals, (3) *formation tests*, which give information about fluid content, pressures, permeability boundaries, and total organic content, and (4) *cores*, which give information about paleontology, palynology, facies, lithology, mineralogy, fluids, and all physical properties. These data are interpreted for information on sedimentology and facies, the depositional environment, erosional hiatus, diagenetic history and fluid movement, and evidence of hydrocarbons. A grid of several wells allows a more complete interpretation than a single well. It must always be recognized that the exploration for crude oil and natural gas involves uncertainty but, on the other hand, the need for crude oil and natural gas continues to increase, which creates a need for more exploration methods to serve as a basis for deciding on the initiation of economic recovery operations.

4.3 EXPLORATION VESSELS

When exploratory drilling is required, a drillship (Chapter 3) might be the vessel of choice. In fact, the drillship can be used as a platform to carry out well maintenance or completion work such as casing and tubing installation, subsea tree installations, and well capping. Drillships are often built to the design specification to meet the requirements necessary for the exploration of (through drilling for) crude oil and natural gas. Drillships have increased in size and capability since they were first employed for the exploration for crude oil and natural gas and the expansion of exploratory drilling seems is assisting in further developments of drillship technology.

Chemical Instrumentation installed on petroleum exploration vessels allows the exploration personnel to search for seeps in the marine environment. In the context of the present chapter, the vessels are equipped with *hydrophone*—a device that will listen to, or detect, the acoustic energy underwater. Hydrophones are usually used below their resonance frequency over a much wider frequency band where they provide uniform output levels. Most hydrophones are based on a piezoelectric transducer that generates electricity when subjected to a pressure change. Such piezoelectric materials, or transducers, can convert a sound signal into an electrical signal since sound is a pressure wave in fluids. Some transducers can also serve as an emitter, but not all have this capability, and may be destroyed if used in such a manner. If all else fails, the exploration company may have no option but to drill a *wildcat well*, which is a small-diameter well drilled in a new area where no other wells exist and generally with little or inconclusive information. Thus many wildcat wells are drilled on instinct, intuition, or a small amount of geology. Many times they are based on photography and experience in a particular area.

Whatever the process of exploration, the basic tool needed for the search for petroleum and natural gas remains: (1) knowledge of the Earth and earth processes of formation, (2) lithology, and (3) structural aspects of the subsea or subterranean formations. Nevertheless, wildcat wells are still drilled but the success rate of wildcat drilling is substantially lower than a well *spudded in* (the process to begin a new well) using all of the geological tools available.

Once a site has been identified as a confined place for the occurrence of crude oil and natural gas, the first stage in the extraction of crude oil and natural is to drill a well into the underground reservoir. Often many wells (*multilateral wells*) will be drilled into the same reservoir, to ensure that the extraction rate will be economically viable. Also, some wells (*secondary wells*) may be used to pump water, steam, acids, or various gas mixtures into the reservoir to raise or maintain the reservoir pressure, and so maintain an economic extraction rate.

REFERENCES

Carragher, P., Hine, G., Leigh-Smith, P., Mayville, J., Nelson, R., Pai, S., Parnum, I., Shone, P., Smith, J., Tischatschke, C., 2003. A new platform for offshore exploration and production. Schlumberger Oilfield Review 25 (4), 40–50.

Constable, S.C., Orange, A.S., Hoversten, M.G., Morrison, H.F., 1998. Marine magnetotellurics for petroleum exploration part 1: a sea-floor system. Geophysics 63, 816–825.

Eidesmo, T., Ellingsrud, S., MacGregor, L.M., Constable, S., Sinha, M.C., Johansen, S., Kong, F.N., Westerdahl, H., 2002. Sea Bed Logging (SBL): A New Method for Remote and Direct Identification of Hydrocarbon Filled Layers in Deepwater Areas. First Break 20, 144–152.

Evans, R.L., Tarits, P., Chave, A.D., White, A., Heinson, G., Filloux, J.H., Toh, H., Seama, N., Utada, H., Booker, J.R., Unsworth, M.J., 1999. Asymmetric electrical structure in the mantle beneath the East Pacific Rise at 17°S. Science 286, 752–756.

Forbes, R.I., 1958. A History of Technology. Oxford University Press, Oxford, United Kingdom.

Haren, R.J., 1984. The Application of Electrical Prospecting Techniques for Crude Oil and Natural Gas Exploration On-Shore. Department of resources and Energy, Canberra, Australia.

Heiland, C.A., 1932. Advances in Techniques and Application of Resistivity and Potential Drop-Ratio Methods. In Oil Prospecting. Reprinted in Early papers of the Society of Exploration Geophysicists. 1947, pp. 420–496.

Karcher, J.C., McDermott, E., 1935. Deep electrical prospecting. AAPG Bulletin 19.

Keller, G.V., 1968. Electrical Prospecting for Oil. Colorado School of Mines, Golden Colorado.

Larue, D.K., Yue, Y., 2003. How stratigraphy influences oil recovery: A comparative reservoir database study concentrating on deepwater reservoirs. The Leading Edge 22 (4), 332–339.

Morrison, H.F., Lee, K.H., Becker, H., 1995. Electrical and Electromagnetic Methods for Reservoir Description and Process Monitoring. Report DOE/DC/95000153 (DE95000153). United States Department of Energy, Washington, DC.

Rust, Jr., W.M., 1938. A historical review of electrical prospecting methods. Geophysics 3, 1–6.

Speight, J.G., 2013. Shale Gas Production Processes. Elsevier-Gulf Professional, Oxford, United Kingdom.

Speight, J.G., 2014a. The Chemistry and Technology of petroleum, 5th Edition Taylor & Francis Group, Boca Raton, Florida.

Speight, J.G., 2014b. Handbook of Petroleum Products Analysis, 2nd Edition John Wiley & Sons Inc., Hoboken, New Jersey.

Unsworth, M.J., 1994. Exploration of mid-ocean ridges with a frequency domain electromagnetic system. Geophysics Journal International 116, 447–467, 1994.

Unsworth, M.J., 2005. New Developments in Conventional Hydrocarbon Exploration with Electromagnetic Methods. CSEG Recorder, April. pp. 34–38.

Wills, J., 2000. Muddied Waters: A Survey of Offshore Oilfield Drilling Wastes and Disposal Techniques to Reduce the Ecological Impact of Sea Dumping. For Ekologicheskaya Vahkta Sakhalina (Sakhalin Environment Watch), Sakhalin, Russia.

CHAPTER FIVE

Drilling Technology and Well Completion

Outline

5.1 INTRODUCTION

The first stage in the extraction of crude oil from an underground reservoir is to drill a well into the reservoir (Speight, 2014). Often many wells (multilateral wells) will be drilled into the same reservoir, to ensure that the extraction rate will be economically viable. Also, some wells (secondary wells) may be used to pump water, steam, acids, or various gas mixtures into the reservoir to raise or maintain the reservoir pressure, and so maintain an economic extraction rate.

In addition to creating drilling rigs that can operate at great water depths, new drilling techniques have evolved, which increase productivity and lower unit costs. The evolution of directional and horizontal drilling to penetrate multiple diverse pay targets is a prime example of technological advancement applied in the offshore. The industry now has the ability to reduce costs by using fewer wells to penetrate producing reservoirs at their optimum locations. Horizontal completions within the formation also extend the reach of each well through crude oil-bearing (and/or natural gas-bearing) formations, thus increasing the flow rates compared with those from simple vertical completions. These advancements can be attributed to several developments. For example, the evolution of retrievable whipstocks allows the driller to exit the cased wells without losing potential production from the existing

Handbook of Offshore Oil and Gas Operations. http://dx.doi.org/10.1016/B978-1-85617-558-6.00005-2

wellbores. Also, top drive systems allow the driller to keep the bit in the side-tracked hole, and mud motor enhancements permit drilling up to 60° per 100-ft-radius holes without articulated systems. In addition, pay zone steering systems are capable of staying within pay zone boundaries.

New innovations in drilling also include multilateral and multibranch wells. A multilateral well has more than one horizontal (or near horizontal) lateral drilled from a single site and connected to a single wellbore. A multibranch well has more than one branch drilled from a single site and connected to a single wellbore. Although not as pervasive in the offshore locale as in the onshore locale because of the necessity of pressure-sealed systems, multilateral and multibranch wells are more important factors in current (and future) offshore development.

Finally, planning for drilling or any offshore operation should include environmental considerations (Chapter 9). In fact, as with any project with the potential to harm the environment, plans must be developed and permits applied for and received before moving any equipment onto the work site. After the plans and permits are obtained, the pre-spud meeting, in addition to discussing well depth, casing points, and rig selection, should cover topics pertinent to the environmental management of the drilling and completion operation. Each site will have particular factors that differ from factors encountered at other sites (*site specificity*) and compliance with the regulation for the site is a must-do. It is also advisable because of public interest in any project that can affect the environment, to engage in public forums and public disclosure of a project should be made as early as possible, even during the time that the project is in the *project pending* stage. Well-presented public disclosure statements released to the potentially affected community may not allay all fears but can help in giving the public an undertaking of what is to be done. In such case, the public may be able to offer viable alternates that assist the project developer. The element of public surprise must (which often is the initiator for many objections) be removed.

5.2 DRILLING

Drilling is the most essential activity in oil and gas recovery. Once a prospect has been identified through the use of geological and exploration techniques (Chapter 2, Chapter 4), it is only through the actual penetration of the formation by the drill bit that the presence of recoverable crude oil and natural gas can be confirmed. The challenging conditions that confront the number of drilling rigs qualified for deep-water operations are limited.

Five rigs capable of drilling in up to 2,500 ft of water were operating in 1995. By 1996, nine were in operation and additional rigs were being upgraded for operations in deep water. Because this set of equipment has expanded more slowly than the demand for drilling in deep water necessitate specialized equipment.

Currently, the major use of offshore structures is for the exploration *and* production of oil and gas—mobile exploratory drilling rigs are used to drill wells to determine the presence or absence (dry hole) of crude oil and/or natural gas at the offshore site. If crude oil or natural gas is present in sufficient quantity to warrant development of the field, the well is plugged until a permanent production platform is in place. During drilling operations, supply vessels continuously support the drilling vessel that has sleeping quarters for the crew and galley facilities, standby boats for safety purposes, a heliport, and are organized in self-contained or tendered configurations.

Drilling offshore wells requires a modified drilling procedure when compared with onshore drilling—offshore wells are drilled by lowering a drill string (consisting of a drill bit, drill collar, and drill pipe) through a conduit (riser) that extends from the drilling rig to the sea floor. The steel drill pipe sections are typically 30 ft in length and weigh approximately 600 lb—as the drilling progresses, additional drill pipe sections are connected at the surface as the well deepens.

For readers unfamiliar with drill bits, the *roller-cone drill bit* usually has three cones with teeth and is designed to break the rock by indention and a gouging action. As the cones roll across the bottom, the teeth press against the formation with enough pressure to exceed the failure strength of the rock at which rock fracture occurs. The lower part of the bit's body supports the roller cones, usually three (but also one, two, or four can be used). Each cone has two or three rows of teeth, which can either be milled from the same block of metal as the bit ("non inserts" bit) or be fabricated from tungsten inserts that are harder and more durable than milled teeth. The external, intermediate, and internal rows of a cone each have a different number of teeth and each tooth is like a chisel, and has a maximum height (penetration depth) and a semi-angle (the angle made by its lateral surface with the bit axis. The wedge of a new tooth can either be sharp or flat and each cone is protected externally by the lobed body of the bit legs. The cones are supported by bearings, which are lubricated and sealed. The bearing axis of the cone forms the cone-journal angle with the horizontal level (or normal to the bit axis). The offset distance is the distance between the cone axis and the drill-bit center (typically measured in millimeters and the offset angle is the angle the cone axis would be rotated to make it to pass through the central bit axis.

Natural diamond bits are constructed with diamonds embedded into a matrix and are used in conventional rotary, turbine, and coring operations. Diamond bits can provide improved drilling rates than roller bits in some formations and all the diamond bit suppliers provide comparison tables between roller bit and diamond bit performance to aid users in bit selection based on economic evaluation. Some of the most important benefits of diamond bits are: (1) bit failure potential is reduced due to there being no moving parts, (2) less drilling effort is required for the shearing cutting action than for to the cracking and grinding action of the roller bit, (3) bit weight is reduced, therefore deviation control is improved, and (4) the low weight and lack of moving parts make them well suited for turbine drilling.

At the sea floor, the riser passes through a system of safety valves (blowout preventer (BOP)), which is used to contain pressures in the well and to prevent a blowout leading to oil leakage. The BOP—similar to the system used in onshore drilling—prevents any oil or gas from seeping out into the water. In the old days of surface drilling, the blowout would be called the *gusher*, which was an environmental disaster. Above the BOP, a *marine riser* extends from the sea floor to the drilling platform above—the marine riser is designed to house the drill bit and drill string and is sufficiently flexible to accommodate the movement of the drilling platform. Strategically placed slip and ball joints in the marine riser allow the subsea well to be unaffected by the pitching and rolling of the drilling platform.

At the surface, a rotary table at the surface turns the drill string and the drill bit teeth penetrate the sea floor sediment and the various rock formations that overly the reservoir while a drilling fluid (often referred to as *drilling mud*) is pumped into the drill pipe from a tank on the surface and the mud flows through perforations in the drill bit. The drilling mud collects the cuttings of rock produced by the drill bit and flows to the surface through the annulus between the well casing and the drill string below the sea bottom (*mud line*) and the riser and the drill string above the mud line. A strainer (filter) is used to remove the cuttings from the drilling mud, which is then recirculated through the mud tank and pumped to the drill string. Discharge of these fluids and cuttings into ocean waters are governed by environmental regulations and protocols (Chapter 10, Chapter 11).

The weight of the mud exerts a pressure greater than the pressure in the rock formations, and, therefore, keeps the well under control. As the drill bit penetrates further into the rock formations, strings of steel pipe (*casing*) are run into the well and cemented into place in order to seal off the walls of the well and maintain the integrity of the well by preventing collapse of the walls.

One of the most important pieces of equipment for offshore drilling is the subsea drilling template, which is used to connect the underwater well site to the drilling platform on the surface of the water. The template consists of an open steel box with multiple holes—the number of holes is dependent on the number of wells to be drilled—that is placed over the well site in an a precise exact position using satellite and GPS technology. The drilling template is secured (cemented) to the sea floor and automatic shutoff valves (BOPs) are attached to the template so that the well can be sealed off if there are problems on the platform or if the drilling rig has to be moved. Cables attach the template to floating platforms and are used to position the drill pipe accurately in the template and wellbore, while allowing for some vertical and horizontal movement of the platform.

The first stages of drilling typically penetrate 1500–3000 ft below the sea floor, after which steel casing is cemented into the hole—the casing helps prevent crude oil or natural gas from escaping into the environment. One of the primary differences between onshore drilling and offshore drilling is that, in offshore drilling, a large pipe (a riser) is used to connect the drilling rig to the seabed. However, the riser is not in place during the drilling of the shallowest part of the hole, but is installed after the casing has been cemented in place. The riser acts as a conduit for the drilling mud and drill cuttings which must be circulated through the well bore and back to the drilling rig so that the cuttings can be removed for disposal.

The drill bit may be a mile or more below the sea bed before crude oil and natural gas resources are reached and the pressure between the reservoir and the well must be controlled by adjusting the flow and weight of the drilling mud. The mud must be sufficiently viscous (heavy) to keep reservoir fluids from entering the borehole during drilling, but not so viscous that the mud penetrates the rock and prevents crude oil and natural gas from reaching the well. The first sign of success is usually an increase in the rate the drill bit penetrates the rock, which is often followed by traces of crude oil or natural gas in the rock cuttings brought up by the drill bit. This stage of drilling must be carefully monitored to prevent releases of crude oil and/or natural gas that might affect the environment.

In some cases, instruments are sent down a wireline and into the well to determine the existence of oil or gas—if crude oil and/or natural gas are present, steel production casing is set in place and is used as the conduit for transporting crude oil and natural gas safely to the surface. The well fluids are typically a mixture of crude oil, natural gas, sand, and brine and the fluids are processed before being sent ashore through a pipeline or transported to shore by a tanker. The degree of processing of the well fluids is dependent upon pipeline requirements (which may limit the amount of sand and brine in the fluids) or tanker regulations.

The wireline logging tools gather data about the thickness of rock layers, porosity, permeability and the composition of the fluids (oil, natural gas, or water) contained in them. These tools can be mounted on the drill string above the bit to send information to the surface continuously during drilling, or they can be lowered into a well after it is drilled. Another instrument—a measurement-while-drilling (MWD) tool—is also used to measure the direction and precise location of the bit while drilling horizontal wells. A common way to determine potential oil or natural gas production is the drill-stem test in which a special tool replaces the bit on the end of the drill string and is lowered into the well. It allows liquids or natural gas from the formation to flow into the empty drill pipe. This gives a good indication of the type and volume of the fluids in the formation, their pressure and rate of flow. Regulations require that the sea floor at any well site must be left in the same condition after drilling as before. This means that the drilling crew plugs the well bore with cement and removes the subsea equipment. A similar procedure is followed when a producing well is no longer economical to operate.

Offshore drilling for oil and natural gas offshore, in some instances hundreds of miles away from the nearest landmass, introduces many challenges to the challenge of onshore drilling. While the actual drilling mechanism used to drill into the sea floor is much the same as can be found on an onshore rig, the sea floor can sometimes be thousands of feet below sea level. Therefore, while with onshore drilling the ground provides a platform from which to drill, at sea an artificial drilling platform must be constructed and used. In fact, drilling in deep and ultra-deep water is one of the main goals for the development of oil fields in new exploration areas.

Underbalanced drilling (UBD), or near balanced drilling with light-weight drilling fluids, has practical applications offshore (Pratt, 2002; Bourgeois, 2003; Santos et al., 2003, 2006). UBD is a technique in which the hydrostatic head of drilling fluid is intentionally designed to be lower than the formation pressure. The hydrostatic head of the fluid may naturally be less than the formation pressure, or it can be induced by adding different substances to the liquid phase of the drilling fluid, such as: (1) natural gas,

(2) nitrogen, (3) air. Whether the underbalanced status is induced or natural, the result may be an influx of formation fluids that must be circulated from the well, and controlled at surface. The technique is useful for infill drilling in depleted fields and in developing lower-permeability reservoirs. UBD is most often used to prevent formation damage since the lightweight drilling fluids are less likely to invade formations, and there is no filter cake or mud cake buildup to impede flow from the reservoir.

There are four main techniques to achieve underbalance, including (1) use of lightweight drilling fluids, (2) gas injection down the drill pipe, (3) gas injection through a parasite string, and (4) foam injection. Using lightweight drilling fluids, such as fresh water, diesel and lease crude, is the simplest way to reduce wellbore pressure. A negative for this approach is that in most reservoirs the pressure in the wellbore cannot be reduced enough to achieve underbalance. The method of injecting gas down the drill pipe involves adding air or nitrogen to the drilling fluid that is pumped directly down the drill pipe. In the offshore environment where drilling challenges are presented, there is a narrow window between the formation pore pressure and fracture gradient (PPFG), which dictates conservative casing programs during conventional drilling. There is also the challenge of flow hazards in shallow water where (1) rig space is limited, (2) the equipment must be small/compact, and (3) drilling gases and other drilling fluid components must be generated or stored on the rig.

In order to drill underbalanced safely in the marine environment, there are different types of rotating control heads for rigs with either surface or subsea BOPs. Returns at the surface need to be controlled by a choke manifold system independent of the rig manifold. Drillers need tools to control and manage equivalent circulating density to prevent downtime. Floating vessels with risers require additional considerations that are not factors in land drilling. The riser is the weakest mechanical component in the blowout prevented stack, and the riser stress must be continually monitored.

Advantages to this technique include improved penetration, decreased amount of gas required, and that the wellbore does not have to be designed specifically for UBD. On the other hand, disadvantages include the risk of overbalance conditions during shut-in. Additionally, there are temperature limits to using foam in UBD, limiting use of the technique to wells measuring less than 12,000 ft deep.

Finally, *dual-gradient drilling* technology (*dual density drilling* technology) is being considered for ultra deepwater wells where the PPFG is particularly narrow and the need to reduce riser stress becomes critical. In this situation, two fluid gradients are maintained in the wellbore using seafloor pumps: (1) seawater in the annulus from the rig floor to the mudline, and (2) a mud column from the mudline to the topside. Mud returns to the surface via an auxiliary line, separate from the conventional riser. Another method of maintaining a dual density column, without a seafloor pump, is to use mud pumps to inject hollow glass spheres into the bottom of the riser to reduce the mud density in the riser to that of seawater.

5.2.1 Drilling Ships and Drilling Rigs

The drilling unit (drilling platform) can take many forms (Chapter 2), depending on the characteristics of the well to be drilled, including the underwater depth of the drilling target. Furthermore, in order to drill successfully the drillship (Chapter 2) must hold a constant position directly above the borehole in the ocean floor. And it must be able to do this in waves, wind, and ocean currents. Because of the great depths, anchors are out of the question and the ship retains its position by means of *dynamic positioning*, which is accomplished by computer controlled thrusters, some of which are mounted on hydraulic pods that retract into the ship when it is underway.

The drilling ship uses satellite navigation systems to find the chosen drill site. When it is in position, transponders are dropped to the seafloor and the thrusters are extended beneath the ship. With the thrusters lowered, the bottom of the ship literally bristles with propellers. Computers on the ship use the transponder signals to activate the various thrusters, which can move the ship forward, backward, or sideways. The ship should also be able to drill relatively unperturbed in high seas. Instead of being connected directly to the drill rig, the pipe string is suspended from a *heave compensator*, which functions like a shock absorber. If the ship rises on a wave, the heave compensator lowers the drill string. More modern drilling ships may also contain laboratories such as those for core handling, sampling, physical properties, chemical properties, thin section preparation, and X-ray photography.

Drilling rigs that use such new technology as top-drive drilling and proposed dual derricks are reducing drilling and completion times. In light of the limited number of vessels available for drilling deep-water wells and the resulting increasing drilling rates for such equipment, shorter operating times are a key advantage expected from dual rig derricks.

In addition to creating drilling rigs that can operate at great water depths, new drilling techniques have evolved, which increase productivity and lower unit costs. The evolution of directional and horizontal drilling to penetrate multiple diverse pay targets is a prime example of technological advancement applied in the offshore. The industry now has the ability to reduce costs by using fewer wells to penetrate producing reservoirs at their optimum locations. Horizontal completions within the formation also extend the reach of each well through crude oil-bearing (and/or natural gas-bearing) formations, thus increasing the flow rates compared with those from simple vertical completions. These advancements can be attributed to several developments. For example, the evolution of retrievable whipstocks allows the driller to exit the cased wells without losing potential production from the existing wellbores. Also, top drive systems allow the driller to keep the bit in the sidetracked hole, and mud motor enhancements permit drilling up to 60° per 100-ft-radius holes without articulated systems. In addition, pay zone steering systems are capable of staying within pay zone boundaries.

New innovations in drilling also include multilateral and multibranch wells. A multilateral well has more than one horizontal (or near horizontal) lateral drilled from a single site and connected to a single wellbore. A multibranch well has more then one branch drilled from a single site and connected to a single wellbore. Although not as pervasive in the offshore as in the onshore because of the necessity of pressure-sealed systems, multilateral and multibranch wells are expected to be more important factors in future offshore development.

5.2.2 Top Drive Drilling

Top drive drilling replaces the kelly method of rotation used in conventional rotary drilling. Using hydraulic or electric motors suspended above the drill pipe enables top drives to rotate and pump continuously while drilling or during the removal of the drill pipe from the hole.

Top drive drilling systems are one of the greatest contributions to the offshore drilling industry. Until 1982, the drill-string was handled and rotated by a kelly joint and a rotary table. Drilling and making connections on offshore rigs had been virtually unchanged for years. Then development of the top drive, which rotates the drill pipe directly and is guided down rails in the derrick, replaced the need for a kelly joint to rotate the drill string. It performs normal hoisting requirements such as tripping and running casing. It also added the ability for drilling with triples, circulating and rotating during tripping, and back-reaming and/or freeing stuck pipe.

Top drive drilling provides a safer drilling operation by reducing the hazards of rotary tongs and spinning chain. In addition, the pipe handling features use hydraulic arms to move drill pipe and drill collars to and from the V-door and monkey board, thereby reducing strenuous work and increasing pipe handling safety. The automatic, driller-operated pipe elevators eliminate accidents caused by drilling crews operating elevators manually during under balanced drilling operations. Well control capability is greatly enhanced because of the ability to screw into the string any point in the derrick to circulate drilling fluids. The remotely operated kelly valve reduces mud spillage when back reaming or breaking off after circulating above the rig floor.

The most important feature of the top drive is the ability to rotate and pump continuously while reaming into or out of the hole. Continuous rotation means substantially reduced friction when removing the string from or tripping back into directional or horizontal wells. In addition, there is less reservoir damage due to reduced usage and subsequent entry of gel/clay particles into the producing formation.

However, the top drive system offers many other benefits, each of which act to increase the performance of the drilling rig and improve the return on investment for the well. For instance, top drives reduce the instances of stuck pipe.

Historically, it has not been considered uncommon to get a drill string stuck in the hole from time to time, and the potential of stuck pipe increases with hole depth and the particular formation being drilled through. Regardless of the depth or type of the formation, drilling with a top drive drastically reduces the instances of stuck pipe. Drilling with 30 m of pipe at a time allows more time for hole conditioning and circulating solids to the surface. Also, because there are fewer connections to be made, the pumps are stopped less often. This results in less circulation time required to achieve uniform distribution of annular cuttings load. All of these factors help keep the bit and the string free to rotate and help prevent sticking.

Besides the stuck pipe that could previously occur while drilling, drill strings can also encounter tight spots when tripping in or out of the hole. If a tight spot is encountered during a trip on a conventional rig, it becomes a major effort to pick up the kelly and begin circulating and rotating the pipe through the trouble zone. However, when tripping on a rig equipped with a top drive, rotation and circulation can be achieved at any point within a matter of seconds. The driller simply needs to set the slips, lower the top drive to engage the drive stem, make up the connection with the top drive pipe-handler, and begin circulation. This feature provides the driller with the added benefit of being able to *back-ream* whenever necessary. Entire sections of the well bore can be *reamed* through without significantly impacting trip times. The result is a conditioned and clean borehole, ensuring a successful casing run.

The ability to quickly and easily connect to the drill string during trips provides benefits that extend beyond just preventing stuck pipe. For instance, consider the situation when a kick is encountered during a trip. On a kelly rig, the crew has little recourse and will find it very difficult to contain the fluids escaping the well without taking drastic measures. In the case where a top drive encounters the same situation, the slips can be set and the top drive connected immediately, thus containing and controlling the well within seconds of a kick being detected. These rapid responses to well kicks have increased the safety of the rig floor and have helped to protect drilling personnel from possible injury.

Other aspects of top drive drilling have led to increased awareness of safety all around the rig floor. When kelly drilling, the rotary table and kelly bushing are spinning rapidly at the rig floor, while the crew is in close proximity. Since top drives eliminate the need for the kelly drive mechanism, and the rotary table is not used to rotate the pipe, the only thing rotating at the drill floor is smooth drill pipe. Also, since the top drive eliminates two out of three drilling connections, the drill crew is less exposed to possible injury; less exposure to possible injury results in less injuries.

While improvements in drilling time and crew safety are well documented, and these features can benefit any drilling rig, certain aspects of top drive drilling have allowed drastic improvements in oil and gas recovery from reservoirs. Enhanced recovery of crude oil and natural gas has been achieved through a combination of extended reach and horizontal drilling programs. Extended reach, or highly deviated wells, increases the horizontal area of a reservoir that can be tapped from a given location. Horizontal completions allow a major increase in the ultimate recovery from a given reservoir. Both of these offers tremendous financial incentive from the operator's perspective, and both of these situations can only be achieve by utilizing a top drive drilling system.

In the case where geological, geographic, or economic factors limit the placement of drilling locations, it may be beneficial to deviate the wells drilled from a given location in order to access certain areas of a reservoir. This is achieved by drilling at angles from 70° to 90° from vertical for extended measured depth. When drilling with a top drive, and taking into account other parameters such as drilling fluid composition, it is now considered commonplace to extend the horizontal reach to several miles.

In fact, the productivity of a conventional well is proportional to the permeability-thickness product. Low productivities result from low values of permeability or formation thickness (or both). This can be compensated for in horizontal wells where the length of the horizontal section is not imposed by nature but chosen. The permeability-length product in horizontal wells plays a role similar to that of the permeability-thickness product of conventional wells. In addition, to increasing productivity, horizontal wells have been shown to increase productivity, to reduce coning tendencies, and to improve recovery by a variety of mechanisms (Sherrard et al., 1987; Wilkerson et al., 1988; Wilson and Willis, 1986).

The long wellbores allow longer completed intervals and therefore increased production rates. In reservoirs overlying an aquifer or located under a gas cap, the increased standoff from the fluid contacts can improve the production rates without causing coning. Additionally, the longer wellbore length serves to reduce the drawdown for a given production rate and thus further reduces coning tendencies. Fractured reservoirs can also benefit from horizontal wells. Long wellbores are likely to intersect more fractures and hence improve both production rate and ultimate recovery. Furthermore, the application of horizontal wells early in a project may allow development with fewer wells because of the larger drainage area of each well. In some fields, the advantages of horizontal drilling may allow development where conventional techniques would be uneconomical.

Most offshore rigs now use top drives—hydraulic or electric motors suspended above the drill string. In some situations, the bit can be turned by a mud motor, a down hole hydraulic drive that is inserted above the bit at the bottom of the drill string. It receives power from the flow of the drilling mud. This technique is often used to drill directional and horizontal wells, which are important to offshore operations. Directional drilling allows a number of wells to be drilled from one location. Horizontal wells can penetrate a long section of the rock formation, providing better contact with the reservoir. This reduces the time it takes to extract crude oil or natural gas from the reservoir and in some cases increases the total amount of product that can be recovered. As drilling technologies and methods have improved, the reach of wells into producing formations continues to increase.

5.2.3 Dual Derricks

The dual derrick system is a relatively recent drilling structure used for deep water drilling operations. Efficiencies exist from allowing dual hook load operations that make this type of structure advantageous (Effenberger et al., 2013).

Most modern drillships have some degree of dual-rig activity (i.e., they have two drilling derricks on one hull) and have the capability to run two riser and two BOP systems with one system drilling and the other completing a well on a subsea template. With this drill-and-complete mode on a multiwell template, efficiency is claimed to increase significantly. For exploration wells, it is possible to run casing with one derrick set and drill with the other, thus reducing total time to complete the operation. Some systems have the capability to produce and store crude oil, thus eliminating the need to flare or burn the produced fluid during well testing.

The dual-activity capability units are, in general, exploration units with the development capability for large-numbered multiwell subsea templates in very deep water. Generally for exploration wells, the greater the depth of water and the shorter the well is, the more commercially attractive the dual-activity units become compare to a standard spread-moored semisubmersible unit.

5.2.4 Directional and Horizontal Drilling

New methods to drill for oil are continually being sought, including directional or horizontal drilling techniques, to reach oil under ecologically sensitive areas, and using lasers to drill oil wells. *Directional drilling* is also used to reach formations and targets not directly below the penetration point or drilling from shore to locations under water (Speight, 2014). A controlled deviation may also be used from a selected depth in an existing hole to attain economy in drilling costs. Various types of tools are used in directional drilling along with instruments to help orient their position and measure the degree and direction of deviation; two such tools are the *whipstock* and the *knuckle joint*. The whipstock is a gradually tapered wedge with a chisel-shaped base that prevents rotation after it has been forced into the bottom of an open hole and used to assure that the bottom of the drill pipe that is oriented in the direction the well is intended to take. As the bit moves down, it is deflected by the taper about 5 from the alignment of the existing hole.

Directional drilling and horizontal drilling is the practice of drilling nonvertical wells. Many prerequisites enabled this technology to become productive. Probably the first requirement was the realization that oil wells are not necessarily vertical and there were several lawsuits in the late 1920s alleging that wells drilled from a rig on one property had crossed the boundary and were penetrating a reservoir on an adjacent property.

Prior experience with rotary drilling had established several principles for the configuration of drilling equipment down hole (bottom hole assembly) that would be prone to drilling a crooked hole in which initial accidental deviations from the vertical would be increased. Counter-experience had also given early directional drillers principles of bottom hole assembly design and drilling practice that would help bring a crooked hole nearer the vertical.

Combined, these survey tools and bottom hole assembly designs made directional drilling possible, but it was perceived as arcane. The next major advance was in the 1970s, when downhole drilling motors (mud motors), driven by the hydraulic power of drilling mud circulated down the drill string, became common. These allowed the bit to be rotated on the bottom of the hole, while most of the drill pipe was held stationary. Including a piece of bent pipe between the stationary drill pipe and the top of the motor allowed the direction of the wellbore to be changed without needing to pull all the drill pipe out and place another whipstock. Coupled with the development of MWD tools, directional drilling became easier. The most recent major advance in directional drilling has been the development of a range of rotary steerable tools that allow three dimensional control of the drill bit without stopping the drill string rotation—these tools have improved the process of drilling highly deviated wells.

To achieve directional drilling, downhole instrumentation is required to deflect the direction of the bit from the drill-string axis. Usually, the directional tool is a downhole motor, either provided with a bending housing (steerable motor) or used with a bending sub above the motor. Another directional technique uses the whipstock, which is a nonsymmetric steel joint forcing the drilling direction. Sometimes this tool is removed after drilling has taken the desired direction.

Directional wells increase the exposed section length through the reservoir by drilling through the reservoir at an angle and allow drilling into the reservoir where vertical access is difficult or not possible—such as when the crude oil reservoir is under a town, under a lake, or lies beneath a difficult-to-drill-formation. Directional drilling also allows more wellheads to be grouped together on one surface location and may not require as many movement of the drillship—there is also less surface area disturbance. For example, on an offshore oil platform, up to about 40 wells can be grouped together—the wells fan out from the platform into the reservoir below. Furthermore, directional drilling allows drilling relief wells to relieve the pressure of a well producing without restraint (blowout) in which another well could be drilled starting at a safe distance away from the blow out, but intersecting the troubled wellbore—this is followed by pumping heavy fluid (kill fluid) is pumped into the relief wellbore to suppress the high pressure in the original wellbore causing the blowout.

Most directional drilling operations follow a well path that is predetermined by engineers and geologists before the drilling commences. When the drilling process is started, periodic surveys are taken with a downhole camera instrument (single shot camera) to provide survey data (inclination and azimuth) of the well bore—the pictures are typically taken at intervals between 30 and 500 ft (commonly 90 ft during active changes of angle or direction) and distances of 200 to 300 ft are more typical while drilling ahead (not making active changes to angle and direction). During critical angle and direction changes, a MWD tool will be added to the drill string to provide continuously updated measurements that may be used for (near) real-time adjustments. The data acquired during the operation indicate whether or not the well is following the planned path and whether or not the orientation of the drilling assembly is causing the well to deviate as planned. Corrections are regularly made by techniques as simple as adjusting rotation speed or the drill string weight (weight on bottom) and stiffness, as well as more complicated and time consuming methods, such as introducing a downhole motor.

Thus, directional drilling is used to: (1) to optimize drilling, as several wells and side tracks—that is, another well drilled using the upper part of the first well can be drilled from the same onshore site or marine platform, (2) to optimize the approach to the target, so that an increased interval of reservoir is crossed by the borehole with sloping or horizontal trajectory—the ultimate result is to improve the subsequent production, (3) to reach targets that are located in zones not accessible by vertical drilling, (4) to cross faults, by choosing a direction close to perpendicular to the fault in order to minimize the drift effects that are induced on drilling by the fault, (5) to drill salt domes from lateral locations, which sometimes can be preferable to vertical drilling because of the problems involved in the drilling of the salt, and (6) to realize side tracking in wells partially damaged—this operation makes it possible to drill a borehole section parallel to an abandoned one.

Until the arrival of modern downhole motors and better tools to measure inclination and azimuth of the hole, directional drilling and horizontal drilling was much slower than vertical drilling due to the need to stop regularly and take time consuming surveys, and due to slower progress in drilling itself (lower rate of penetration). These disadvantages have shrunk over time as downhole motors became more efficient and semi-continuous surveying became possible.

A disadvantage of wells with a high inclination was that prevention of sand influx into the well was less reliable and needed higher effort. Again, this disadvantage has diminished such that, provided sand control is adequate planned, it is possible to carry it out reliably.

5.2.5 Multilateral Drilling Technology

During the 1990s, the technology of drilling branches in the reservoir was developed and technology enables the drilling of more than two wells (commonly three to six or up to eight production bores) out from a *mother bore* in the production zones of interest. In certain circumstances, the technology has provided access to new reserves and allowed large reductions in overall well costs.

A *combination well* is a well drilled for two purposes: (1) production from a thin oil zone and (2) gas production from a gas cap. The horizontal well section is suited to extracting oil from a thin oil layer, which in the illustrated case has a gas cap above and a water zone below. Production from such a well can better be optimized with a low reservoir draw down, to avoid provoking the water to cone or gas to cusp into the wellbore prematurely. After production of the oil zone, gas can be produced through the higher perforations in order to avoid drilling a new well for the purpose.

5.3 SALT DEPOSITS

Technology has provided access to areas that were either technically or economically inaccessible owing to major challenges, such as deposits located in very deep water or located below salt formations. While the major additions to production and reserves in the Gulf of Mexico have occurred in deep waters, work in refining the discovery and recovery of oil and gas deposits in subsalt formations must be noted as another promising area of potential supplies.

Eighty-five percent of the continental shelf in the Gulf of Mexico, including both shallow- and deep-water areas, is covered by salt deposits, which comprises an extensive area for potential crude oil and natural gas development. Phillips Petroleum achieved the first subsalt commercial development in the Gulf of Mexico with its Mahogany platform. This platform, which was set in August 1996, showed that commercial prospects could be found below salt (in this case below a 4000 ft salt sheet).

The subsalt accumulations can be found in structural traps below salt sheets or sills. The first fields under salt were found by directional wells drilling below salt overhangs extending out from salt domes. Experience in field development close to salt-covered areas indicated that not all salt features were simple dome-shaped features or solid sheets. Often the salt structure was the result of flows from salt deposits that extended horizontally over sedimentary formations that could contain oil. The salt then acts as an impermeable barrier that entraps the crude oil and natural gas in accumulations that may be commercially viable prospects.

The identification of structures below salt sheets was the first problem to overcome in the development of subsalt prospects, as the salt layers pose great difficulty in geophysical analysis. The unclear results did not provide strong support for investing in expensive exploratory drilling. The advent of high-speed parallel processing, pre- and post-stack processing techniques, and 3-D grid design helped potential reservoir resolution and identification of prospects.

Industry activity in subsalt prospect development has been encouraged also by improvements in drilling and casing techniques in salt formations. Drilling through and below salt columns presents unique challenges to the drilling and completion of wells. The drilling of these wells requires special planning and techniques. Special strings of casing strategically placed are paramount to successful drilling and producing wells.

The highly sophisticated technology available to firms for offshore operations does not necessarily assure success in their endeavors, and the subsalt prospects illustrate this point. The initial enthusiasm after the Mahogany project was followed by a string of disappointments in the pursuit of subsalt prospects. After a relative lull in activity industry-wide, Anadarko announced a major subsalt discovery in shallow water that should contain at least 140 million barrels of oil equivalent (BOE), with reasonable potential of exceeding 200 million BOE. Successes of this magnitude should rekindle interest in meeting the challenge posed by salt formations.

Subsalt development has also been slowed because the majority of prospects have been leased or recovery from the subsalt is delayed by production activities elsewhere on a given lease. Subsalt operations apparently will be more a factor in the future as flows from leases presently dedicated to other production decline and the leases approach the end of their lease terms, which will promote additional development to assure continuation of lease rights.

5.4 WELL COMPLETION

Once the final depth has been reached, the well is completed to allow oil to flow into the casing in a controlled manner.

The well consists of a wellhead, which supports the well casing in the ground, and a pod (*submerged Christmas tree, wet tree*), which contains valves to control the flow and to shut off the flow in the case of an emergency or a leak in the riser. This pod Subsea wells are expensive, but not as expensive in deepwater as placing a platform at the site. If a subsea well ceases to produce, or if its rate of production falls below economic limits, it is necessary to bring in a mobile drilling unit to remove the tree and perform the workover. This can be an extremely expensive operation and if the outcome of the workover is in doubt, the operator may choose to abandon the well instead. Because of this, much of the oil and gas in reservoirs produced through subsea trees may be left behind. Subsea wells may also result in lower reservoir recovery simply because of the physics of their operation. The chokes and valves placed in a subsea tree result in a pressure drop in the flow of oil or gas. When the well formation drops below a certain threshold, production ceases to flow. The difference in cut-off pressure between a subsea well and a surface well can be as much as 1000 psi versus 100 psi.

First, a *perforating gun* is lowered into the well to the production depth. The gun has explosive charges to create holes in the casing through which oil can flow. After the casing has been perforated, a small-diameter pipe (*tubing*) is run into the hole as a conduit for oil and gas to flow up the well and a *packer* is run down the outside of the tubing. When the packer is set at the production level, it is expanded to form a seal around the outside of the tubing. Finally, a multivalve structure (the *Christmas tree*; Figure 5.1) is installed at the top of the tubing and cemented to the top of the casing. The Christmas tree allows them to control the flow of oil from the well.

The average rate of production from deep-water wells has increased as completion technology, tubing size, and production facility efficiencies have advanced. Less expensive and more productive wells can be achieved with extended reach, horizontal and multilateral wells. Higher rate completions are possible using larger tubing (5-in. or more) and high-rate gravel packs. Initial rates from Shell's Auger Platform were about 12,000 barrels of oil per day per well. These flow rates, while very impressive, have been eclipsed by a well at BP's Troika project on Green Canyon Block 244, which produced 31,000 barrels of oil on January 4, 1998.

Another area of development for completion technology involves subsea well completions that are connected by pipeline to a platform that may be miles away. The use of previously installed platform infrastructure as central producing and processing centers for new fields allows oil and gas recovery from fields that would be uneconomic if their development required their own platform and facilities. Old platforms above and on the continental slope have extended their useful life by processing deep water

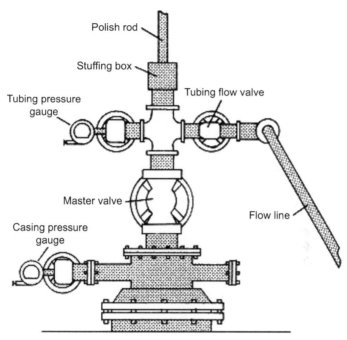

Figure 5.1 The Christmas tree: A collection of control valves at the wellhead.

fields. A prime example of this innovation is the Mensa field, which gathers gas at a local manifold and then ships the gas by pipeline to the West Delta 143 platform 68 miles up the continental shelf.

The exploitation of deep water deposits has benefited from technological development directed at virtually all aspects of operation. Profitability is enhanced with any new equipment or innovation that either increases productivity, lowers costs, improves reliability, or accelerates project development (hence increasing the present value of expected returns). In addition to the major developments already discussed, other areas of interest for technological improvement include more reliable oil subsea systems (which include remotely operated vehicle systems), bundled pipeline installations of 5 miles or more that can be towed to locations, improved pipeline connections to floating and subsea completions, composite materials used in valving, and other construction materials.

The advantages of adopting improved technology in deep water projects are seen in a number of ways. For example, well flow rates for the Ursa project are 150% more than those for the Auger project just a few years earlier. The economic advantages from these developments are substantial as the unit capital costs were almost halved between the two projects. The incidence of dry holes incurred in exploration also has declined with direct reduction in project costs. The number of successful wells as a fraction of total wells has increased dramatically, which reflects the benefits of improvements in 3-D seismic and other techniques. Lastly, aggressive innovation has improved project development by accelerating the process from initial stages to the point of first production. Rapid development requires not only improvements in project management, but also better processes to allow construction of new facilities designed for the particular location in a timely fashion. Project development time had ranged up to 5 years for all offshore projects previously. More recent field development has been conducted in much less time, with the period from discovery to first production ranging between 6 and 18 months. Experience with deep-10 water construction and operations has enabled development to proceed much faster, with time from discovery to production declining from 10 years to just over 2 years by 1996. Accelerated development enhances project economics significantly by reducing the carrying cost of early capital investment, and by increasing the present value of the revenue stream. Design improvements between the Auger and Mars projects allowed Shell to cut the construction period to 9 months with a saving of $120 million.

The completion of offshore wells for offshore petroleum production involves two steps (1) tubular casing lines the length of the well bore to ensure safe control of crude oil and natural gas, (2) natural gas flows to the surface under its own pressure, but oil may need to be coaxed to the surface. In fact, well completion for producing crude oil and natural gas from offshore locations is very similar to the process used on dry land and similar principles may be employed, but generally in a smaller area. Crude oil and natural gas wells are prepared for production through a process called completion. In the first step, a production casing is cemented into the well bore.

The casing—tubular steel pipe connected by threads and couplings—lines the total length of the well bore to ensure safe control of the crude oil and natural gas, prevent water entering the well bore and keep the rock formations from falling or collapsing into the well bore. Once the cement has set, the production tubing can be put in place. Production tubing is steel pipe that is smaller in diameter than the production casing. Production tubing was traditionally made up of joined sections of pipe, similar to the string of pipe used for drilling, but most offshore wells today use coiled tubing, a continuous, high-pressure-rated hollow steel cylinder. Production tubing is lowered into the casing and hangs from a sea floor installation called the wellhead. The wellhead has remotely operated valves and chokes that allow it to regulate the flow of oil and gas. The casing is then perforated to allow crude oil and natural gas to flow into the well. This is done with a perforating gun, an arrow of shaped explosive charges that is lowered into the well. An electrical impulse fires the charges that perforate the casing, surrounding cement, and reservoir rock.

Natural gas flows to the surface under its own pressure, but oil may need to be coaxed to the surface by pumping in the later stages of a well's lifespan. In many crude oil and natural gas wells, the formation must also be stimulated by physical or chemical means. This procedure creates channels beyond the perforations that allow crude oil and natural gas to flow back to the well. The two most common stimulation methods are acidizing and fracturing (also known as "fracking"). Acidizing involves injecting acids under pressure through the production tubing and perforations and into the formation.

Fracking involves pumping a fluid, such as a water gel, down the hole under sufficient pressure to create cracks in the formation. Whether these techniques are used depends on a number of factors, including potential environmental impacts as well as the geology of the reservoir. Subsea pipelines can be used to connect multiple offshore wells to processing and transportation facilities. Subsea pipelines were used for the Cohasset-Panuke project and play a key role in the Sable Offshore Energy Project, which also uses a major subsea pipeline from the production area to onshore facilities.

Once the final depth has been reached, the well is completed to allow oil to flow into the casing in a controlled manner. First, a *perforating gun* is lowered into the well to the production depth. The gun has explosive charges to create holes in the casing through which oil can flow. After the casing has been perforated, a small-diameter pipe (*tubing*) is run into the hole as a conduit for oil and gas to flow up the well and a *packer* is run down the outside of the tubing. When the packer is set at the production level, it is expanded to form a seal around the outside of the tubing. Finally, a multivalve structure (the *Christmas tree*; Figure 5.1) is installed at the top of the tubing and cemented to the top of the casing. The Christmas tree allows them to control the flow of oil from the well.

Tight formations are occasionally encountered and it becomes necessary to encourage flow. Several methods are used, one of which involves setting off small explosions to fracture the rock. If the formation is mainly limestone, hydrochloric acid is sent down the hole to produce channels in the rock. The acid is inhibited to protect the steel casing. In sandstone, the preferred method is hydraulic fracturing. In this technique, a fluid with a viscosity high enough to hold coarse sand in suspension is pumped at very high pressure into the formation, fracturing the rock. The grains of sand remain, helping to hold the cracks open.

REFERENCES

Bourgeois, G.D., 2003. Under-Balanced Drilling Experience in a Shallow Clastic Oil Field, Offshore Sabah, South China Sea. Proceedings. SPE Asia Pacific Oil & Gas Conference in Jakarta, April. Society of Petroleum Engineers, Richardson, Texas.

Effenberger, M., Ross, G., Martinez, J., Beck, R., Botros, T., Jesudasen, S., et al., 2013. Wind Testing of a Dual Derrick. Paper No. OTC 24262. Proceedings. Offshore Technology Conference, Houston, Texas. May 6–9.

Pratt, C.A., 2002. Underbalanced Drilling: The Past, The Present, and the Future. Proceedings. SPE/AFTP Section Meeting, Paris, France. September 16. Society of Petroleum Engineers, Richardson, Texas.

Santos, H., Leuchtenberg, C., Shayegi, S., 2003. Micro-Flux Control: The Next Generation in Drilling Process for Ultra-deepwater. Proceedings. Offshore Technology Conference, Houston, May 5–8. Society of Petroleum Engineers, Richardson, Texas.

Santos, H., Reid, P., McCaskill, J., Kinder, J., Kozicz, J., 2006. Deepwater Drilling Made More Efficient and Cost Effective; Using the Microflux Control Methods and an Ultralow Invasion Fluid to Open the Mud-Weight Window. SPE Paper OTC 17818. Proceedings. Offshore Technology Conference, Houston, Texas. May 1–4. Society of Petroleum Engineers, Richardson, Texas.

Sherrard, D.W., Brice, B.W., MacDonald, D.G., 1987. Application of horizontal wells at Prudhoe Bay. Journal of Petroleum Technology 39 (11), 1417–1425.

Speight, J.G., 2014. The Chemistry and Technology of Petroleum, 5th Edition CRC Press, Taylor & Francis Group, Boca Raton, Florida.

Wilkerson, J.P., Smith, J.H., Stagg, T.O., Walters, D.A., 1988. Horizontal drilling techniques at Prudhoe Bay, Alaska. Journal of Petroleum Technology 40 (11), 1445–1451.

Wilson, R.C., Willis, D.N., 1986. Successful High Angle Drilling in the Statfjord Field. Paper No. SPE 15465. Proceedings. SPE Annual Technical Conference and Exhibition, New Orleans, Louisiana. October 5--8. Society of Petroleum Engineers, Richardson, Texas.

CHAPTER SIX

Production

Outline

6.1 INTRODUCTION

Crude oil and natural gas accumulate over geological time in porous underground rock formations called reservoirs that are at varying depths in the crust of the Earth and, in many cases, elaborate equipment is required to recover (produce) the crude oil and natural gas (Chapter 1, Chapter 3) (Speight and Ozum, 2002; Parkash, 2003; Hsu and Robinson, 2006; Gary et al., 2007; Speight, 2011a, 2014a). The oil is usually found trapped in a layer of porous sandstone, which lies just beneath a dome-shaped anticline (Figure 6.1) or folded layer of some nonporous rock such as limestone. In other formations, the oil is trapped at a fault, or break in the layers of the crust (Speight, 2014a). Whether it is onshore or offshore, the anatomy of a reservoir is complex and is *site specific* (*reservoir specific*), both microscopically and macroscopically. Because of the various types of accumulations and

Handbook of Offshore Oil and Gas Operations. http://dx.doi.org/10.1016/B978-1-85617-558-6.00006-4

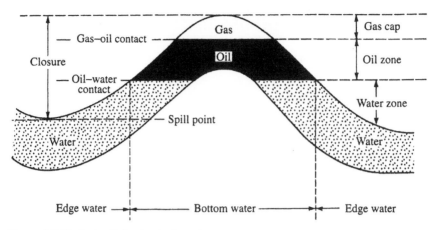

Figure 6.1 Typical anticlinal petroleum trap.

the existence of wide ranges of both rock and fluid properties, reservoirs respond differently and must be treated individually.

Generally, crude oil reservoirs exist with an overlying *gas cap*, in communication with aquifers, or both. The crude oil resides together with natural gas and water in very small holes (pore spaces) and fractures in the reservoir rock—the concept of a reservoir being a swimming-pool type structure is not geologically viable. The size, shape, and degree of interconnection of the pores are not consistent and vary considerably within the reservoir rock. Typically, because of density segregation, the crude oil layer in the sandstone is usually underlain by sandstone saturated with mineral-containing water (brine). The oil is released from the reservoir by release of the reservoir pressure, which is accomplished by drilling a well and puncturing the limestone layer on either side of the dome (Figure 6.1). If the peak of the formation is penetrated (due to an incomplete survey leading to poor knowledge of the reservoir position and structure and the haste to produce crude oil), only the gas is obtained. For the same reason—incomplete survey and haste to produce crude oil—if the penetration is made too far from the center, only brine water is obtained. Generally, the crude oil (conventional crude oil) and natural gas in a reservoir is usually under high great pressure that it flows naturally—the pressure forces the crude oil out of the reservoir. With time—sooner rather than later if the anatomy of the reservoir is not known and the drilling has not been well planned—the reservoir pressure later diminishes so that the oil must be pumped from the well. If this happens, natural gas or water is sometimes pumped into the well to replace the oil that is withdrawn (*reservoir repressurizing*).

Modern offshore crude oil and natural gas production activities take place in waters of the coastal nations of the earth—and this includes lacustrine activities as well as the oceanic activities. Offshore operators are no longer confided to wharves or jetties that jut into the ocean but, instead, wells are drilled from modern steel or concrete structures which are, in many cases, movable and many can float while being moved, and often while drilling in waters over 7500 ft deep and often more than 200 miles from the shore. Offshore development concepts include fixed platforms, floating platforms, subsea wells, as well as a variety of other units (Chapter 3).

Subsea production systems (subsea wells) typically lie directly on the seafloor. In the shallower waters of the Arctic, however, ice keels or icebergs pose the potential risk of colliding with the equipment. In these areas, the subsea equipment may be placed in a hole dredged into the seafloor so that the ice will pass over it without causing any damage. In deeper water depths, pipeline and subsea equipment may be safely placed directly on the seafloor, below the potential threat of ice keels.

The functions of offshore production facilities are very much the same as those described for land operations (Speight, 2014a) that must be able to accommodate the location and types of the reservoir sediments (Chapter 2) as well as the types of crude oil to be recovered (Chapter 1) (Speight, 2009, 2014a). In fact, the offshore production platform is a gathering station where crude oil and natural gas are collected and processed before transport to an onshore facility. Such activities reflect the similarity of the offshore facilities to the onshore facilities but that is where the similarity ends. However, the design and layout of the offshore facilities are very different from the onshore facilities: (1) a platform has to be installed above sea level before drilling and process facilities can be placed off shore, (2) there are no utilities offshore—all light, water, power, and living quarters have to be installed to support the platform operations, (3) weight and space restrictions make platform based storage tanks nonviable, so alternative storage methods or facilities for immediate transport of the crude oil and natural gas have to be employed.

Offshore oil and gas production is more challenging than land-based installations due to the remote and harsher environment and a large part of the innovation in the offshore petroleum and gas exploration and recovery (production) sector concerns overcoming these challenges, including the need to provide very large production facilities. Production and drilling facilities may be very large, such as the Troll A platform standing on a depth of approximately 1000 ft (Figure 6.2). Another type of offshore platform may float with a mooring system to maintain it on location. While a floating system may be lower cost in deeper waters than a fixed platform, the

Figure 6.2 Production and drilling complex. *Source: Figure reproduced from El-Reedy, 2012, fig. 1.7.*

dynamic nature of the platforms introduces many challenges for the drilling and production facilities.

The ocean depth can add hundreds of feet or more to the fluid column which increases the equivalent circulating density and downhole pressures in drilling wells, as well as the energy needed to lift produced fluids for separation on the platform. As a result the current trend is to conduct more of the production operations subsea by (1) separating water from oil and reinjecting the water rather than pumping it up to a platform or (2) flowing to onshore, with no installations visible above the sea. Subsea installations help to exploit resources at progressively deeper waters—locations which had been inaccessible—and overcome challenges posed by sea ice such as in the Barents Sea. One such challenge in shallower environments is seabed gouging by drifting ice features—the means of protecting offshore installations against ice action includes burial in the seabed.

Offshore manned facilities also present logistics and human resources challenges—and offshore oil platform is a small community in itself with cafeteria, sleeping quarters, management and other support functions. Also, in many cases staff members are transported by helicopter for a 2-week shift and supplies and waste are transported by ship—the supply deliveries need to be carefully planned because storage space on the platform is limited. In fact, subsea facilities are also easier to expand, with new separators or different modules for different oil types, and are not limited by the fixed floor space of an above-water installation.

The success of offshore exploration and production during the past four decades can be attributed, in large part, to technological advances (Chapter 4). Innovative technologies, such as new offshore production systems, three-dimensional (3-D) seismic surveys, and improved drilling and completion techniques, have improved the economics of offshore activities and enabled development to occur in deeper, more remote environments. Progress in offshore technology is exemplified by advances in production platforms, which provide a base for operations, drilling, and then production, if necessary. For many years, the standard method for offshore development was to utilize a fixed structure based on the sea bottom, such as an artificial island or man-made platform. Use of this approach in ever-deeper waters is hindered by technical difficulties and economic disadvantages that grow dramatically with water depth.

The focus of this chapter is the major developments in exploration, drilling, completion, and production technology. It also briefly discusses subsalt deposits, which comprise an additional area of promising application for the new technologies. Since 85% of the continental shelf in the Gulf of Mexico is covered by salt deposits, the potential for hydrocarbon development may be quite large.

This chapter, for the most part, deals with those recovery methods that are applied to offshore crude oil as well as to the recovery of onshore crude oil and, in some cases, to recovery of heavy oil from reservoirs. Many of the methods used for onshore oil recovery have not yet been applied to offshore crude oil but that may change as reservoir energy diminishes and more drastic recovery methods are necessary. Methods (other than mining methods that are not applicable to offshore locations) that are being proposed for the recovery of bitumen from tar deposits are presented elsewhere (Speight, 2009, 2014a). Such methods include steam stimulation (SAGD) and variants thereof and there may, in the future, be opportunities to apply combustion methods to undersea reservoirs and deposits.

6.2 SUBSEA PRODUCTION

Subsea production facilities are installations that are entirely submerged under the sea and such installations are often used to produce crude oil and natural gas from locations that cannot easily be reached by directional drilling from an existing platform and where a separate platform would be unnecessary and uneconomic.

Mature crude oil and natural gas fields are areas in which crude oil and natural gas have been produced for a number of years, existing production tends to decline. Remaining accumulations are often too small to justify stand-alone offshore platform developments so subsea production systems offer an alternate development option by tying wells located at the seabed to the existing platforms and pipelines as well as to onshore processing facilities and terminals—thus, crude oil and natural gas accumulations that would be uneconomic as stand-alone developments can be developed economically (Amin et al., 2005).

6.2.1 Subsea Facilities

Briefly and by way of introduction to subsea production of crude oil and natural gas, in order for the crude oil and natural gas to reach a processing facility, production from the reservoir will flow through connectors (jumpers), manifolds, flowlines and (for surface platforms) risers designed to withstand pressures and temperatures prevalent at the subsea depths. Indeed, fluid produced from a deep-water reservoir experiences significant change in pressure and temperature as it moves from pore space to production riser. As a result, artificial lift may be required inside the wellbore to produce the fluids to the subsea wellhead, or tree.

The typical high pressure (HP) wellhead at the sea bottom with the Christmas tree and choke connects to a production riser (offshore) or gathering line (onshore) brings the well flow into the manifolds. As the reservoir is produced, wells may fall in pressure and become low pressure (LP) wells. This line may include several check valves. The choke, master and wing valves are relatively slow, therefore in the case of production shutdown, the pressure on the first sectioning valve closed will rise to the maximum wellhead pressure before these valves can close.

Short pipeline distances are not a problem, but longer distances may cause a multiphase well flow to separate and form severe slugs—plugs of liquid with gas in between—traveling in the pipeline. Severe slugging may upset the separation process and cause overpressure safety shutdowns. Slugging might also occur in the well as described earlier. Slugging can be controlled manually by adjusting the choke, or by automatic slug controls—areas of heavy condensate might form in the pipelines. At HP, these plugs may freeze at normal sea temperature, for example, if production is shut down or with long offsets. This can be prevented by injecting ethylene glycol—glycol injection is not used at all facilities.

Check valves allow each well to be routed into one or more of several manifold lines. There will be at least one for each process train plus additional manifolds for test and balancing purposes. The test manifold allows one or more wells to be routed to the test separator and, since there is only one process train, the HP and LP manifolds allow groups of HP and LP wells to be taken to the first and second stage separators, respectively (Figure 6.3). The chokes are set to reduce the wellhead flow and pressure to the desired HP and law pressure, respectively.

Production wells are free-flowing or lifted—a free flowing crude oil well has downhole pressure to reach suitable wellhead production pressure and maintain an acceptable well-flow. If the formation pressure is too low, and lacking sufficient pressure to encourage fluid flow, artificial lift using a pump becomes the operative manner of crude oil production. Onshore facilities typically use a horsehead pump (Figure 6.4) and an offshore facility may use the same or a derivation of this types of pump. In these cases, a gas lift system or electrical submersible pump (ESP) will be employed. In terms of the temperature, deep water is cold and temperatures at the sea floor may be on the order of $-40°C$ ($-40°F$) seafloor these

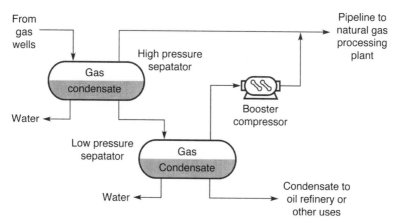

Figure 6.3 Schematic of high-pressure and low-pressure separation.

Figure 6.4 A horsehead pump.

temperatures must be accommodated beyond the subsea tree where fluids enters the manifold. The change in fluid temperature between the tree and the manifold will depend on the thermal management strategy of the production system. Some systems use electrically heated flowlines, whereas other systems may use foam-insulated pipe or the flow line may be buried beneath the seafloor.

In addition, high-viscosity crude oil (heavy crude oil) may be difficult to move and factors such as in-pipe corrosion, hydrate formation, organics deposition (wax deposition or deposition of asphaltene constituents) can be sufficiently severe to impede flow to surface processing facilities (Speight, 2014a). However, flow-related problems are influenced by the chemical composition of the produced fluid(s) as well as by the temperature and pressure as the fluids are moved from the reservoir to the production platform. These issues can be addressed and the effects diminished by thorough testing of the crude oil and natural gas which will allow the producer to anticipate and manage conditions that affect hydraulic performance of the production system (Speight, 2001; Amin et al., 2005; Speight, 2009, 2014a).

Typically, the main components of subsea systems are: (1) the *subsea well system*, which consists of the downhole completion system, the subsea tree system, and the drilling rig/vessel system, (2) the *subsea production system*, which consists of the subsea control system, the support and protection structures and the manifolds, intervention systems, and subsea processing systems, and (4) the *subsea pipeline system*, which consists of the subsea flow lines, the ancillary systems (including the umbilicals, which may include control cables as well as lines for injection of fluids at the wellhead), the riser system, and the subsea pipelines.

Typically, the wells are initially drilled into the reservoir are exploration and production wells. However, at some stage of the operation, the wells may be divided into production and injection wells. By definition, the former are for production of oil and gas while injection wells are drilled to inject gas or water into the reservoir. The purpose of injection is to maintain overall and hydrostatic reservoir pressure and force the oil toward the production wells. When injected water reaches the production well (*water breakthrough*), logging instruments (often based on radioactive isotopes added to injection water) are used to detect breakthrough. Injection wells are fundamentally the same as production wellheads—the difference being the direction of flow and therefore mounting of some directional components such as the choke.

Various types of subsea production systems are available and the most basic subsea satellite is a single *subsea wellhead* with *subsea tree*, which monitors and controls the production of a subsea well. The subsea tree is connected to the wellhead of a completed well and is also connected to a production facility by a series of pipelines and umbilical lines. Subsea trees have been used since the 1950s and use a design that reflects the structure and use of the above-ground (onshore) *Christmas tree* (Chapter 2).

Thus, a subsea tree (also called a *wet tree*) is, like the onshore Christmas tree, a series of valves and gauges located at the wellhead. A control module situated at the subsea tree allows the facility to be remotely operated and water injection, chemical injection, and gas lift can be provided from the host facility. However, there are various types of subsea tree designs and each type is rated for water depth, temperature, pressure, and the expected flow of crude oil and natural gas. For example, the *dual bore subsea tree* was the first tree to include an annulus bore for troubleshooting, well servicing, and well conversion operations, especially in the often-hostile environment of the North Sea. These trees can now be specified with guideline or guidelineless position elements for production or injection well applications.

On the other hand, the *standard configurable tree* is usually used in shallow waters measuring up to 3000 ft deep. *High-pressure high-temperature trees* are designed for pressures up to 16,500 psi and temperatures ranging from -33 to

175°C (−27 to 345°F) and are capable of use in unfriendly environments, such as the North Sea. Other subsea trees include *horizontal trees, mudline suspension trees, monobore trees*, and *large bore trees*. A control module, usually situated on the subsea tree, allows the production platform to remotely operate the subsea facility. These single satellites are commonly used to develop small reservoirs near to a large field and can also be used to provide additional production from, or peripheral water injection support to, a field which could not adequately be covered by drilling additional wells from the platform. Furthermore, an exploration or appraisal well, if successful, can be converted to a subsea producer if hydrocarbons are discovered. In this case the initial well design would have to allow for any proposed conversion.

When the number of wells grows in number, a *subsea production template* is generally recommended for use with groups of six or more wells. The *template* has the same function as a manifold, but is different with regard to the location of wells. On templates, the wells themselves are located directly below the structural frame, which protects and connects them. Hence, subsea templates are structures that allow several wells to be drilled and completed from one location. The templates are fabricated from large tubular members and incorporate a receptacle for each well and a three or four point leveling system. Moreover, it is possible to place part of the template over existing exploration wells and tie these wells collectively individually into the template production system.

After onshore construction, the template is transported offshore to the crude oil production site and lowered to the seabed using a crane barge or, if small enough, the template can be lowered beneath a semi-submersible rig. Prior to drilling the first well, piles are driven into the sea bed to hold the template in place and, as the first well is being drilled, the template is connected to the host facility with flowlines, umbilicals, and risers. A *chemical injection umbilical* will also typically be tied to the template or subsea facility and connected to a distribution manifold. As soon as the subsea tree on the first well has been commissioned, production can commence and the drilling rig can then be moved to another template slot after which drilling the next well is commenced.

If one or more clusters of single wells are required, an *underwater manifold system* can be deployed and used as a focal point to connect each subsea well. Subsea manifolds are typically fabricated are usually made from tubular steel parts (similar to a template), and include such components as pipes, chokes, valves, and flow meters—the valves and chokes are usually remotely operated. The subsea trees sit on the seabed around the main manifold (compared to the template) but only one set of pipelines and umbilicals (as with the template) are required to connect the manifold to the host facility. It is common for subsea tree control systems to be mounted on the manifold and not on the individual trees—flow lines and umbilicals connect each well to the manifold. A set of pipelines, umbilicals, and risers is required to convey crude oil and natural gas from the manifold to the surface host facility. It is common for subsea trees to be mounted directly at the manifolds, and not on the individual subsea wells (Jahn et al., 1998). A complex manifold will generally have a set of dedicated subsea control modules—in this way, the manifold valves and monitoring flow line sensors can be controlled.

Since subsea systems are remote from their host production location, which may be located on a platform or onshore, they require remote control and surveillance. Modern subsea components are controlled via *umbilicals*, which may consist of the following types of cables and lines: (1) electric cables, which are used for powering the control system and some actuating components, and for communication—signals are sent back to the surface to provide information on the status of valves and chokes, flow rate measurements, pressure measurements, and temperature measurements, (2) hydraulic cables, which are used to operate actuating components such as valves, (3) chemical lines, which are used for transporting substances that are to be injected into the production pipeline, such as monoethylene glycol (MEG), methanol, scale inhibitors, and emulsion breakers, and (4) optical cables, which are a modern alternative to electric cables for communications with the subsea equipment.

The umbilicals connect the subsea control distribution unit and the surface (or onshore) control station. The umbilicals transfer information about the status of subsea equipment such as temperature measurements, pressure measurements, choke settings, and flow rates back to the surface. Based on this information, the surface control station will return signals that will initiate appropriate corrective measures if the status of the subsea measurements is unsatisfactory.

Subsea technology is continuing to advance as the offshore oil industry seeks to produce oil and gas from ever increasing water depths using modern offshore platforms, pipelines, and subsea systems (Chapter 2). As the oil and gas industry has matured and many existing fields have reached the end of their life, the facilities used to develop these fields have to be decommissioned in an environmentally friendly manner (Chapter 9). On the other hand the active life of the equipment may be extended—an example of active life extension is the reuse of existing oil and gas separation and transportation systems on the matured/expired field. Also, many new fields, particularly wet gas or high gas–oil ratio (GOR) fields, are not economically viable if the field has to be exploited with its dedicated gas and condensate separation and transportation systems (Amin and Waterson, 2002).

The operational feasibility of the subsea system is high if either (1) the total production of the field is routed to nearby offshore oil and gas facilities which can provide the necessary processing requirements or (2) the total production of crude oil and natural gas can be transported directly to shore without processing. In both of these cases, total production has to be transported via multiphase flow lines that carry the produced fluids (gas/crude oil/water). Furthermore, the current trend for crude oil natural and gas exploration and recovery is for more of the production to be achieved with subsea facilities, such as facilities to separate sand from the fluids oil and reinject the sand rather than pump it to the platform, or even to pumping the sand onshore, with no installations visible above the sea.

Subsea installations further the goal of developing crude oil and natural gas resources which have previously been considered to be inaccessible because of the depth of the water coverage. These systems can range in complexity from (1) a single satellite well with a flow line linked to a fixed platform or an onshore installation, or (2) to several wells on a template or clustered around a manifold. Furthermore, these systems can be used to develop reservoirs, or parts of reservoirs, which require drilling of the wells from more than one location. Deep water conditions, or even ultra-deep water conditions, can also be developed by means of a subsea production system, especially where traditional surface facilities might be either technically unfeasible or uneconomical due to the water depth. However, the development of subsea crude oil and natural gas fields requires that the system must be sufficiently reliable enough to prevent damage to the safeguard the environment and make the exploitation of the resource economically feasible.

6.2.2 Shallow Water Facilities

Many petroleum and natural gas reservoirs are located in areas where the typical land-based drilling rigs cannot be used. In shallow inland waters, lakes, or wetland areas, a *drilling platform* and other drilling-recovery equipment may be mounted on a barge, which is then be floated into position and then secured in a stable position on the bottom of the waterway— typically such operations are restricted to water depths of less than 50 ft. The actual drilling platform can be raised above the water on masts if necessary—reminiscent of the construction of the homes of the stone-age lake dwellers in Europe. Once the platform is in place, drilling and other operations on the well make use of an opening through the barge hull.

Shallow-water production went through three peaks, culminating in 1997, with each peak of production progressively lower than the previous peak due to the expansion of drilling territory and development of more advanced drilling and production methods for crude oil and natural gas. From 1997 through 2003, shallow-water production declined and in the years 2004 and 2005, hurricanes Ivan, Katrina and Rita caused extensive damage to Gulf of Mexico oil production facilities that reduced production, particularly in 2005. In terms of real numbers, before landfall, Hurricane Katrina destroyed 46 oil and gas platforms and four drilling rigs. Four weeks later, Hurricane Rita came charging through the Gulf and destroyed a further 63 platforms and one drilling rig. Interior Fortunately by early 2007, thanks to the efforts of the crude oil and natural gas industry, crude oil and natural gas production was well on the way reaching to pre-Ivan-Katrina-Rita levels. However, the offshore crude oil and natural gas industry is always wary of hurricanes and the damage that can be caused by these natural phenomena.

While shallow-water oil production in the central and western Gulf of Mexico is in decline, other areas in the Gulf of Mexico are opening up to exploration and production but it is still believed that the amount of oil and gas in new areas is not sufficient in that region of the Gulf of Mexico to make a significant impact on shallow-water production. Drilling activities in the Orinoco Delta (Venezuela) and the Gulf of Paria (a w waterway spanning the space between the island of Trinidad and Venezuela) are performed primarily with shallow water jack-ups, swamp barges, and conventional shallow water drilling barges resting upon a flat carrier barge. The most common diving tasks are survey, sand bagging, piles recovery, pipeline routing, inspections, anchor recovery, and pipeline and cable lay. Underwater surveys are performed prior to the positioning of the drilling unit, with the job conducted from a small diving vessel.

6.2.3 Deep Water Facilities

Deep water is arbitrarily typically defined to cover the water depth greater than 1000 ft. The term *ultra-deep water* is often arbitrarily used for water depths in excess 5000 ft, but this is subject to interpretation and re-evaluation.

Generally, *deepwater* refers to fields located offshore in significant depths of water—typically in excess of 1000 ft but there is no clear decisions or definition of the depth that constitutes deepwater. Basically, at any given point in time, routinely practiced technologies are considered conventional offshore technologies, while state-of-the-art technologies that stretch the industry's production capabilities are considered deepwater. Sometimes the term *ultra-deepwater* is used to describe the water depths at which exploration is currently taking place, but for which available production technology is only just feasible or has just become viable. Moreover, fields are now being developed in water depths in excess of 6500 ft in the Gulf of Mexico, offshore West Africa, and offshore Brazil are at least deepwater production, if not ultra-deepwater production. It is estimated that 40% of undiscovered deepwater resources are at a water depth of between 6500 and 10,000 ft and 30% at between 10,000 and 13,000 ft. Beyond 13,000 ft water depth, the potential for oil and gas resources is largely unknown and, currently—speculative.

However, deepwater operations (which represent a significant resource potential, pose major technical and engineering challenges and they involve very high costs, which are only affordable for very prolific reservoirs. The trend to deepwater and ultra-deepwater resources can be expected to continue, enabling access to even deeper waters and, more important, reducing the cost of drilling and producing within current frontiers. In fact, advances in technology have rendered smaller accumulations within shallower water worthy of development (i.e., economical) as the deepwater limit moved further offshore. The discoveries of large fields under deeper water have justified the development of infrastructure on which smaller fields can then piggyback.

Exploration and production activities for crude oil and natural gas have increasingly turned to deepwater prospects, for example, in water that is more than 1650 ft deep. These accumulations have previously been out of reach for conventional platform developments, but the use of floating production facilities and subsea technology has made exploitation in such areas possible. Such water depth presents major challenges to production and water depth dominates all process, design, and economic considerations (Amin et al., 2005). The major deepwater areas in the world are offshore Brazil, West Africa, and in the Gulf of Mexico—for example, oil production takes place at water depths of 10,000 ft offshore Brazil. Although river deltas are promising areas for exploration, the sea bottom outside the large river deltas is likely to consist of very soft materials which can lead to instability issues. The constant transport of mud with large rivers may cause mudflows or seafloor slides which will influence the possible location of wells, templates, and pipeline routing.

Bottom-supported steel jackets and concrete platforms are impractical in deep water from a technical and economic point of view and floating moored structures are preferred. In deep and especially ultra-deep open waters over continental shelves, drilling is done from free-floating platforms or from platforms made to rest on the bottom. Floating rigs are most often used for exploratory drilling, while bottom-resting platforms are usually associated with well drilling in an established field. One type of floating rig is the drill ship—an oceangoing vessel with a derrick mounted in the middle, over an opening for the drilling operation (Chapter 3)—which is usually held in position by six or more anchors, although some vessels are capable of precise maneuvering with directional thrust propellers. Even so, drill ships will roll and pitch from wave action, making the drilling difficult and a more stable platform is preferred using semisubmersible vessels. In these vessels (Chapter 3), buoyancy is achieved by the underwater hull while the operational platform is held well above the surface on slender supports. Normal wave action affects the platforms very little and these units are also kept in place during drilling by either anchors or precise maneuvering.

In deeper, more open waters over continental shelves, drilling is achieved from free-floating platforms or from platforms made to rest on the bottom. Floating rigs are most often used for exploratory drilling, while bottom-resting platforms are usually associated with the drilling of wells in an established field. One type of floating rig is the drill ship, which is an oceangoing vessel with a derrick mounted in the middle, over an opening for the drilling operation. The ship is usually held in position by six or more anchors, although some vessels are capable of precise maneuvering with directional thrust propellers. Even so, drill ships will roll and pitch from wave action, making the drilling difficult. A more stable platform is obtained with semisubmersible vessels. In these vessels, buoyancy is afforded by a hull that is entirely underwater, while the operational platform is held well above the surface on slender supports. Normal wave action affects the platforms very little. These vessels are also kept in place during drilling by either anchors or precise maneuvering.

Fixed platforms, which rest on the seafloor, are very stable, although they cannot drill in water as deep as floating platforms can. The most popular type is called a *jack-up rig*. This is a floating (but not self-propelled) platform with legs that can be lifted high off the seafloor while the platform is towed to the drilling site. There the legs are cranked downward by a rack-and-pinion gearing system until they encounter the seafloor and actually raise the platform 10–20 m above the surface. The bottoms of the legs are usually fastened to the seafloor with pilings. Other types of bottom-setting platforms may rest on rigid steel or concrete bases that are constructed onshore to the correct height. After being towed to the drilling site, flotation tanks built into the base are flooded, and the base sinks to the ocean floor. Storage tanks for produced oil may be built into the underwater base section. In some platforms, the legs have been replaced by cables fastened to the seafloor. The platform is pulled down on the cables so that its buoyancy creates a tension in the cables that holds it firmly in place. These platforms typically operate in water up to 350 ft deep.

Fixed platforms, which rest on the seafloor, are stable although drilling in deep water than can be achieved by floating platforms is not always possible. The most popular type of platform is called a *jack-up rig*, which is a floating (but not self-propelled) platform with legs that can be lifted high off the seafloor while the platform is towed to the drilling site (Chapter 3). Once on site, the legs are cranked downward by a rack-and-pinion gearing system until they encounter the seafloor and actually raise the platform 30–65 ft above the surface of the sea. The bottoms of the legs are usually fastened to the seafloor with pilings—other types of bottom-setting platforms may rest on rigid steel or concrete bases that are constructed onshore to the correct height—in some platforms, storage tanks for produced oil may be built into the underwater base section. After being towed to the drilling site, flotation tanks built into the base of such platforms are flooded, after which the base sinks to the ocean floor. In some platforms (which typically operate in water up to 300 ft deep), the legs are replaced by cables that are secured to the seafloor and the buoyancy of the platform creates a tension in the cables that serves to hold the platform firmly in place.

For both fixed and floating platforms, the drill pipe must still transmit both rotary power and drilling mud to the drill bit and, in addition, the mud must be returned to the platform for recirculation rather than discharged and cause environmental damage (Chapter 10) (Wills, 2000). In order to accomplish these functions through seawater, an outer casing (the *riser*) must extend from the seafloor to the platform. Also, a guidance system (usually consisting of cables fastened to the seafloor) must be in place to allow equipment and tools from the surface to enter the well bore. In the case of floating platforms, there will inevitably be some motion of the platform relative to the seafloor, hence the requirement for flexibility of the system.

The economic oil production from deep-water reservoirs relies on a group of new production technologies including mainly: (1) long horizontal or multilateral wells (producing with high power electric submersible pumps, hydraulic pumps or submarine multiphase pumps) to compensate the decrease in productivity caused by the high oil viscosity), (2) efficient heat management systems, and (3) compact oil–water separation systems. The proliferation of deepwater development projects will continue to grow, as long as the technology and financial incentives make the ventures increasingly possible. The potential for oil and gas in deep-water areas has been realized by the development of the 3-D and subsalt geophysical technologies. Deepwater operations vary significantly compared to conventional operations in shallow waters—there are different environmental conditions—but successful results show a very lucrative potential, particularly in the long-run. The opportunity to produce oil and gas at much higher rates exists is where the interest lies in these operations.

Floating production systems offer a practical method for developing medium and deepwater oil and gas reservoirs. As exploration and production heads into deeper waters, floating production facilities become increasingly more economical than fixed or guyed-tower and tension-leg platforms (Manning and Thompson, 1995). Floating production systems had been tried on the UK Continental Shelf in the Argyll field in 1975 and used successfully for oil production in other regions, such as Brazil, West Africa, North-western Australia and Southeast Asia (Chianis, 2003).

Flexible risers are essential components of floating production systems and are used to transport production fluids from the seabed to the production vessels where the riser must deal with substantial vessel motion, can be viewed as an established technology in this respect (Patel and Seyed, 1995). A flexible riser can be designed into different shapes to meet such demands. For deep-water exceeding 5000 ft, the free-standing configuration is new and have many advantages compared with other riser forms. Consequently, it is increasingly important that proper attention is paid to the design and selection of the riser system early in the design loop (Hatton and Howells, 1996). A considerable part of flexible riser system design is the determination of configuration parameters so that the riser can safely sustain the loadings for which it is to be designed. A well-designed riser configuration is safe and provides compliancy to motion of the surface vessel in a suitable manner by minimizing the station-keeping requirements for the vessel. Furthermore, while there is an engineering-oriented selection process for the choice of riser—depending on the use and site conditions—knowledge of the reservoir and local environmental conditions are equally important. The reservoir conditions are the basis for defining the production parameters, which, in turn, is the main input for establishing the subsea layout (definition of the number of subsea manifolds as well as the position of the surface unit, and the pipeline route. In addition, environmental conditions are the major reason for selecting the surface unit concept (the riser interface with the surface unit) are the vessel motion and the riser hang-off, which provides (is to provide a structural support between the riser or umbilical and the outer J-tube).

In general, the nature of the water current gradients in a particular developments, dictates the riser shape, the type, and relative position of any risers. The free hanging configuration is often considered to be the simplest riser (Hatton and Howells, 1996) but an issue is that if there are any significant first order wave motions at the vessel connection, the amplitude of dynamic tension is transferred directly to the seabed which inevitably leads to compression at the riser touchdown point. Furthermore, the free-hanging catenary riser (riser in a catenary shape—the shape that a hanging cable or chain assumes) is not very compliant to vessel motions, where riser top tension increases rapidly with far vessel offset, and large vessel offset motions result in correspondingly large and undesirable motions of the riser/seabed touchdown point. The challenge is even more significant when steel catenary risers are considered for harsh environments, where extreme and long-term environmental conditions are amongst the most severe in the world, causing the risers to be highly dynamic and fatigue sensitive (Hatton and Howells, 1996; API, 2002).

In flow systems, such as risers, multiphase flow transport can create challenges which can significantly change design requirements (Amin and Waterson, 2002). Hence, one of the objectives of the system design is *flow assurance*, that is, the transmission system must operate in a safe, efficient, and reliable manner throughout design life. The term *flow assurance* covers the whole range of possible flow problems in pipelines that include both multiphase flow and fluid related effects (such as gas hydrate formation, wax deposition, asphaltene precipitation, corrosion, scaling, and severe slugging).

The avoidance or remediation of these problems is essential to flow assurance that enables optimization of the production system for the range of expected conditions including start-up, shutdown, and turndown. However, as production systems go deeper into the ocean, flow assurance becomes a major issue for the offshore production systems and traditional approaches are inappropriate for deepwater production systems due to extreme distances, depths, or temperature. Maintaining fluid flow is one of the most critical challenges of the deepwater context, where temperatures, HPs, and poorly consolidated reservoirs can all lead to blockages in production lines. To find solutions, it is necessary to innovations related to multiphase flow hydrate prevention solutions.

Pipeline–riser systems transport multiphase flow from satellite wells to a central production platform via a single pipeline. However, in a single multiphase pipeline segregated flow of liquid and gas may cause problems—the actual velocity of the gas phase is faster than the actual liquid velocity and the liquid phase has the tendency to accumulate in the dips and inclined pipe sections causing irregular flow behavior. As a result, large volumes of liquid (slugs, riser-induced slugs, hydrodynamic slugs) may flow through the pipeline. Furthermore, operational changes, such as startup and production increase, can create large liquid slugs, severe slugging (Hunt, 1996). For this system liquid slugs will accumulate in the riser and the pipeline, blocking the flow passage for the gas flow., which results in a compression and pressure build-up in the gas phase that will eventually push the liquid slug up the riser and a large liquid volume be produced into the separator, that causes severe consequences eventually leading to shut down (Schmidt et al., 1985; Mokhatab, 2007).

6.2.4 Arctic Environments

Development of subsea crude oil and natural gas reservoirs in the Arctic may experience considerable downtime during drilling, which may be restricted by ice conditions during large parts of the year. In addition, use of drill ships in combination with ice feature management may extend the drilling season and the need for the ship to be free of the potential ice field before the sea goes into a winter freeze condition must be incorporated into any drilling plans. Freedom from the potential ice fields is a particular necessity for jack-up rigs, which generally have low tolerances for impacts from ice—operational limitations with considerable production downtime must be anticipated, especially if well maintenance is required and, after disconnecting production facilities due to ice restrictions, the reconnection time may be long.

Subsea technology in Arctic conditions is presently used at the Terra Nova and White Rose fields on the Grand Banks off the eastern coast of Canada, in 400 ft deep water. Oil production wells were predrilled from a semi-submersible drilling platform and wellheads (with the necessary production manifolds) were placed in excavations in the seafloor, for protection from icebergs. A network of more than 25 miles of flexible flow lines is used to transfer hydrocarbons to and from the wells—produced natural gas (and any other gases) are separated from the crude oil and reinjected into the reservoir for possible future extraction. At the Terra Nova location, drifting icebergs can occur without any pack ice cover. The field installations rely on the capability to disconnect the riser system should one fail to tow away a threatening iceberg. In contrast, the nearby Hibernia development relies on a large concrete gravity structure for protection from any wayward icebergs.

In contrast to ice bergs, pack ice is an extensive ice cover which moves with water currents and such conditions require somewhat different solutions than those prescribed for icebergs in open water at Terra Nova. Drifting pack ice may require the use of propulsion to head against the ice drift direction, allowing the vessel to withstand the forces from drifting ice features, which may be supplemented with icebreaker assistance and the capability of disconnecting and temporarily leaving the site, should the ice situation be unmanageable.

In shallow Arctic waters where drifting ice is a hazard for fixed platforms, artificial islands are constructed. Onshore in Arctic areas, permafrost makes drilling difficult because melting around and under the drill site makes the ground unstable. Here, too, artificial islands are built up with rock or gravel. Also, in these waters, waters where drifting ice is a hazard for fixed platforms, artificial islands are constructed. Onshore in Arctic areas, permafrost makes drilling difficult because melting around and under the drill site makes the ground unstable. Here, too, artificial islands are built up with rock or gravel. Oil production has a long history in the Gulf of Mexico with production starting in the shallow waters off Texas and Louisiana in the late 1940s. The pinnacle of shallow-water (drilling and recovery oil from depths less than 1000 ft) production occurred in the early 1970s when the production rate exceeded one million barrels per day.

Deepwater Arctic natural gas and gas condensate fields tend to be several hundred kilometers from shore, in water approximately 1000 ft deep, and exposed to occasional ice and drifting icebergs. Due to the distance from shore, such a development may be unsuited to a full well stream transfer to shore. It may be advantageous to use support vessel(s) at the site for control, MEG injection, and possibly separation purposes and compression.

6.2.5 Subsea Processing

Subsea processing involves a series of technologies of which some have been implemented, but which are still under development, which may provide the following functions: (1) gas separation, (2) liquid separation, (3) gas compression, (4) liquid pressure boosting, (5) desanding, (6) gas treatment, and (7) injection into reservoirs. These processing techniques are expected to yield a number of advantages, notably a significant potential for improved oil recovery (IOR) particularly in deep water, which can enable new development, improve the utilization of existing infrastructure, enable longer tie-backs from subsea to shore solutions, and avoid discharges into the sea. In fact, technology advances in improved oil may offer the most promising methods for future offshore oil recovery.

Several subsea processing solutions have been implemented or are planned for the Norwegian offshore, representing important steps forward in technology development. They include subsea separation (Tordis field), well-stream compression (Åsgard field), and water injection (Tyrihans field).

6.2.6 Subsea Intervention

Most wells require some intervention during their lifetime, in order to extend the life of the wells. These include (Christie et al., 1999): (1) installation or service of subsea control valves, (2) change of gas-lift valves, (3) production logging, (4) change of damaged and/or broken tubing, (5) removal of scale and deposited wax or asphaltene constituents, (6) perforation of new sections in the reservoir, (7) cementation of existing perforations to shut off water flow, and application of chemical treatments (such as *acidization*).

Traditionally, interventions have been accomplished by the use of a drilling rig and a marine riser with diver assistance. However, returning a drilling rig to a subsea well to perform intervention tasks is a very expensive option. Therefore, the industry has investigated more cost-effective methods for subsea intervention. Some light intervention tasks are performed by *remotely operated vehicles* (ROVs). A typical intervention task involving a ROV would be visual inspection of subsea equipment, valve operations, installation operations (wire handling and tie-ins), nondestructive testing (crack testing), cleaning, maintenance, and repair.

Light well intervention (LWI) technology is a riser-less technology that involves lowering intervention equipment directly into a pressurized well by means of a wireline paid out from a dynamically positioned vessel, usually a ship—a key element is a lubricator system used to ensure a tight seal. Recognizing that wireline intervention may cut the cost by as much as 50% in comparison with conventional rig operations, several qualification campaigns have been carried out. As a result, riser-less well intervention has become part of the standard subsea tool kit. Further suggestions involve the use of composite cables and coiled tubing operations as well as increasing the lubrication length to extend the capabilities of wireline tools (Zittel et al., 2006).

6.3 RECOVERY

If the underground pressure in the oil reservoir is sufficient and the temperature is sufficiently high to maintain the liquids in the as fluid state, the crude oil will be forced to the surface under this pressure (primary

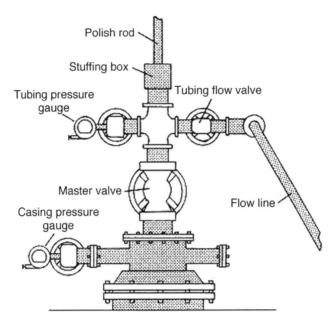

Figure 6.5 The Christmas tree: A collection of control valves at the wellhead.

recovery)—the pressure and temperature are often referred to collectively as the reservoir energy. Natural gas (associated natural gas) is often present with crude oil, which also supplies needed underground pressure for primary recovery. In this situation, it is sufficient to place an arrangement of valves—the Christmas tree (Figure 6.5) on the well head to connect the well to a pipeline network for storage and processing.

Recovery, as the term is applied in the petroleum industry, is the production of crude oil and/or natural gas from a reservoir. There are several methods by which this can be achieved that range from recovery due to reservoir energy (i.e., the pressure and temperature within the reservoir is sufficient to enable the crude oil (with any natural gas)) to flow from the reservoir, through the well, and to the surface without assistance. On the other hand, *enhanced oil recovery* (EOR, also known as *enhanced oil production—EOP*—as well as IOR) methods require that energy must be added to the reservoir to produce the crude oil. However, the effect of the method on the oil and on the reservoir must be considered before application and whether or not the method will diminish the reservoir energy too quickly leaving much of the oil to be recovered by application of enhanced recovery methods.

Thus, once the well is completed, the flow of oil into the well is commenced. If the reservoir consists of limestone, acid may be is pumped down the well (acidizing; Section 3.4) to improve the flow of oil to the well. For sandstone reservoir rock, a specially blended fluid containing *proppants* (sand, walnut shells, aluminum pellets) is pumped down the well to improve the oil flow through the channels—the pressure from this fluid creates small fractures in the sandstone and the proppants hold these fractures open. Once the oil is flowing, the oil rig is removed from the site and production equipment is set up to extract the oil from the well.

Briefly, a proppant is a solid material, typically treated sand or manufactured ceramic materials, designed to keep an induced hydraulic fracture open, during or following a fracturing treatment. The proppant is added to a the fluid used for hydraulic fracturing (*fracking*) which may vary in composition depending on the type of fracturing used, and can be gel, foam, or water-based. Radioactive tracer isotopes may be included in the fracking fluid to determine the injection profile and location of fractures created by hydraulic fracturing (Reis, 1976). Except for diesel-based additive fracturing fluids, noted by the US Environmental Protection Agency to have a higher proportion of volatile organic compounds and carcinogenic benzene-toluene-ethylbenzene-xylene (BTEX) (Figure 6.6), use of fracking fluids is monitored and will be subject to future legislation.

A well is always carefully controlled in its flush stage of production to prevent the potentially dangerous and wasteful *gusher*. This is actually dangerous condition, and is (hopefully) prevented by the blowout preventer and the pressure of the drilling mud. In most wells, acidizing or fracturing the well starts the oil flow. For a newly opened formation and under ideal conditions the proportions of gas may be so high that the crude oil is, in fact, a solution of liquid in the natural gas that leaves the reservoir rock so efficiently that a core sample will not show any obvious oil content—an

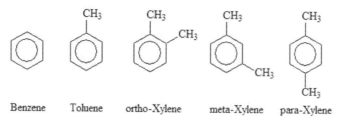

Figure 6.6 Structure of benzene, toluene, and the isomeric xylenes – ethylbenzene ($C_6H_5C_2H_5$,) is a homolog of toluene.

indication of this situation is the wellhead production of a high ratio of natural gas to crude oil—this ratio may be zero for fields in which the rock pressure has been dissipated.

Drilling does not end when production commences and continues after a field enters production. Extension wells must be drilled to define the boundaries of the crude oil pool. In-field wells are necessary to increase recovery rates, and service wells are used to reopen wells that have become clogged. Additionally, wells are often drilled at the same location but to different depths, to test other geological structures for the presence of crude oil. Thus, production of crude oil and natural gas commences when a well is first opened is usually by natural flow forced by the pressure of the gas or fluids that are contained within the deposit (*primary production, primary recovery*) (Speight, 2009, 2014a). There are several means that serve to drive the petroleum fluids from the formation, through the well, and to the surface, and these methods are classified as either natural or applied flow.

6.3.1 Primary Recovery

If the underground pressure in the oil reservoir is sufficient, the crude oil is forced to the surface under this pressure (primary recovery—natural methods). Gaseous fuels or natural gas are usually present, which also supplies needed underground pressure. In this situation it is sufficient to place a complex arrangement of valves (the Christmas tree) at the well head to connect the well to a pipeline network for storage and processing.

Thus, *primary oil production* (*primary oil recovery*) is the first method of producing oil from a well and depends upon natural reservoir energy to drive the oil through the complex pore network to producing wells. If the pressure on the fluid in the reservoir (reservoir energy) is great enough, the oil flows into the well and up to the surface. Such driving energy may be derived from liquid expansion and evolution of dissolved gases from the oil as reservoir pressure is lowered during production, expansion of free gas, or a gas cap, influx of natural water, gravity, or combinations of these effects. In fact, crude oil moves out of the reservoir into the well by one or more of primary production three processes. These processes are: *dissolved gas drive, gas cap drive,* and *water drive*. Early recognition of the type of drive involved is essential to the efficient development of an oil field.

In *dissolved gas drive*, the propulsive force is the gas in solution in the oil, which tends to come out of solution because of the pressure release at the point of penetration of a well. Dissolved gas drive is the least efficient type of natural drive as it is difficult to control the GOR; the bottom-hole pressure drops rapidly, and the total eventual recovery of petroleum from the reservoir may be less than 20%.

If gas overlies the oil beneath the top of the trap, it is compressed and can be utilized (*gas cap drive*) to drive the oil into wells situated at the bottom of the oil-bearing zone. By producing oil only from below the gas cap, it is possible to maintain a high GOR in the reservoir until almost the very end of the life of the pool. If, however, the oil deposit is not systematically developed so that bypassing of the gas occurs, an undue proportion of oil is left behind. The usually recovery of petroleum from a reservoir in a gas cap field is 40%–50%.

Usually the gas in a gas cap (*associated natural gas*) contains methane and other hydrocarbons that may be separated out by compressing the gas. A well-known example is *natural gasoline* that was formerly referred to as *casinghead gasoline* or *natural gas gasoline*. However at HPs, such as those existing in the deeper fields, the density of the gas increases and the density of the oil decreases until they form a single phase in the reservoir. These are the so-called retrograde condensate pools because a decrease (instead of an increase) in pressure brings about condensation of the liquid hydrocarbons. When this reservoir fluid is brought to the surface and the condensate is removed, a large volume of residual gas remains. The modern practice is to cycle this gas by compressing it and inject it back into the reservoir, thus maintaining adequate pressure within the gas cap, and condensation in the reservoir is prevented. Such condensation prevents recovery of the oil, for the low percentage of liquid saturation in the reservoir precludes effective flow.

The most efficient propulsive force in driving crude oil into a well is natural *water drive*, in which the pressure of the water forces the lighter recoverable oil out of the reservoir into the producing wells. In anticlinal reservoirs (Figure 6.1), the structurally lowest wells around the flanks of the dome are the first to come into water. Then the oil–water contact plane moves upward until only the wells at the top of the anticline are still producing oil; eventually these also must be abandoned as the water displaces the oil.

In the water drive process, it is essential that the removal rate be adjusted so that the water moves up evenly as space is made available for it by the removal of the hydrocarbons. An appreciable decline in bottom-hole pressure is necessary to provide the pressure gradient required to cause water influx. The pressure differential needed depends on the reservoir permeability; the greater the permeability, the less the difference in pressure necessary. The recovery of crude oil from the reservoir in properly operated water drive pools may be as high as 80%. The force behind the water drive may be hydrostatic pressure, the expansion of the reservoir water, or a combination of both. Water drive is also used in certain submarine fields.

Gravity drive is an important factor when oil columns of several thousands of feet exist, as they do in some North American fields. Furthermore, the last bit of recoverable oil is produced in many pools by gravity drainage of the reservoir. Another source of energy during the early stages of withdrawal from a reservoir containing under-saturated oil is the expansion of that oil as the pressure reduction brings the oil to the bubble point (the pressure and temperature at which the gas starts to come out of solution).

For primary recovery operations, no pumping equipment is required. If the reservoir energy is not sufficient to force the oil to the surface, then the well must be pumped. In either case, nothing is added to the reservoir to increase or maintain the reservoir energy or to sweep the oil toward the well. The rate of production from a flowing well tends to decline as the natural reservoir energy is expended. When a flowing well is no longer producing at an efficient rate, a pump is installed.

The recovery efficiency for primary production is generally low when liquid expansion and solution gas evolution are the driving mechanisms. Much higher recoveries are associated with reservoirs with water and gas cap drives and with reservoirs in which gravity effectively promotes drainage of the oil from the rock pores. The overall recovery efficiency is related to how the reservoir is delineated by production wells. Thus for maximum recovery by primary recovery it is often preferable to sink several wells into a reservoir, thereby bringing about recovery by a combination of the methods outlined here.

6.3.2 Secondary Recovery

Over the lifetime of the reservoir the pressure will fall and, at some point, there will be insufficient underground pressure to force the crude oil to the surface. *Secondary oil recovery* uses various techniques to aid in recovering oil from depleted or LP reservoirs. Other secondary recovery techniques increase the reservoir pressure by water injection, natural gas reinjection and gas lift, which injects air, carbon dioxide or some other nonreactive gas into the reservoir. In addition to the use of pumps on the surface (*balanced-beam pump, horse head pump, sucker rod pump*), submerged pumps (ESPs) are also used to provide mechanical lift to the fluids in the reservoir. Other secondary recovery techniques increase the reservoir pressure by water injection and gas injection, which injects air or some other gas into the reservoir are used.

The most commonly recognized oil-well pump is the reciprocating or plunger pumping equipment, which is easily recognized by the *horsehead* beam pumping jacks. A pump barrel is lowered into the well on a string of 6 in. (inner diameter) steel rods known as sucker rods. The up-and-down movement of the sucker rods forces the oil up the tubing to the surface. A walking beam powered by a nearby engine may supply this vertical movement, or it may be brought about through the use of a pump jack, which is connected to a central power source by means of pull rods. Electrically powered centrifugal pumps and submersible pumps (both pump and motor are in the well at the bottom of the tubing) have proven their production capabilities in numerous crude oil production applications.

There are also *secondary oil recovery* operations that involve the injection of water (*waterflood*) or gas (*gasflood*) into the reservoir in which separate wells are used for injection and production. The injected fluids maintain reservoir pressure or repressure the reservoir after primary depletion and displace a portion of the remaining crude oil to production wells. In fact, the first method recommended for improving the recovery of oil was probably the reinjection of natural gas, and there are indications that gas injection was utilized for this purpose before (Speight, 2014a and references cited therein). In secondary recovery, the injected fluid must dislodge the oil and propel it toward the production wells. Reservoir energy must also be increased to displace the oil. Using techniques such as gas and water injection, there is no change in the state of the crude oil. Similarly, there is no change in the state of the crude oil during miscible fluid displacement technologies.

The success of secondary recovery processes depends on the mechanism by which the injected fluid displaces the oil (displacement efficiency) and on the volume of the reservoir that the injected fluid enters (conformance or sweep efficiency). In most proposed secondary projects, water does both these things more effectively than gas. It must be decided if the use of gas offers any economic advantages because of availability and relative ease of injection. In reservoirs with high permeability and high vertical span, the injection of gas may result in high recovery factors as a result of gravity segregation, as described in a later section. However, if the reservoir lacks either adequate vertical permeability or the possibility for gravity segregation, a frontal drive similar to that used for water injection can be used (dispersed gas injection). Thus, dispersed gas injection is anticipated to be more effective in reservoirs that are relatively thin and have little dip. Injection into the top of the formation (or into the gas cap) is more successful in reservoirs with higher vertical permeability (200 md or more) and enough vertical relief to allow the gas cap to displace the oil downward.

During the withdrawal of fluids from a well, it is usual practice to maintain pressures in the reservoir at or near the original levels by pumping either gas or water into the reservoir as the hydrocarbons are withdrawn. This practice has the advantage of retarding the decline in the production of individual wells and considerably increasing the ultimate yield. It also may bring about the conservation of gas that otherwise would be wasted, and the disposal of brines that otherwise might pollute surface and near-surface potable waters.

Water injection is still predominantly a secondary recovery process (*waterflood*) but if some channels in the reservoir are larger than others and the water tends to flow freely through these, bypassing smaller passages where the oil remains, a partial solution to this problem is possible by *miscible fluid flooding*. In this process, liquid butane and propane are pumped into the ground under considerable pressure, dissolving the oil and carrying it out of the smaller passages; additional pressure is obtained by using natural gas.

6.3.3 Enhanced Oil Recovery

Together, primary recovery and secondary recovery allow 25%–35% (v/v) of the crude oil to be recovered from the reservoir. Primary (or conventional) recovery can leave as much as 70% of the petroleum in the reservoir. Such effects as microscopic trapping and by-passing are the more obvious reasons for the low recovery. There are two main objectives in secondary crude oil production. One objective is to supplement the depleted reservoir energy pressure, and the second objective is to sweep the crude oil from the injection well toward and into the production well. In fact, secondary oil recovery involves the introduction of energy into a reservoir to produce more oil. For example, the addition of materials to reduce the interfacial tension of the oil results in a higher recovery of oil.

Thus, conventional primary and secondary recovery processes are ultimately expected to produce about one-third of the original oil discovered, although recoveries from individual reservoirs can range from less than 5% to as high as 80% of the original oil in place. This broad range of recovery efficiency is a result of variations in the properties of the specific rock and fluids involved from reservoir to reservoir as well as the kind and level of energy that drives the oil to producing wells, where it is captured (Speight, 2009, 2014a).

Conventional oil production methods may be unsuccessful because the management of the reservoir was poor or because reservoir heterogeneity has prevented the recovery of crude oil in an economical manner. Reservoir heterogeneity, such as fractures and faults, can cause reservoirs to drain inefficiently by conventional methods. Also, highly cemented or shale zones can produce barriers to the flow of fluids in reservoirs and lead to high residual oil saturation. Reservoirs containing crude oils with low API gravity often cannot be produced efficiently without application of EOR methods because of the high viscosity of the crude oil. In some cases, the reservoir pressure was depleted prematurely by poor reservoir management practices to create reservoirs with low energy and high oil saturation.

Tertiary recovery is started before secondary oil recovery techniques are no longer enough to sustain production. For example, thermally enhanced oil recovery methods are recovery methods in which the crude oil is heated to reduce the viscosity, which is an important factor that must be taken into account when heavy oil is recovered from a reservoir (Speight, 2009, 2014a). In fact, certain reservoir types, such as those with very viscous crude oils and some low-permeability carbonate (limestone, dolomite, or chert) reservoirs, respond poorly to conventional secondary recovery techniques. In these reservoirs it is desirable to initiate EOR operations as early as possible. However, each reservoir has unique fluid and rock properties, and specific chemical systems must be designed for each individual application. The chemicals used, their concentrations in the slugs, and the slug sizes depend upon the specific properties of the fluids and the rocks involved and upon economic considerations.

Enhanced oil recovery processes use *chemical* methods (which also include *fluid phase behavior methods)* or *thermal methods* to reduce or eliminate the capillary forces that trap oil within pores, to thin the oil or otherwise improve its mobility or to alter the mobility of the displacing fluids. In some cases, the effects of gravity forces, which ordinarily cause vertical segregation of fluids of different densities, can be minimized or even used to advantage. The various processes differ considerably in complexity, the physical mechanisms responsible for oil recovery, and the amount of experience that has been derived from field application. The degree to which the EOR methods are applicable in the future will depend on development of improved process technology. It will also depend on improved understanding of fluid chemistry, phase behavior, and physical properties; and on the accuracy of geology and reservoir engineering in characterizing the physical nature of individual reservoirs.

There are continuing efforts to evaluate the potential for EOR in offshore fields, most of them are at the early stages of evaluation or may not be economically attractive with the current technology. As a result, it is expected that commercial applications of EOR methods will likely not be regularly applied for several years. Constraints related to surface facilities constraints and environmental regulations (e.g., chemical additives for EOR also represent major hurdles for large EOR applications in offshore fields. It is more than likely that waterflooding and gas injection and their combined processes (e.g., WAG: water assisted gas injection) will continue to support offshore production in the near term.

6.3.3.1 Chemical Methods

Chemical methods include polymer flooding, surfactant (micellar or polymer and microemulsion) flooding, and alkaline flood processes. *Polymer flooding* (*Polymer augmented waterflooding*) is waterflooding in which organic polymers are injected with the water to improve horizontal and vertical sweep efficiency. The process is conceptually simple and inexpensive, and its commercial use is increasing despite relatively small potential incremental oil production. Surfactant flooding is complex and requires detailed laboratory testing to support field project design. As demonstrated by field tests, it has excellent potential for improving the recovery of low-viscosity to moderate-viscosity oil. Surfactant flooding is expensive and has been used in few large-scale projects. Alkaline flooding has been used only in those reservoirs containing specific types of high-acid-number crude oils.

Microemulsion flooding (*micellar/emulsion flooding*) refers to a fluid injection process in which a stable solution of oil, water, and one or more surfactants along with electrolytes of salts is injected into the formation and is displaced by a mobility buffer solution (Reed and Healy, 1977; Dreher and Gogarty, 1979). Injecting water in turn displaces the mobility buffer. Depending on the reservoir environment, a pre-flood may or may not be used. The microemulsion is the key to the process. Oil and water are displaced ahead of the microemulsion slug, and a stabilized oil and water bank develops. The displacement mechanism is the same under secondary and tertiary recovery conditions. In the secondary case, water is the primary produced fluid until the oil bank reaches the well.

Conventional waterflooding can often be improved by the addition of polymers (*polymer flooding*) to injection water to improve the mobility ratio between the injected and in-place fluids. The polymer solution affects the relative flow rates of oil and water and sweeps a larger fraction of the reservoir than water alone, thus contacting more of the oil and moving it to production wells. Polymers currently in use are produced both synthetically (polyacrylamides) and biologically (polysaccharides). The polymers may also be cross-linked *in situ* to form highly viscous fluids that will divert the subsequently injected water into different reservoir strata.

Polymer flooding has its greatest utility in heterogeneous reservoirs and those that contain moderately viscous oils. Oil reservoirs with adverse waterflood mobility ratios have a potential for increased oil recovery through better horizontal sweep efficiency. Heterogeneous reservoirs may respond favorably as a result of improved vertical sweep efficiency. Because the microscopic displacement efficiency is not affected, the increase in recovery over waterflood will likely be modest and limited to the extent that sweep efficiency is improved, but the incremental cost is also moderate. Currently, polymer flooding is being used in a significant number of commercial field projects. The process may be used to recover oils of higher viscosity than those for which a surfactant flood might be considered.

Surfactant flooding is a multiple-slug process involving the addition of surface-active chemicals to water. These chemicals reduce the capillary forces that trap the oil in the pores of the rock. The surfactant slug displaces the majority of the oil from the reservoir volume contacted, forming a flowing oil–water bank that is propagated ahead of the surfactant slug. The principal factors that influence the surfactant slug design are interfacial properties, slug mobility in relation to the mobility of the oil–water bank, the persistence of acceptable slug properties, and slug integrity in the reservoir.

Alkaline flooding adds inorganic alkaline chemicals, such as sodium hydroxide, sodium carbonate, or sodium orthosilicate, to the water to enhance oil recovery by one or more of the following mechanisms: interfacial tension reduction, spontaneous emulsification, or wettability alteration (Morrow, 1996). These mechanisms rely on the *in situ* formation of surfactants during the neutralization of petroleum acids in the crude oil by the alkaline chemicals in the displacing fluids. Although emulsification in alkaline flooding processes decreases injection fluid mobility to a certain degree, emulsification alone may not provide adequate sweep efficiency. Sometimes polymer is included as an ancillary mobility control chemical in an alkaline waterflood to augment any mobility ratio improvements due to alkaline-generated emulsions.

Miscible fluid displacement (*miscible displacement*) is an oil displacement process in which an alcohol, a refined hydrocarbon, a condensed petroleum gas, carbon dioxide, liquefied natural gas, or even exhaust gas is injected into an oil reservoir, at pressure levels such that the injected gas or alcohol and reservoir oil are miscible; the process may include the concurrent, alternating, or subsequent injection of water.

The procedures for miscible displacement are the same in each case and involve the injection of a slug of solvent that is miscible with the reservoir oil followed by injection of either a liquid or a gas to sweep up any remaining solvent. It must be recognized that the miscible *slug* of solvent becomes enriched with oil as it passes through the reservoir and its composition changes, thereby reducing the effective scavenging action. However, changes in the composition of the fluid can also lead to wax deposition as well as deposition of asphaltene constituents. Therefore, caution is advised.

Other parameters affecting the miscible displacement process are reservoir length, injection rate, porosity, and permeability of reservoir matrix, size and mobility ratio of miscible phases, gravitational effects, and chemical reactions. Miscible floods using carbon dioxide, nitrogen, or hydrocarbons as miscible solvents have their greatest potential for enhanced recovery of low-viscosity oils. Commercial hydrocarbon-miscible floods have been operated since the 1950s, but carbon dioxide-miscible flooding on a large scale is relatively recent and is expected to make the most significant contribution to miscible enhanced recovery in the future.

Carbon dioxide is capable of displacing many crude oils, thus permitting recovery of most of the oil from the reservoir rock that is contacted (*carbon dioxide-miscible flooding*). The carbon dioxide is not initially miscible with the oil. However, as the carbon dioxide contacts the *in situ* crude oil, it extracts some of the hydrocarbon constituents of the crude oil into the carbon dioxide and carbon dioxide is also dissolved in the oil. Miscibility is achieved at the displacement front when no interfaces exist between the hydrocarbon-enriched carbon dioxide mixture and the carbon dioxide-enriched oil. Thus, by a *dynamic (multiple-contact)* process involving interphase mass transfer, miscible displacement overcomes the capillary forces that otherwise trap oil in pores of the rock.

In some applications, particularly in carbonate (limestone, dolomite, and chert/fine-grained quartz) reservoirs where it is likely to be used most frequently, carbon dioxide may prematurely break through to producing wells. When this occurs, remedial action using mechanical controls in injection and production wells may be taken to reduce carbon dioxide production. However, substantial carbon dioxide production is considered normal. Generally this produced carbon dioxide is reinjected, often after processing to recover valuable light hydrocarbons.

For some reservoirs, miscibility between the carbon dioxide and the oil cannot be achieved and is dependent upon the oil properties. However, carbon dioxide can still be used to recover additional oil. The carbon dioxide swells crude oils, thus increasing the volume of pore space occupied by the oil and reducing the quantity of oil trapped in the pores. It also reduces the oil viscosity. Both effects improve the mobility of the oil. Carbon dioxide-immiscible flooding has been demonstrated in both pilot and commercial projects, but overall it is expected to make a relatively small contribution to EOR.

The solution GOR for carbonated crude oil should be measured in the normal way and plotted as GOR in volume per volume versus pressure. The greater the solubility of carbon dioxide in the oil, the larger is the increase in the solution GOR. In fact, the increase in the GOR usually parallels the increase in the oil formation volume factor due to swelling. It should be noted that the gas in any GOR experiment is not carbon dioxide but contains hydrocarbons that have vaporized from the liquid phase. Consequently, whether the GOR is measured in a pressure–volume–temperature cell or from a slim tube experiment, compositional analysis must be carried out to obtain the composition of the gas as well as that of the equilibrium liquid phase. If actual measured values are not available, the correlation developed for crude oil containing dissolved gases can be used but give only approximate values at best. Since the density of pure gases is a function of pressure and temperature, for crude oil saturated with gases, the density in the mixing zone must be specified as a function of pressure and mixing zone composition.

6.3.3.2 Thermal Methods

Thermal methods for oil recovery have found most use when the oil in the reservoir has a high viscosity. For example, heavy oil is usually highly viscous (hence the use of the adjective *heavy*), with a viscosity ranging from approximately 100 centipoises to several million centipoises at the reservoir conditions. In addition, oil viscosity is also a function of temperature and API gravity (Speight, 2000 and references cited therein). *Thermal EOR processes* add heat to the reservoir to reduce oil viscosity and/or to vaporize the oil. In both instances, the oil is made more mobile so that it can be more effectively driven to producing wells. In addition to adding heat, these processes provide a driving force (pressure) to move oil to producing wells.

Thermal recovery methods include cyclic steam injection, steam flooding, and *in situ* combustion. The steam processes are the most advanced of all

EOR methods in terms of field experience and thus have the least uncertainty in estimating performance, provided that a good reservoir description is available. Steam processes are most often applied in reservoirs containing viscous oils and tars, usually in place of rather than following secondary or primary methods. Commercial application of steam processes has been underway since the early 1960s. *In situ* combustion has been field tested under a wide variety of reservoir conditions, but few projects have proven economical and advanced to commercial scale.

Steam drive injection (steam injection) has been commercially applied since the early 1960s. The process occurs in two steps: (1) steam stimulation of production wells, that is, direct steam stimulation, and (2) steam drive by steam injection to increase production from other wells (indirect steam stimulation).

When there is some natural reservoir energy, steam stimulation normally precedes steam drive. In steam stimulation, heat is applied to the reservoir by the injection of high-quality steam into the produce well. This cyclic process (*huff and puff process, steam soak process*) uses the same well for both injection and production. The period of steam injection is followed by production of reduced viscosity oil and condensed steam (water). One mechanism that aids production of the oil is the flashing of hot water (originally condensed from steam injected under HP) back to steam as pressure is lowered when a well is put back on production.

Cyclic steam injection is the alternating injection of steam and production of oil with condensed steam from the same well or wells. The process involves three stages: (1) injection, during which a measured amount of steam is introduced into the reservoir, (2) the soak period, which requires that the well be shut in for a period of time – usually on the order of usually several days—to allow uniform heat distribution to reduce the viscosity of the oil or, alternatively, to raise the reservoir temperature above the pour point of the oil, and (3) production, in which the mobile oil is produced through the same well. The cycle is repeated until the production of crude oil diminishes to a point of no returns.

Cyclic steam injection is used extensively in heavy-oil reservoirs, tar sand deposits, and in some cases to improve injectivity prior to steam flooding or *in situ* combustion operations. Cyclic steam injection is also called *steam soak* or the *huff 'n' puff* method.

In situ combustion is normally applied to reservoirs containing low-gravity oil but has been tested over perhaps the widest spectrum of conditions of any EOR process. In the process, heat is generated within the reservoir by injecting air and burning part of the crude oil. This reduces the oil viscosity and partially vaporizes the oil in place, and the oil is driven out of the reservoir by a combination of steam, hot water, and gas drive. *Forward combustion* involves movement of the hot front in the same direction as the injected air. *Reverse combustion* involves movement of the hot front opposite to the direction of the injected air.

The relatively small portion of the oil that remains after these displacement mechanisms have acted becomes the fuel for the *in situ* combustion process. Production is obtained from wells offsetting the injection locations. In some applications, the efficiency of the total *in situ* combustion operation can be improved by alternating water and air injection. The injected water tends to improve the utilization of heat by transferring heat from the rock behind the combustion zone to the rock immediately ahead of the combustion zone.

The performance of *in situ* combustion is predominantly determined by the four following factors: (1) the quantity of oil that initially resides in the rock to be burned, (2) the quantity of air required to burn the portion of the oil that fuels the process, (3) the distance to which vigorous combustion can be sustained against heat losses, and (4) the mobility of the air or combustion product gases.

Using combustion to stimulate oil production is regarded as attractive for deep reservoirs and, in contrast to steam injection, usually involves no loss of heat. The duration of the combustion may be short (<30 days) or more prolonged (approximately 90 days), depending upon requirements. In addition, backflow of the oil through the hot zone must be prevented or coking occurs.

6.3.4 Acidizing

In the *acidizing* process, a reservoir formation is stimulated by pumping a solution containing reactive acid to improve the permeability and enhance production of a well. In sandstone formations, the acids help enlarge the pores, while in carbonate formations, the acids dissolve the entire matrix. The process can be divided into two categories: (1) matrix acidizing, which is typically used in sandstone formations and the acid is pumped into a well at LPs, dissolving sediments and mud solids, increasing the permeability of the rock, enlarging the natural pores, and stimulating the flow of oil and gas, (2) fracture acidizing, which is typically used in carbonate formations, involves pumping acid at higher pressures—lower pressures than those used during hydraulic fracturing—and the acids fracture the rock, allowing for the flow of crude oil and natural gas.

The acidizing process is typically used in aging wells that are in the final stages of production. The process primarily uses hydrochloric and hydrofluoric (HF) acids at highly diluted concentrations, between 1 and 15% (v/v) and can also be used (at concentrations of acid) to dissolve oil bearing shale. However, the acids are dangerous and strict process control is necessary—for example, HF acid (HF) can corrode glass, steel, and rock and actions are required to prevent the acid from dissolving the well casing, which is intended to keep oil and chemicals from contaminating the surrounding rock or water. In addition, HF acid can cause severe burns to the skin and eyes, and can damage lungs in ways that are not immediately noticeable. If absorbed through the skin, even in minute amounts, and left untreated, it can cause death. HF acid is a liquid at low temperatures, but at 19°C (67°F), the acid becomes a dense vapor cloud that hovers near the ground and does not dissipate.

6.3.5 Offshore EOR

Enhanced oil recovery in offshore fields is constrained not only by reservoir anatomy and lithology but also by surface facilities and environmental regulations and (depending upon the location) will more than likely be limited compared with that for onshore fields. As a result, the main production strategy of offshore fields has been pressure maintenance by gas and water injection.

Despite environmental conditions in the area (e.g., hurricane seasons, another nontechnical issues), oil production from deep waters in the Gulf of Mexico is expected to continue to increase in the future. An increase in oil production will be associated with new projects, the announcement of numerous deepwater discoveries, improved reservoir characterization, recent drilling successes, and new federal incentives for the development of deep gas resources rather than EOR projects. Mexico is another example of offshore reservoir production that is supported by gas injection, bottom water drive, and/or water injection.

Although there are several initiatives to evaluate the potential for EOR projects in offshore fields, most of them are at the early stages of evaluation or may not be economically attractive with the current technology. Therefore, it is expected that commercial applications of EOR methods will require further evaluation. Surface facility constraints and environmental regulations (e.g., chemical additives for EOR) also represent major hurdles for applications of EOR projects in offshore fields.

Considering the price volatility of energy markets, the risk associated with this type of project is high, reducing the probability of the implementation of EOR projects. Therefore, waterflooding and gas injection and the combined processes (e.g., water-assisted gas injection, WAG) will continue to support offshore production in the near term.

6.4 RISKS AND ENVIRONMENTAL EFFECTS

The nature of the offshore operations—extraction of volatile substances sometimes under extreme pressure in a hostile environment—has risk and frequent accidents and tragedies occur (Chapter 10).

Oil spills are an ever-present environmental risk of offshore oil production projects. Further effects are the leaching of heavy metals that accumulate in buoyancy tanks into water; and risks associated with their disposal. There has been concern expressed at the practice of partially demolishing offshore rigs to the point that ships can traverse across the site.

6.5 CRUDE OIL AND NATURAL GAS TREATMENT

Typically, crude oil and natural gas often need cleaning before refining by removal of undesirable constituents such as high-sulfur, high-nitrogen, and high-aromatics (such as polynuclear aromatic components and acidic components (Speight, 2014a, 2014b). A controlled visbreaking treatment would *clean up* such crude oils by removing these undesirable constituents (which, if not removed, would cause problems further down the refinery sequence) as coke or sediment (Radovanović and Speight, 2011). However, there always remains the matter of dealing with any waste products from the visbreaker.

In addition to taking preventative measure for the refinery to process these feedstocks without serious deleterious effects on the equipment, refiners will need to develop programs for detailed and immediate feedstock evaluation so that they can understand the qualities of a crude oil very quickly and it can be valued appropriately and management of the crude processing can be planned meticulously.

The fluids from the well contain a mixture of oil, gas, sand, and brine water. This mixture is processed on the platform for (at least) water and solids (Figure 6.7) so that the fluid meets pipeline specification for fluid transportation; the well fluid may also be transported to shore by a tanker (Chapter 7). In terms of offshore on-platform refining, the way lies open for the future

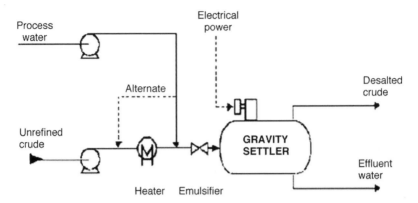

Figure 6.7 Schematic of water removal from and desalting of crude oil. http://www.osha.gov/dts/osta/otm/otm_iv/otm_iv_2.html. *Source: OSHA Technical Manual, Section IV, Chapter 2: Petroleum Refining Processes.*

installation of simple visbreaking equipment to prepare heavier oils to meet the specifications for transportation by pipeline (Bai and Bai, 2005).

In offshore crude oil and natural gas production operations, the dry completion wells on the main field center feed directly into production manifolds, while outlying wellhead towers and subsea installations feed via multiphase pipelines back to the production risers. Risers—the system that connects a pipeline to the topside structure—and allows transport of the well fluids to the surface structure. For floating or structures, this involves a way to take up weight and movement but for heavy crude oil and in Arctic areas where low temperatures cause influence the crude oil properties, diluents and heating may be needed to reduce viscosity and allow fluid flow.

For wells that are predominantly gas-producing wells, the produced gas (sometimes low-boiling condensate) can be taken directly to natural gas treatment and/or compression stations (Figure 6.3, Figure 6.8), whichever is the most appropriate choice. More often, the well gives a combination of natural gas, crude oil, and water plus various contaminants which must be separated and processed. The production separators come in many forms and designs, with the classical variant being the gravity separator although membrane technology is becoming increasingly popular (Baker, 2002; Mokhatab et al., 2006; Speight, 2014a). To accomplish gravity separation, the well flow is fed into a horizontal separator vessel (Figure 6.9)—the retention period is typically 5 min—which allows the natural gas constituents to separate from the water, which settle to the bottom with any crude oil constituents taken out at a mid-point of the separator. Variations in the

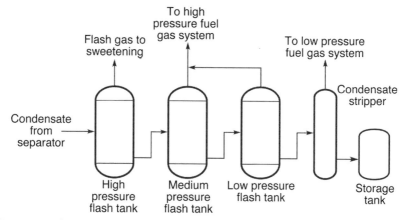

Figure 6.8 Schematic of high-pressure and low-pressure separation.

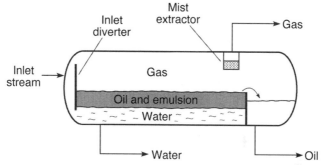

Figure 6.9 A gas–oil–water separator.

pressure inside the vessel allow different separations to be accomplished (Figure 6.10). In some cases, crude oil-associated constituents (natural gas liquids) are also separated and sent transported to a refinery for use in a petrochemical plant or as a source of energy as a blend stock for gasoline.

Most surface platforms do not allow storage of natural gas—the gas may be reinjected into the well to encourage oil production—but crude oil is often stored before loading on a vessel, such as a shuttle tanker taking oil to a larger tanker terminal, or direct to a crude carrier. Offshore production facilities without a direct pipeline connection generally rely on crude storage in the base or hull, to allow a shuttle tanker to offload on an as needed or scheduled basis—if the amount of crude oil produced is stable. A larger production complex may have an associated tank farm terminal allowing the storage of different grades of crude to take up changes in demand, such as delays in transport.

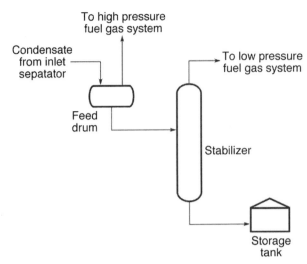

Figure 6.10 A high-pressure fuel gas separation system.

Finally, a metering station allows offshore production operators to monitor and manage the crude oil natural gas exported from the production platform. The metering station will be fitted with meters that not only measure the crude oil and/or natural gas flow but also *do not* impede the crude oil as it flows through the pipeline. As it is typical for onshore metering stations, the offshore metered volume represents a transfer of ownership from a production division to another division within the company (*custody transfer metering*).

REFERENCES

Amin, R., Waterson, A., 2002. The Challenges of Multiphase Flow. Proceedings. 14th International Oil & Gas Industry Exhibition and Conference, OSEA 2002. Singapore Exhibition Services PTE Ltd., Singapore, October–November.

Amin, A., Riding, M., Shepler, R., Smedstad, E., Ratulowski, J., 2005. Subsea Development from Pore to Process. Schlumberger Oilfield Review, Spring, pp. 4–17.

API, 2008. Recommended Practice for Flexible Pipe, 4th Edition American Petroleum Institute, Washington, DC, ANSI/API Recommended Practice 17B.

Bai, Y., Bai, Q., 2005. Subsea Pipelines and Risers, 2nd Edition Elsevier BV, Amsterdam, Netherlands.

Baker, R.W., 2002. Future Directions of Membrane Gas Separation Technology. Industrial Engineering Chemistry Research 41, 1393–1411.

Chianis, J.W., 2003. Deepwater Dry-Tree Units for Southeast Asia. PetroMin 29 (8), 30–43.

Christie, A., Kishino, A., Cromb, J., Hensley, R., Kent, E., McBeath, B., Stewart, H., Vidal, A., Koot, L., 1999. Subsea Solutions. Schlumberger Oilfield Review, Winter, pp. 2–19.

El-Reedy, M.A., 2012. Offshore structures: Design, Constructions, and Maintenance. Gulf Professional Publishers, Elsevier, Boston, Massachusetts.

Gary, J.G., Handwerk, G.E., Kaiser, M.J., 2007. Petroleum Refining: Technology and Economics, 5th Edition CRC Press, Taylor & Francis Group, Boca Raton, Florida.

Hatton, S.A., Howells, H., 1996. Catenary Hybrid Risers for Deepwater Locations Worldwide. Proceedings. Advances in Riser Technologies Conference, Aberdeen, United Kingdom, June.

Hsu, C.S., Robinson, P.R. (Eds.), 2006. Practical Advances in Petroleum Processing Volume 1 and Volume 2. Springer Science, New York.

Hunt, A., 1996. Fluid properties determine flowline blockage potential. Oil & Gas Journal 94 (29), 62–66.

Jahn, F., Cook, M., Graham, M., 1998. Hydrocarbon Exploration and Production. Elsevier, Amsterdam, Netherlands.

Manning, F.S., Thompson, R.E., 1995. Oil Field Processing, Volume 2: Crude Oil. PennWell Publishing Company, Tulsa, Oklahoma.

Mokhatab, S., Poe, W.A., Speight, J.G., 2006. Handbook of Natural Gas Transmission and Processing. Elsevier, Amsterdam, Netherlands.

Mokhatab, S., 2007. Severe slugging in a catenary-shaped riser: experimental and simulation studies. Petroleum Science and Technology 25, 719–740.

Parkash, S., 2003. Refining Processes Handbook. Elsevier-Gulf Professional Publishing Company, Burlington, Massachusetts.

Patel, M.H., Seyed, F.B., 1995. Review of flexible riser modeling and analysis techniques. Engineering Structures 17 (4), 293–304.

Radovanović, L., Speight, J.G., 2011. Visbreaking: A Technology of the Future. Proceedings. First International Conference – Process Technology and Environmental Protection (PTEP 2011). University of Novi Sad, Technical Faculty "Mihajlo Pupin," Zrenjanin, Republic of Serbia. December 7. pp. 335–338.

Reis, J.C., 1976. Environmental Control in Petroleum Engineering. Gulf Professional Publishers, Houston, Texas.

Schmidt, Z., Brill, J.P., Beggs, H.D., 1980. Experimental study of severe slugging in a two-phase flow pipeline riser-pipe system. SPE Journal 20, 407–414.

Speight, J.G., 2001. Handbook of Petroleum Analysis. John Wiley & Sons Inc., New York.

Speight, J.G., Ozum, B., 2002. Petroleum Refining Processes. Marcel Dekker Inc., New York.

Speight, J.G., 2009. Enhanced Recovery Methods for Heavy Oil and Tar Sands. Gulf Publishing Company, Houston, Texas, 2009.

Speight, J.G., 2014a. The Chemistry and Technology of Petroleum, 5th Edition CRC Press, Taylor & Francis Group, Boca Raton, Florida.

Speight, J.G., 2014b. High Acid Crudes. Elsevier-Gulf Professional Publishing, Oxford, United Kingdom.

Wills, J.W.G., 2000. Muddied Waters—A Survey of Offshore Oilfield Drilling Wastes and Disposal Techniques to Reduce the Ecological Impact of Sea Dumping. Ekologicheskaya Vahkta Sakhalina, Yuzhno-Sakhalinsk, Russia.

Wolfson, A., Van Blaricom, G., Davis, N., Lewbe, G.S., 1979. The marine life of an offshore oil platform. Marine Ecology Progress Series 1, 81–89.

Zittel, R.J., Beliveau, D., O'Sullivan, T., Mohanty, R., 2006. Reservoir Crude Oil Viscosity Estimation from Wireline NMR Measurement. Paper No. SPE 101689. Proceedings. SPE Annual Technical Conference, San Antonio, Texas. September 24–27. Society of Petroleum Engineers, Richardson, Texas.

CHAPTER SEVEN

Transportation

Outline

7.1 INTRODUCTION

Most crude oil and natural gas fields are at a considerable distance from the refineries that convert crude oil into usable products, and therefore the oil must be transported in pipelines and tankers (Speight, 2014a). However, most crude oil needs some form of treatment near the reservoir before it can be carried to considerable distances through the pipelines or in the tankers. Railroad cars and motor vehicles are also used to a large extent for the transportation of petroleum products.

Fluids produced from a well are seldom pure crude oil. In fact, the oil often contains quantities of gas, saltwater, or even sand. Processing facilities on offshore platforms remove unwanted fluids from crude oil before it is stored and shipped. Thus, separation must be achieved before transportation. Separation and cleaning usually take place at a central facility that collects the oil produced from several wells. Gas can be separated conveniently at the wellhead. When the pressure of the gas in the crude oil as it comes out at the surface is not too great, a simple flow tank fitted with baffles can be used to separate the gas from the oil at atmospheric pressure. If a considerable amount of gas is present, particularly if the crude oil is under considerable pressure, a series of flow tanks is necessary. The natural gas itself may contain as impurities one or more nonhydrocarbon substances. The most abundant of these impurities is hydrogen sulfide, which imparts a noticeable odor to the gas. A small amount of this compound is considered advantageous as it gives an indication of leaks and where they occur, as mentioned.

Another step that needs to be taken in the preparation of crude oil for transportation is the removal of excessive quantities of water. Crude oil at the wellhead usually contains emulsified water in proportions that may reach amounts approaching 80%–90%. It is generally required that crude oil to be transported by pipeline contain substantially less water than may appear in the crude at the wellhead. In fact, water contents from 0.5% to 2.0% have been specified as the maximum tolerable amount in a crude oil to be moved by pipeline. It is therefore necessary to remove the excess water from the crude oil before transportation.

In an emulsion, the globules of one phase are usually surrounded by a thin film of an emulsifying agent that prevents them from congregating into large droplets. In the case of an oil–water emulsion, the emulsifying agent may be part of the heavier (asphaltic) more polar constituents. The film may be broken mechanically, electrically, or by the use of demulsifying agents and the proportion of water in the oil reduced to the specified amounts, thereby tendering the crude oil suitable for transportation.

The transportation of crude oil may be further simplified by blending crude oils from several wells, thereby homogenizing the feedstock to the refinery. It is usual practice, however, to blend crude oils of similar characteristics although fluctuations in the properties of the individual crude oils may cause significant variations in the properties of the blend over a period of time. However, the technique of blending several crude oils before transportation, or even after transportation but before refining, may eliminate the frequent need to change the processing conditions that would perhaps be required to process each of the crude oils individually.

The arrival of large quantities of petroleum oil at import and refining centers has brought about the need for storage facilities. The usual form of crude oil storage is the collection of large cylindrical steel storage tanks (*tank farm*) that are a familiar sight at most refineries and shipping terminals. The tanks vary in size, but some are capable of holding up to 950,000 barrels of oil. Crude oil may also be stored in such geological features as salt domes. The domes have been previously leached or hollowed out into huge underground caves, such as those used by the US Strategic Petroleum Reserve in Louisiana and Texas. Other underground storage facilities include disused coal mines and artificial caverns. Natural gas is, on occasion, stored in old reservoirs from which the gas has been recovered. The gas is pumped under pressure into the reservoir at times of low gas demand so that it can be retrieved later to meet peak demand.

7.2 PIPELINES

The transportation of crude oil (which replaced whale oil as the illuminant of choice in the 19th century) has had an interesting history in which the system has evolved to the modern-day system of transportation. Early producers shipped crude oil to market in earthenware vessels aboard slow-moving barges—on the other hand, the use of natural gas in antiquity is somewhat less well documented, although historical records indicate that the use of natural gas (for other than religious purposes) dates back to about 250 AD when it was used as a fuel in China. The gas was obtained from shallow wells and was distributed through a piping system constructed from hollow bamboo stems. Since then, the need to move increasingly large quantities of crude oil and natural gas has evolved considerably. As containers (for liquids) evolved, the 42-gallon oil barrel became the (and remains) United States standard for measurement. The unit originated in the 1860s—at the time there was no standard measure for oil—and producers simply used whiskey and molasses barrels or whatever was handy until customers demanded something more uniform (and with less impurity from the prior contents of the barrels). The most common size cask capacity (42 US gallons, approximately 35 UK or Imperial gallons) became the agreed-upon standard. As US oil companies became active in other countries, many of these countries also adopted that 42 US gallon barrel as the standard unit of measure of crude oil. For much of international trade, however, the common unit is the metric ton (2.204 pounds), which is approximately 7 US barrels.

For the purposes of this text, the focus is on offshore transportation using pipelines and sea-going tankers. Thus, when the crude oil and natural gas have been separated into liquid and gas and any other non-refinery or superfluous material (such as water) removed, the crude oil and natural gas for the offshore facility are sent to onshore refineries and/or to gas processing plants on land. Most offshore oil and gas production is transported by pipelines to onshore facilities, which require that pipelines be laid from the offshore facility to an onshore center and from thence the crude oil and natural gas are sent to a refinery or gas processing facility.

The technique for laying pipelines under water in the current sense is not an old technology and had its beginnings in England during World War II. Initiated by the necessity of the war-time need for fuels, steel tubes were welded together and coiled around floating drums. One end of the pipe was fixed to a terminal point and as the floating drums were towed across the English Channel, the pipe was unreeled from the drum. Once in securely place (or at least as secure as it could be under the circumstances), the pipeline connected fuel supply depots in England with distribution points in Europe to support the Allied march across France and into Germany.

Pipeline designs vary depending on what they are transporting—crude oil, natural gas, or refined products—and their function. The world's longest underwater pipeline is the Langeled project, which was completed in 2007 to transport natural gas approximately 750 miles from Norway to England. Moreover, more than a third of the worldwide growth in drilling is expected to come from offshore and technological advances in pipeline construction and safety are accelerating with most efficient and safety the goals.

A subsea pipeline (offshore pipeline, subsea pipeline, marine pipeline) is a pipeline that is laid on the seabed or below the seabed inside specially constructed trench (Brown, 2006; Gerwick, 2007; Bai and Bai, 2010; Dean, 2010; Palmer and Been, 2011). While subsea pipelines are used primarily to carry oil or gas, transportation of water by pipeline is often of equal importance. At this point, it is worthy of note that a distinction is sometimes made between a *flow line* and a *pipeline*—the former is an *intra-field* pipeline, insofar as it is used to connect subsea wellheads, manifolds, and the platform *within* a particular development field. The latter type of pipeline, sometimes referred to as an *export pipeline*, is used to bring the resource to a shore-based center. Sizeable pipeline construction projects need to take into account a large number of factors, such as the offshore ecology, geo-hazards and environmental loading—they are often undertaken by multidisciplinary, international teams.

7.2.1 Installation

Pipelines can be installed by a number of methods depending on the site conditions. Barges and other types of vessels are often used to construct and lay the pipelines—some barges are equipped to weld the pipes together and then lower them to the seabed one pipe or joint at a time as the vessel slowly moves ahead. Each weld is inspected using X-ray and/or ultrasonic techniques to ensure a proper weld and the integrity of the pipeline system. For burying the pipelines, specialized equipment is used to dig a trench and then cover the pipeline—the methods vary depending on the depth of the trench and the water depth as well as strength of the seafloor soils. In the most obvious option, the trench may be dug before the line is laid down or, as another option, the trench can be dug around a pipeline that has been laid on the seabed.

Originally, with reference to the shore-based pipelines, there were two principal technological trends during that period that increased pipeline use: (1) more widespread use was made of large diameter pipe, and (2) more efficient diesel pumps became available for installation at stations along the line, replacing steam-driven models, as well as other developments such as (3) the introduction of welded rather than screw-in couplings, (4) the use of high carbon steel to replace lap-welded pipe, and (5) the replacement of diesel equipment by electrically powered pumps. The increased use of subsea pipelines has served to advance the technology even further.

Pipelines may be used to transport different types of crude oil (batch transportation) and when the different batches (rather than a blend of two or more crude oils) must be kept separated and mixing prevented, slugs of kerosene, water, or occasionally inflatable rubber balls can be used to separate the batches. However, there also occasions when the batches of the different crude oils are transported through the pipelines without such separators. The properties of each crude oil may be such that mixing, other than the formation of a narrow interface, is prevented.

One of the early, and critical, tasks in a subsea pipeline planning exercise is the route selection of the most appropriate route—which is not always the shortest or most convenient route. Route selection must to consider a variety of issues, some of a political nature, but most issues deal with (1) geo-hazards, (2) physical factors along the prospective route, and (3) other uses of the seabed in the area considered. This task begins with a fact-finding exercise—often a standard desk-paper study—that includes a survey of geological maps, bathymetry, subsea topography, aerial and satellite photography, as well as information from the various navigation authorities.

However, the primary physical factor to be considered in subsea pipeline construction is the state of the seabed—whether it is smooth (relatively flat) or uneven (corrugated, with high points and low points). If the seabed is uneven, the pipeline will include free spans when it connects two high points, leaving the section in between without any physical support and, if an unsupported section is too long, the bending stress exerted onto it (due to its weight, including the weight of the crude oil inside the pipeline) may be excessive leading to serious damage. In addition, vibration from sea current-induced vortexes may also become an issue and corrective measures for unsupported pipeline spans include seabed leveling and post-installation support, such as the construction of a berm (a raised small hill of sand between the high points) or sand completely filling (infilling) the area between the high points and below the pipeline. The strength of the seabed is another significant parameter—on the one hand, if the sand (or soil) is not strong enough, the pipeline may sink into to an extent where inspection, maintenance procedures, and prospective tie-ins (connections with other pipelines) become difficult to carry out. On the other hand and at the other extreme, if the seabed is rocky, it is often difficult and expensive to trench and, at high points, abrasion and damage may occur to the external coating of the pipeline. In an ideal world, the sand-soil should be such as to allow the pipe to settle into it to some extent, thereby providing the pipeline a measure of some lateral stability.

There are several other physical factors that should be given consideration prior to building a pipeline and these are: (1) *seabed mobility*, which is related to the presence of features that move with time, such that a pipeline that was supported by the crest of one such feature during construction may be in a trough later during the operational life of the pipeline—the evolution of these features is difficult to predict so it is preferable to avoid the areas where they are known to exist, (2) *subsea landslides*, which result from high sedimentation rates and occur on steeper slopes and can also be triggered by earthquakes—when the soil around the pipe is subjected to a slide, especially if the resulting displacement is at high angle to the line, the pipe within it can incur severe bending and consequent tensile failure, (3) *high currents*, which are troublesome insofar as these currents hinder pipe laying operations—in shallow seas, tidal currents may be strong in a straight between two islands and it may be preferable to lay the pipeline on another route even if the alternative route is a longer path, (4) *waves*, which in shallow water can also be problematic (some observers relate the strength of the current to wave activity) for pipeline laying operations and to pipeline stability because of the scouring action of the water, and (5) *ice-related issues*, which occur in freezing waters where the under-water part of floating ice may, in shallower waters, come into contact with the seabed and the pipeline and cause damage to the pipeline.

In terms of the pipeline size, the diameter of the pipe must be sufficient to allow the maximum volume to pass with the minimal resistance for optimal efficiency. The velocity must also be maintained at a sufficiently high rate to keep the pipe free of corrosion or debris that can plug the pipe.

The pipes are usually made of high-quality carbon steel that is produced to specific standards, tested and quality checked from its raw steel state to the finished product—pipes used for transporting corrosive fluids (high acid crude oil may be included in this category) must contain special corrosion resistant alloys. Generally, the pipes or joints are manufactured in approximately lengths on the order of 40 ft. For the transportation of heavy crude oil (high-viscosity oil), the pipeline may need to be heated to maintain oil flow and, depending on the distance the oil needs to travel, the pipeline may need to be insulated or equipped with additional pumps or heating stations to maintain flow.

In general, many of the pipeline systems available use pipe material up to 48 in. in diameter (although lately larger diameter pipe has become more favorable) and sections of pipe may be up to 40 ft long. In fact, the walls of the pipe vary from 3 in. in diameter up to 72 in. in diameter for gas pipe lines to 72 in. in diameter and also for high capacity fluid lines—the wall thickness of the pipeline typically ranges from 0.39 to 3.0 in.—the pipe can also be designed for fluids at high temperature and pressure. The structure is often shielded against external corrosion by coatings such as an asphalt derivative or epoxy material, supplemented as well as protection by cathodic protection with sacrificial anodes (Chapter 8) (Speight, 2014b). Concrete or fiberglass wrapping provides further protection against abrasion—the use of a concrete coating is also useful to compensate for the negative buoyancy of the pipeline, especially when it carries a low density fluid, such as crude oil.

When the manufacture of the pipeline is complete, the pipe is coated to protect it from corrosion and may have special insulation applied before it is shipped to the installation site. Typically, the inside wall of the pipeline is not coated when used for transporting crude oil but when it carries seawater or corrosive substances (such as hydrogen sulfide which may not have been completely removed from natural gas), the pipeline can be coated with epoxy, polyurethane, or polyethylene materials—it can also be cement-lined. In the shipment of crude oil and/or natural gas (or any other type of liquids and gases) where any amount of leakage is unacceptable and the pipelines may experience an internal pressure on the order of 1500 psi, the pipe segments are joined by full penetration welds—mechanical joints are also used.

Several systems can be used—for a subsea pipeline, the choice in favor of any one of them is based on the following factors: physical and environmental conditions (such as currents and wave regime), availability of equipment and costs, water depth, pipeline length and diameter, constraints tied to the presence of other lines and structures along the route. These systems are generally divided into four broad categories: (1) *pull-tow system*, (2) *S-lay system*, (3) *J-lay system*, and (4) *reel-lay system*.

In the *pull-tow* system, the subsea pipeline is assembled onshore and then towed to location. Assembly is done either parallel or perpendicular to the shoreline—in the former case, the full line can be built prior to tow out and installation. A significant advantage with the pull/tow system is that pre-testing and inspection of the line are performed onshore, not at sea. As for the towing procedures, a number of configurations can be used, which may be categorized as follows: *surface tow*, *near-surface tow*, *mid-depth tow*, and *off-bottom tow*.

In the *surface-tow system*, the pipeline remains at the surface of the water during tow, and is then sunk into position at site. The line has to be buoyant, which can be achieved by use attaching buoyancy units to the pipeline. The surface tow system is not appropriate for rough seas and is vulnerable to lateral currents. In the *near-surface tow system*, the pipeline remains just below the water surface to reduce any effects of wave action—the buoyancy units used to maintain the line at the desired level may be affected by rough seas. In the *mid-depth tow system*, the pipeline is not buoyant and the line is suspended in a catenary between two towing vessels. The shape of that catenary (the *sag*) is a balance between the line's weight, the tension applied to it by the vessels and hydrodynamic lift on the chains and the amount of allowable sag is limited by the depth of the water. The *off-bottom tow system*, is similar to the mid-depth tow, but (using chains dragging on the seabed as a guide) the line is maintained within three-to-seven feet or more from the sea bottom. Finally, in the *bottom tow system*, the pipeline is dragged onto the bottom—the line is not affected by waves and currents—and if rough seas develop after the commencement of the towing operation the line can be abandoned and recovered when the wave action diminishes. The challenges with this type of system relate to the need for an abrasion-resistant coating and interaction with other subsea objects (such as other pipelines, reefs, and boulder) and, as a result, the bottom tow system is commonly used for river crossings and crossings between shores.

In the *S-lay system*, the pipeline assembly is done at the installation site, on board a vessel that has all the equipment required for joining the pipe segments: pipe handling conveyors, welding stations, X-ray equipment, joint-coating and module. The *S* notation refers to the shape of the pipeline as it is laid onto the seabed. The pipeline leaves the vessel at the stern or bow from a supporting structure called a *stinger* that guides the downward motion of the pipe and controls the convex-upward curve (the *over bend*). As it continues toward the seabed, the pipe has a convex-downward curve (the *sag bend*) before coming into contact with the seabed (*touch down point*). The sag bend is controlled by a tension applied from the vessel (via *tensioners*) in response to the pipeline's submerged weight. The pipeline configuration is monitored so that it will not get damaged by excessive bending. This on-site pipeline assembly approach, referred to as *lay-barge* construction, is known for its versatility and self-contained nature—despite the high costs associated with this vessel's deployment, it is efficient and requires relatively little external support. But it may have to contend with severe sea states—these adversely affect operations such as pipe transfer from supply boats, anchor-handling and pipe welding. Recent developments in lay-barge design include *dynamic positioning* and the *J-lay system*.

In areas where the water is very deep, the S-lay system may not be appropriate because the pipeline leaves the stinger to go almost straight down. To avoid sharp bending at the end of it and to mitigate excessive sag bending, the tension in the pipeline would have to be high. Doing so would interfere with the vessel's positioning, and the tensioner could damage the pipeline. A particularly long stinger could be used, but this is also objectionable since that structure would be adversely affected by winds and currents. The *J-lay system*, one of the latest generations of lay-barge, is better suited for deep water environments. In this system, the pipeline leaves the vessel on a nearly vertical ramp (or tower). There is no over bend—only a sag bend of catenary nature (hence the *J* notation), such that the tension can be reduced. The pipeline is also less exposed to wave action as it enters the water. However, unlike the *S-lay system*, where pipe welding can be done simultaneously at several locations along the vessel deck's length, the *J-lay system* can only accommodate one welding station.

In the *reel-lay system*, the pipeline is assembled *onshore* and is spooled onto a large drum typically approximately 66 ft) by 20 ft in size, mounted on board a purpose-built vessel. The vessel then goes out to location to lay the pipeline. Onshore facilities to assemble the pipeline have inherent advantages: they are not affected by the weather or the sea state and are less expensive than seaborne operations. Pipeline supply can be coordinated: while one line is being laid at sea, another one can be spooled onshore. A single reel can have enough capacity for a full length flow line. The reel-lay system, however, can only handle lower diameter pipelines—up to approximately 16 in. Also, the kind of steel making up the pipes must be able to undergo the required amount of plastic deformation as it is bent to proper curvature (by a spiral J-tube) when reeled around the drum, and straightened back (by a straightener) during the layout operations at the installation site.

A subsea pipeline may be laid inside a trench as a means of safeguarding it against marine activity (such as boat such as anchors) and trawling activities. This may also be required in shore approaches to protect the pipeline against currents and wave action. Trenching can be done prior to pipeline lay (*pre-trenching*), or afterward by seabed removal from below the pipeline (*post-trenching*). In the latter case, the trenching device rides on top of, or straddles, the pipeline. A buried pipeline has better protection than a pipeline in an open trench—burying is generally accomplished by covering the structure with rocks quarried from a nearby shoreline or by covering the pipeline with the soil excavated from the seabed during trenching (backfilling). A significant drawback to burial is the difficulty in locating a leak should it arise, and for the ensuing repairing operations.

Because of its lower density natural gas is much more expensive to ship than crude oil. Most natural gas moves by pipeline, but in the late 1960s tanker shipment of cryogenically liquefied natural gas began, particularly from the producing nations in the Pacific to Japan. Special alloys are required to prevent the tanks from becoming brittle at the low temperatures ($-161°C$, $-258°F$) required to keep the gas liquid. Thus, the means by which *natural gas* is transported depends upon several factors: (1) the physical characteristics of the gas to be transported, whether in the gaseous or the liquid phase, (2) the distance over which the gas will be moved, (3) features such as the geological and geographic characteristics of the terrain, including land and sea operations, (4) the complexity of the distribution systems, and (5) environmental regulations that are relevant to the mode of transportation. In the last case, such factors as the possibility of pipeline rupture as well as the effect of the pipeline itself on the ecosystems need to be addressed. In general, and aside from any economic factors, it is possible to construct and put in place a system capable of transporting natural gas in the gaseous or liquid phase that allows system flexibility.

There are many such pipeline systems throughout the world and the United States. However, natural gas pipeline companies must meet environmental and legal standards. Economic standards are also a necessity: it would be extremely foolhardy (and economic suicide!) if a company were to construct several pipelines when one such system would suffice. Construction of a pipeline system involves not only environmental and legal considerations but also compliance with the regulations of the local, state, and/or federal authorities.

The gas pressure in long-distance pipelines may vary up to 5000 psi, but pressures up to 1500 psi are more usual. To complete the pipeline, it is necessary to install a variety of valves and regulators that can be opened or closed to adjust the flow of gas. The system must also be capable of shut down of any section in which an unexpected rupture may be caused by natural events (such as weather) or even by unnatural events (such as sabotage). Most of the valves or regulators in the pipeline system can now be operated by remote control, so that in the event of a rupture the system can be closed down. This is especially valuable where it may take a repair crew considerable time to reach the site of the breakdown.

The trend in recent years has been to expand the pipeline system into marine environments where the pipeline is actually under a body of water. This has arisen mainly because of the tendency for petroleum and natural gas companies to expand their exploration programs to the sea. Lines are now laid in marine locations where depths exceed 500 ft and cover distances of several hundred miles to the shore. Excellent examples of such operations include the drilling operations in the Texas gulf and in the North Sea.

One early concern with the laying of pipelines under a body of water arose because of the buoyancy of the pipe and the subsequent need to place the pipe in a permanent position on the floor of the lake bed or sea bed. In such instances, the negative buoyancy of the pipe can be overcome by the use of a weighted coating (e.g., concrete) on the pipe. Other factors, such as laying the pipe without too much stress which would otherwise induce a delayed rupture), as well as the anchoring and positioning of the pipe on the seabed, are major issues that need to be addressed.

The final phase of pipeline installation is to conduct a hydrostatic test, a check on system integrity while under pressure. Water is pumped into the pipeline and then pressurized to a minimum 1.25 times its design pressure for an extended period of time. Once this final testing has determined that the system is sound, the water is removed and the pipeline is set for operation.

7.2.2 Maintenance

In addition to prolonging the in-service life of the pipeline, a maintenance program must allow maximum flow of oil through the pipeline system. In order to assure flow, the two most common procedures for flow maintenance (internal maintenance) are (1) chemical treatment and (2) mechanical cleaning using pigs and bother pro = grams (which are often used together) can prevent pipeline flow problems.

Chemicals (such as pour-point depressants, flow improvers, corrosion inhibitors, biocides, and gas hydrate-prevention products) used in treating oil and gas pipelines are often applied using pigs to enhance performance and efficiency. Pigs are also used to remove paraffin deposits (as well as remove sand, chalk, rust, and scale deposits), apply corrosion inhibitors, clean deposits from the line, and remove accumulations of water. Water is the source of several problems in oil and gas pipelines in that it allows corrosion to occur and bacterial growth, which generates hydrogen sulfide, causes corrosion, and produce plugging slimes and solids as separate phases in the oil. If not foreign non-oil bodies are not removed, the end result is corrosion of the pipeline (Chapter 9) (Speight, 2014b).

7.2.2.1 Chemical Treatment

In terms of chemical treatment, there is a need to removed paraffin wax that is deposited from waxy crude oil (paraffin fouling). Paraffin-treating compounds are used to (1) reduce the viscosity of an oil as it cools while traversing a pipeline to maintain full flow capacity in the pipeline, (2) minimize paraffin deposition on the walls of the pipe, which causes diameter reduction (*arteriosclerosis of the pipeline!*) and a need for increased pumping requirements to maintain the flow rate, (3) prevent the development of high gel strength in crude oil which adversely affects so that the pipeline restarting after the pipeline is shut-in and cools to low ambient temperatures such as are found deepwater pipelines, and (4) minimize plugging of instrumentation and metering equipment.

The predominant types of paraffin-control compounds are (1) crystal modifiers and (2) dispersants. *Crystal modifiers* (pour-point depressants, flow improvers) distort the growth and three-dimensional shape of paraffin crystals and the paraffin deposits occur as small, round particles rather than acicular (needle-like) crystals—the latter types of crystals interlock and form gels which significantly increase oil viscosity. Thus, crystal modifiers change the crystal shape which is less likely to attach to the walls of the pipe or to other wax crystals. In addition, the crystal size remains small and the crystals are less prone to sedimentation and agglomeration.

Dispersants are surface-active compounds (surfactants) which alter the surface energy of paraffin crystals and thereby change the energy of the interaction (interfacial energy) the paraffin crystal and the oil and adversely affects the likelihood of the crystals to deposit on pipeline walls—the crystals are also less likely to interact with each other—and remain dispersed in the oil.

Crystal modifiers are typically added (continuously) at a temperature above the cloud point of the oil to be effective. The cloud point of the oil is that temperature at which the oil becomes *cloudy* due to separation of paraffin crystals and represents the solubility limit of paraffin in the oil (Speight, 2014a, 2014c). once deposited form the oil, paraffin are difficult (if not imposable) to dissolve in the oil when the oil is below the cloud point and the deposited paraffin material (wax) must be removed either by addition of solvent-dispersant chemicals, or mechanical or thermal methods—generally, mechanical methods are preferred. If hot oil is added in an attempt to dissolve the paraffin deposits this is only a partial measure and transfer of paraffin fouling to another part of the pipeline system—the paraffin deposits are likely to re-deposit as the oil cools again.

7.2.2.2 Use of Pigs

Pigs are commonly used (called *pigging*) to remove paraffin deposits from the internal surfaces of pipelines—the function of the pig (as a scraper) is to (1) scrape the adhered wax from the pipe wall and (2) remove the deposits from the pipeline. The interaction of a surface of the pig against the pipeline wall causes a shearing or scraping effect after which the deposit debris is suspended into the oil flow ahead of the pig and carried out of the line. The ability of a pig to remove wax is not necessarily the tight sealing capability but mainly the cutting, scraping, or pushing characteristics of the pig.

In most onshore and offshore pipelines, a *pig* is a standard device in pipeline transport and pig is used (1) to test for hydrostatic pressure, (2) to check for dents and crimps on the sidewalls inside the pipe, and (3) to conduct periodic cleaning and minor repairs. Because of the potential for mixing and the equally high potential for phase separation during the mixing, it is often necessary to pass cylindrical steel cleaners (*pigs*) through the pipelines, between pumping stations, to maintain the pipeline clear of deposits that are the result of phase separation of crude oil constituents.

There are many instances where either a chemical-treatment program or a pigging program should be sufficient to control paraffin deposition or remove paraffin deposits in a pipeline. However, in actual pipeline operating conditions, neither program is a sure method of fouling mitigation—this is often the case in pipelines that transport high-cloud point crude oil or the flow velocity is low and the crude oil has a high content of asphaltene constituents (Speight, 2014a). Thus, it is often necessary (if not essential) to combine chemical treatment with pig use. In such a combined program, pigs should be run periodically to scrape-off accumulated paraffin deposits on the walls of the pipe which the chemical program has not been able to prevent—this leads to reduced chemical consumption, as the goal is no longer complete prevention of deposits through use of the chemicals. In addition, the wax should be removed in controllable amounts by pig use and once pigs have removed all of the possible wall-adhering wax, chemicals can be used to treat the remaining paraffins in the oil.

7.2.2.3 Corrosion Control

Corrosion is an extremely serious problem associated with offshore structures, especially in pipelines (Chapter 9) (Speight, 2014b). Most corrosion programs involve chemical treatment with corrosion inhibitors, which form a protective layer on the walls of the pipe and which are available in various basic types: (1) oil-soluble water-dispersible chemicals, (2) water-soluble, limited-solubility chemicals, and (3) volatile chemicals, each of which is designed for specific pipeline conditions. The chemical inhibitors can be applied in a batch procedure to form a protective film that may last for several months or the chemicals can be continuously metered into the pipeline in low concentrations leading the coating the pipeline wall with thin film that is maintained over time.

Corrosion inhibitors are cationic surfactant chemicals that chemically bond to any negatively charged surface and include such inhibitors as metals, corrosion products such as iron carbonate, iron sulfide, iron oxide, and sand with clay. If deposits of dirt, corrosion products, and bacteria, are inside the pipeline, the chemical inhibitors may be prevented from contacting the pipeline walls that are already coated with dirt and/or corrosion products—pipelines should be as clean when applying corrosion inhibitor for which cleaning by application of a pigging program may be suitable.

If there is stratification of liquids (laminar flow) in the pipeline this may have an adverse effect on corrosion-inhibitor treatment and flow patterns must be taken into consideration applying corrosion inhibitors. For example, when multiphase conditions exist, liquids stratify along the bottom of the pipe and water will form a separate layer beneath the crude oil. Under these conditions, the corrosion inhibitor may not be in sufficient contact with the upper walls of the pipe, leaving a significant portion of the pipeline surface unprotected. Furthermore, if wet gas is carried in the pipeline, condensation of water and hydrocarbons caused by cooling occurs over the entire internal surface of the pipe. The condensed liquids descend to the base of the pipeline and collect in low spots and also in sections that are inclined in upward slope (*liquid hold-up*), which causes increases in pressure drop through the line as well as adversely affecting corrosion-inhibitor treatment–it becomes difficult to treat both the liquids and the exposed pipe wall.

Water is a source of several problems in oil and gas pipelines in that it allows corrosion to occur and bacteria to grow. Frequent pigging is advised to keep accumulated water and other liquids to a minimum.

7.3 TANKERS

The volume of international trade in oil increased as a result of world economic growth. The largest oil consumers are the most heavily industrialized countries such as the United States Western Europe and Japan. OECD countries account for about 75% of global crude oil imports. Since oil consumption and production do not happen in the same places, international oil trade is a necessity to compensate the imbalances between supply and demand. Unlike most other countries, a major portion of OPEC's oil is traded in international markets.

Tank trucks are used for both lock and intermediate hauling from manufacturing and distance hauling from manufacturing and terminal points to individual domestic, commercial, and industrial consumers that maintain storage tanks on their premises. Because of costs, most bulk deliveries by truck fall within a radius of 300 miles. Seagoing tankers (Table 7.1, Figure 7.1), on the other hand, can be sent to any destination where a port can accommodate them and can be shifted to different routes according to need.

Since the first oil tanker began shipping oil in 1878 in the Caspian Sea, the capacity of the world's maritime tanker fleet has grown substantially. As of 2005, about 2.4 billion tons of petroleum was shipped by maritime transportation, which is roughly 62% of all the petroleum produced. The remaining 38% is either using pipelines (dominantly), trains or trucks. Crude oil alone accounted for 1.86 billion tons. The dominant modes of petroleum transportation are complimentary, notably when the origins or destinations are landlocked or when the distance can be reduced by the use of land routes. The maritime circulation of petroleum follows a set of maritime routes between regions where it is been extracted and regions where it is been refined and consumed. More than 100 million tons of oil is shipped each day by tankers. About half the petroleum shipped is loaded in the Middle East and then shipped to Japan, the United States and Europe. Tankers bound to Japan are using the Strait of Malacca while tankers bound

Table 7.1 Descriptions of Various Oil Tankers

Class*	Length**	Beam**	Draft++	Overview
Tanker*	205 m 675 ft	29 m 95 ft	16 m	Less than 50,000 deadweight tons, mainly used for transportation of refined products (gasoline, gasoil).
Aframax	245 m 800 ft	34 m 110 ft	20 m 65 ft	Approximately 80,000 deadweight tons (American Freight Rate Association)—mid size vessels. Typically engage in long to medium haul oil trades from West Africa to the North Sea to the East Coast and Gulf Coast of the United States. Alternatively, Aframax vessels typically engage in medium to short haul oil trades and can carry cargos of 80,000 to 120,000 dwt.
Suezmax	285 m 940 ft	45 m 145 ft	23 m 75 ft	Between 125,000 and 180,000 deadweight tons, originally the maximum capacity of the Suez Canal—mis-size vessels. Typically engage in long to medium haul oil trades from West Africa to the North Sea to the East Coast and Gulf Coast of the U.S.
VLCC	350 m 1150 ft	55 m 180 ft	28 m 90 ft	Very large crude carrier. Up to around 300,000 deadweight tons of crude oil. Like the ULCC, these tankers also transport oil in long-haul trades mainly from the Arabian Gulf to Western Europe and the United States via the Cape of Good Hope and Asia.
ULCC	415 m 1350 ft	63 m	35 m 115 ft	Ultra Large Crude Carrier. Capacity exceeding 300,000 deadweight tons. Largest vessel of the tanker fleet—the largest examples have a deadweight of over 550,000 deadweight tons. They typically transport oil in long-haul trades mainly from the Arabian Gulf to Western Europe and the United States via the Cape of Good Hope and Asia.

* Panamax and Handysize tankers are the smallest vessels in the world fleet and typically trade in short haul business and can transport cargos of 80,000 dwt to as little as 10,000 dwt.
** Approximate.

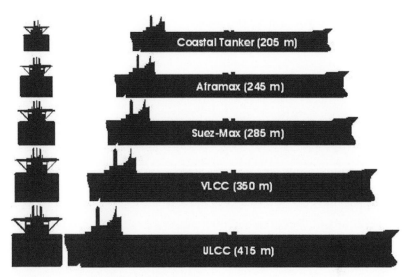

Figure 7.1 Profiles of various tanker sizes.

to Europe and the United States will either use the Suez Canal or the Cape of Good Hope, pending the tanker's size and its specific the destination. International oil trade is often correlated with oil prices, as it is the case for the United States.

The world tanker fleet capacity (excluding tankers owned or chartered on long-term basis for military use by governments) was about 280 million deadweight tons in 2002. There are roughly 3500 tankers available on the international oil transportation market. The cost of hiring a tanker is known as the charter rate. It varies according to the size and characteristics of the tanker, its origin, destination and the availability of ships, although larger ships are preferred due to the economies of scale. In addition, due to environmental and security considerations, single-hulled tankers are gradually phased out to be replaced by double-hulled tankers.

Tankers transport crude oil from their points of production to their points of consumption, which are typically oil refineries. The main clients within the industry include oil companies, oil traders, large oil consumers, petroleum product producers, and government agencies. The contracts by which crude oil is transported include spot charters, time charters and "bareboat" charters.

The pricing of crude oil transportation services occurs in a highly competitive global tanker charter market. A broker is usually involved in the deal and acts as an intermediary between the vessels owner and the charterer. The major hubs of shipping are located in New York, London, Oslo, Singapore and Tokyo.

7.3.1 Types of Tankers

The oil tanker fleet in divided into five major categories based on size and (hence) carrying capacity (Table 7.1). In order to benefit from economies of scale charterers typically charter the largest possible vessel that can be accommodated in their arrival and discharge ports.

The common rule is that the volume that can be carried in a tanker increases as a function of the cube of its length. For instance, a ULCC is about twice the length of a coastal tanker (415 m versus 205 m), but can carry about eight times the volume (50,000 tons versus 400,000 tons). Because of their huge mass, tankers have a large inertia, making them very difficult to steer. A loaded supertanker could take as much as 3 km and 15 min to come to a full stop and has a turning diameter of about 2 km.

Double-hulled, double-bottomed shuttle tankers are designed to transport crude oil from the offshore platforms to a transshipment terminal. Varying quantities of saltwater and natural gas are usually produced along with crude oil. Offshore production platforms include processing facilities that remove these fluids from the crude oil before it is stored and shipped. The saltwater is reinjected into the producing formation to maintain pressure in the reservoir. The natural gas, often called solution gas, is used for heat and power on the platform, and the remainder is reinjected into the reservoir. The Hibernia project initially flared natural gas while injection wells were being drilled, and a small amount of natural gas is always flared to maintain an ignition source in case production operations are interrupted and gas must be diverted to the flare stack. New technologies such as compressed natural gas tankers are being investigated as a way to deliver the natural gas eventually to onshore markets. Both Hibernia and Terra Nova include crude oil storage facilities onboard the production platforms. Double-hulled, double-bottomed shuttle tankers, specially designed for use in the waters off Newfoundland and Labrador, carry the crude oil from the platforms directly to market or to a trans-shipment terminal at Whiffen Head, Newfoundland and Labrador, for storage and transshipment to tankers that carry the oil to refineries.

7.3.2 Tanker Demand and Supply

Tanker demand is expressed in "ton-miles" and is measured as the product of (a) the amount of oil transported in tankers, multiplied by (b) the distance over which this oil is transported. Tonnage of oil shipped is primarily a function of global oil consumption, which is driven by economic activity as well as the long-term impact of oil prices on the location and related volume of oil production. Tonnage of oil shipped is also influenced by transportation alternatives such as pipelines.

The distance over which oil is transported is the more variable element of the ton-mile demand equation. It is determined by seaborne trading and distribution patterns, which are principally influenced by the locations of production and the optimal economic distribution of the production to destinations for refining and consumption. Seaborne trading patterns are also periodically influenced by geo-political events that divert tankers from normal trading patterns, as well as by inter-regional oil trading activity created by oil supply and demand imbalances.

The United States is the leading importer of crude oil in the world. Since 1995, U.S. demand for crude oil has risen in the aggregate by 6.8%, whereas U.S. crude oil production has decreased by 11.6% during the same period. Driven by the imbalance of supply and demand, U.S. crude oil imports have increased by 25.6% from 1995 to 2003.

Tanker supply increases with the deliveries of new buildings and decreases with the scrapping of older vessels. Typically new buildings are delivered 18–36 months after they are ordered. Every two and a half years oil tankers undergo a class survey, which with time becomes progressively more expensive. If the number of new buildings delivered stays below the number of older tankers scrapped, the demand for modern tonnage will increase, as may the rates they command.

7.3.3 Consolidation

The seaborne crude oil transportation business is highly fragmented and is generally provided by two types of operators: independent ship owners and captive fleets of privately and state owned oil companies. Within the industry, independent owners account for approximately 80.4% of the tanker capacity, and the top ten owners account for 26.4% of the world tanker fleet. The continued concern among the oil companies to secure safe modern tonnage by dealing with large trusted owners has greatly influenced the continued consolidation within the industry. Additionally, the drive toward consolidation has provided the larger owners with leverage to better control operating costs by taking advantage of economies of scale. Through consolidation of the mid-size tanker market, General Maritime will be able to create a sector specific focus. It intends to have a versatile fleet of similar ships that will be able to cater to a broad range of charterers.

7.3.4 Focus on Safety

Environmental protection has been a major focus of the tanker industry over the past years. Regulations such as OPA 90 and IMO have caused tanker owners to take extra care in the maintenance of their vessels and plan ahead to the time their vessels will no longer be allowed to trade. Oil disasters such as the Exxon Valdez in 1989 the Erika in 1999 and the Prestige in 2002 have forced charterers to exercise extreme caution in hiring only the most modern and well-maintained vessels to trade within U.S. waters.

With major oil companies seeking modern double hull vessels, the demand has increased for these ships, thus prompting higher charter rates. With a modern fleet of Aframax and Suezmax ships well bellow the industry's average age, General Maritime is well positioned to take advantage of the higher rates being offered by the charterers.

The seagoing tanker fleets that are owned, or used, by the world's oil companies are also responsible for the movement of a considerable portion of the world's crude oil. In fact, seagoing tankers form one of the most characteristic features associated with the transportation of petroleum. Many of these ships are of such a size that there are few ports that can handle them. Instead these large ships (VLCCs, very large crude carriers, and ULCCs, ultra large crude carriers) spend their time sailing the seas between different points, filling up and off-loading without ever entering port. Special loading jetties, artificial islands, or large buoys moored far offshore have been developed to load or off-load these tankers.

In general, the larger the tanker the lower its unit cost of transportation. As a result, the size of tankers during the 1960s and 1970s grew steadily. During the 1930s and 1940s, the average size of tankers was about 12,000 deadweight ton (dwt = the number of tons of cargo, stores, and fuel that a ship can usually carry). By the 1950s a 33,000 dwt tanker was considered standard size. At present, the VLCC and ULCC classes involve tankers of over 200,000 and 300,000 dwt. A few tankers in the 500,000 dwt range also exist, and some have even exceeded this size.

The cargo space in the tankers is usually divided into two, three, or four rows of cargo tanks by longitudinal bulkheads. These are further divided into individual tanks (from 25 to 40 ft long in large vessels) by transverse bulkheads. Access to the tanks is through oil-tight hatches on the deck, cargo being loaded or discharged by means of the ship's own pumps, which may have a capacity in excess of 4000 ton/h. There has lately been some discussion about the wisdom of building such tankers with double hulls.

In theory, the single-hull tanker has a better chance of staying afloat but will spill some of the cargo into the sea.

Over the past two decades, it has also become evident that crude oil tankers are usually shorter lived than most other cargo ships. Crude oil cargoes can deposit corrosive sludge on the bottom of the hold, and gasoline cargoes can also have a corrosive effect on the steel of the tanks. As a result, a tanker may last only 12 years instead of the 20 year life of a cargo ship, although protective coatings have been developed that help to withstand the corrosive effects of petroleum products.

Natural gas is also transported by seagoing vessels. The gas is either transported under pressure at ambient temperatures (e.g., propane and butanes) or at atmospheric pressure but with the cargo under refrigeration (e.g., liquefied petroleum gas). For safety reasons, petroleum tankers are constructed with several independent tanks so that rupture of one tank will not necessarily drain the whole ship, unless it is a severe bow-to-stern (or stern-to-bow) rupture. Similarly, gas tankers also contain several separate tanks.

Natural gas presents different transportation requirements problems. Before World War II its use was limited by the difficulty in transporting it over long distances. The gas found in oil fields was frequently burned off; and unassociated (dry) gas was usually abandoned. After the war, new steel alloys permitted the laying of large-diameter pipes for gas transport in the United States. The discovery of the Groningen field in the Netherlands in the early 1960s and the exploitation of huge deposits in Soviet Siberia in the 1970s and 1980s led to a similar expansion of pipelines and natural gas use in Europe.

REFERENCES

Bai, Y., Bai, Q., 2010. Subsea Engineering Handbook. Gulf Professional Publishing, New York.

Brown, R.J., 2006. Past, Present, and Future Towing of Pipelines and Risers. Proceedings. 38th Offshore Technology Conference (OTC), Houston, Texas.

Dean, E.T.R., 2010. Offshore Geotechnical Engineering—Principles and Practice. Thomas Telford, Reston, Virginia.

Gerwick, B.C., 2007. Construction of Marine and Offshore Structures. CRC Press, Taylor & Francis Group, Boca Raton, Florida.

Palmer, A.C., Been, K., 2011. Pipeline geohazards for Arctic conditions. In: McCarron, W.O. (Ed.), Deepwater Foundations and Pipeline Geomechanics. J. Ross Publishing, Fort Lauderdale, Florida.

Palmer, A.C., King, R.A., 2008. Subsea Pipeline Engineering, 2nd Edition PennWell, Tulsa, Oklahoma.

Speight, J.G., 2014a. The Chemistry and Technology of Petroleum, 5th Edition CRC Press, Taylor & Francis Group, Boca Raton, Florida.

Speight, J.G., 2014b. Oil and Gas Corrosion Prevention. Gulf Professional Publishing, Elsevier, Oxford, United Kingdom.

Speight, J.G., 2014c. Handbook of Petroleum Product Analysis, 2nd Edition John Wiley & Sons Inc., Hoboken, New Jersey.

Corrosion

Outline

Handbook of Offshore Oil and Gas Operations. http://dx.doi.org/10.1016/B978-1-85617-558-6.00008-8

8.1 INTRODUCTION

Corrosion in the refining industry has been known for almost 100 years (Speight, 2014a,b); several and the harsh environments in which many platforms operate are also causing a considerable number of corrosion-related issues. In fact, the offshore environment is highly corrosive and special stresses put additional pressure on the coating systems used for protection of the steel structures (Braestrup et al., 2005; Guo, 2005; Palmer and King, 2008). Offshore maintenance is difficult and the cost is exorbitantly high; it is therefore very important to select quality coating systems and to ensure that they are applied correctly and under the right conditions.

Floating offshore production systems such as *floating production, storage and offloading* facilities (FPSO facilities), SPAR (originally designated as a *single point articulated riser*, now generically applied cylindrical floating structures), *tensioned-leg platforms*, and *deep draft caisson vessels* (DDCV) present a number of corrosion control challenges. Corrosion control of such structures is necessary because of (1) the use of high strength materials on these structures, (2) the need to control weight and reduce corrosion allowance, (3) the contact of the various parts of the structures, (4) a high number of non-welded connections, and (5) difficulties in inspection and monitoring (Ashworth, 1994).

The corrosion control systems for these types of structures should be specified with caution and a high degree of conservatism. The operator must also be aware of in-service inspection requirements and allow appropriate monitoring and inspection aids to be specified during construction.

The most common method used for the protection of steel in offshore environments is the use of various types of coatings and for the immersion zone, coatings combined with cathodic protection (CP).

For proper performance of coating systems and thus durable corrosion protection for such extreme conditions, the fundamental parameters in coating selection must be respected and application that are each critical to achieve the required result. These parameters include the (1) type and condition of the substrate—usually steel, (2) environment and possible additional stresses, (3) quality of any coatings employed, and (4) quality control, such as application of the coating and the inspection protocols.

An important aspect of corrosion that is often not recognized as much as other causes (Speight, Chapter 3) is crude oil quality, can be described in terms of the types and quality of processes required to manufacture a wide range of products. Crude oil value—to a production operator and to a refinery—is based on the expected yield and value of the products value, less the operating costs expected to be incurred to achieve the desired yield. Ensuring that the quality of crude oil received is equivalent to the purchased quality (value acquired is equal to value expected) is one of the greatest challenges facing the industry today. However, there is a deficiency in the equation balance occurs when a crude oil contains corrosive constituents— such as the high acid crude oils (Speight, 2014a,b,c)—and the simple equation shown below takes on a new meaning:

$$Quality = yield\ of\ products + value\ of\ products$$

Furthermore, difficulties in minimizing differences between purchased quality and received quality are significantly higher when multiple crude oils are processed as a blend, and the complexity of the crude delivery system increases. Shipping crude oil through multiple pipelines and redistribution storage tanks—a reality faced by most producers and refiners—results in the delivered crude oil being a composite of the many crude oils encountered *en route*. Thus, the resultant composite blend may vary significantly from the expected purchased quality, and the sources of quality problems are much more difficult to estimate.

The crude assay determines how the crude is represented in optimization models. Inaccurate crude oil characteristics will distort the perceived value and lead to nonoptimal crude slates. One of the first things in crude selection is to ensure that the potentially acceptable crude oil is accurately represented by the analytical data (Speight, 2014a,c). An accurate crude assay (property interpolation/extrapolation accuracy, updated test history) for each potential crude oil to be considered acceptable should represent the predicted quality of the crude, if purchased. Most assay databases are dated or incomplete and, typically, the majority of the crude oil assays are more than 2 years old and do not include analysis for, or indications of, corrosive constituents (Speight, 2014a,b,c).

Crude oil quality problems occur regardless of whether the dominant crude slate is comprised of domestic crude delivered from an offshore location by pipeline or via a waterborne delivery system. In both cases, using simple categories such as gravity or sulfur do not provide an accurate measure of the value of a particular crude oil value, and monitoring only gravity and sulfur does not provide adequate safeguards for the integrity of the crude oil while it is in transit. These easy-to-use field test methods are inadequate, but fortunately do not need to be relied upon for refinability as well as corrosive properties. More sophisticated analyses (with an analysis for constituents likely to cause corrosion and correlated to crude oil acceptance) can provide a comprehensive estimate of crude oil quality. This analysis needs to give weight to quality consistency, where appropriate, as well as to improved yield, reduced operating expenses and the *fit* to processing units.

In addition, a combination of aging plants, greater fluid corrosiveness and tightening of health, safety, security, and environment requirements has made corrosion management a key consideration for producers and refiners. The prevention of corrosion/erosion through live monitoring provides asset and integrity managers with a real-time picture of how their facility is coping with the high demands placed upon it by corrosive fluids. This information can assist in risk management and auditing. Furthermore, steel pipework and vessels are always at risk of corrosion or erosion (Figures 8.1– 8.3). Unless monitored, there is a risk of failure, which may impact the safety of workers and the environment. The financial costs of operational interruption, repairs, and reputational damage must also be considered.

As oil and gas operators produce and process ever more corrosive or erosive hydrocarbon streams, the demands on plant metallurgy steadily increase. Permanently installed sensor systems can deliver a continuous picture of asset condition over time, at a comparable cost to that of a single manual inspection. This picture can be correlated with process conditions that may be causing corrosion or erosion, and strategies to minimize corrosion, such as inhibitor use (Sastri, 1998; Knag, 2005). With such knowledge, the asset manager can move beyond merely knowing whether corrosion or erosion is occurring, to understanding why and at what rate. This understanding enables operators to make better-informed decisions.

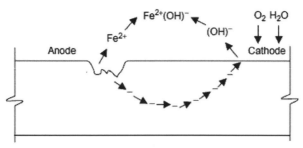

Figure 8.1 Corrosion process on a steel surface. *Source: Figure reproduced from El-Reedy, 2012, Fig. 6.1.)*

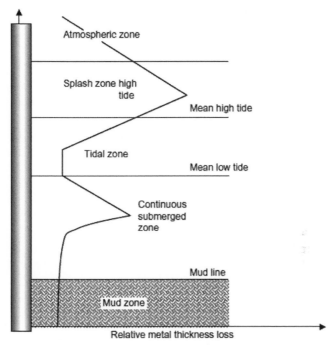

Figure 8.2 Corrosion process on a steel piling surface. *Source: Figure reproduced from El-Reedy, 2012, Fig. 6.6.)*

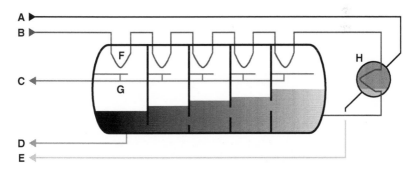

Figure 8.3 Schematic of a multistage flash desalinator.
A: Steam inlet
B: Seawater inlet
C: Potable water outlet
D: Waste outlet
E: Steam outlet
F: Heat exchange area
G: Condensation and collection area
H: Brine heater

Furthermore, in addition to these equipment reliability-monitoring systems, analytical tests of process streams are vital to processing high-acid crude oils. The monitoring of TAN numbers and other relevant properties of is of high importance. TAN numbers should be routinely measured in different process streams to optimize dosing of inhibitors (Sastri, 1998; Knag, 2005). Tests involving potentiometric titration are normally used for TAN measurement. Elements—such as trace metals—should be monitored with inductively coupled plasma (ICP) mass spectrometry or ICP optical emission spectrometry instruments. These machines use ICP for elemental analysis.

Some metals in process streams are measured from a corrosion viewpoint. Vanadium (V) and nickel (Ni) are measured to monitor the content of the heavy metals in VGO streams that go to the FCC or to the HC, as these are poisonous to FCC and HC catalysts. Sometimes they also impact product yields. Additionally, P content in process streams must be monitored – inhibitors for high-TAN crude are P-based compounds, and there are certain restrictions on the allowable limit of phosphorus in the catalyst of cracker units.

8.2 OFFSHORE CORROSION

Corrosion is typically identified by the appearance of rust on the steel surface of the offshore structure). The chemical reactions driving the corrosion process are caused by chloride attack—corrosion of steel occurs by an anodic reaction (Speight, 2014b). The offshore structure is exposed to saline water during its whole lifetime and the effect of the sea water on the integrity of the materials of construction is an important aspect of the in-service life of an offshore structure.

Sea water is the most corrosive of the natural environments that materials have to withstand, but it is much less corrosive than many environments encountered in industry, such as mineral acids, as are used in various processes (Chapter 2, Chapter 4) (Rowlands, 1994). This situation has important implications for desalination materials in that there is a wide range of corrosion resistant alloys readily available, which have been developed for the chemical and process industries, and which are resistant to sea water.

Corrosion due to the presence of extreme corrosive elements can be classified under several headings (Speight, 2014b, and references cited therein):
- Uniform corrosion, which occurs as a decrease in metal thickness per unit of time or uniform deposit of corrosion products on the surface of the metal (sometimes also known as general corrosion).

- Galvanic corrosion, which results from contact between two different materials in a conducting, corrosive environment. Galvanic corrosion may result in the very rapid deterioration of the least resistant of the two materials leading to a fatal failure. Avoiding the mixing of different materials, for example on tubes and fittings or valves, is the most common method of minimizing the problem.

- Crevice corrosion, which is an electromechanical oxidation reduction process. It occurs within localized volumes of stagnant trapped solution trapped in pockets, corners or beneath a shield of some description. The corrosive process is greatly accelerated if chlorine, sulfide, or bromide ions are present in the electrolyte solution. Crevice corrosion is considered far more dangerous than uniform corrosion as the rate at which it acts can be up to 100 times higher.

- Pitting corrosion, which is characterized by deep, narrow holes that can penetrate inward extremely rapidly while the remainder of the surface stays intact. Perforation of a component can occur in a few days with no appreciable reduction in weight of the overall structure. Stainless steels are particularly sensitive to pitting corrosion in seawater environments.

- Intergranular corrosion, which progresses along the grain boundaries of an alloy and can result in the catastrophic failure of equipment, especially if tensile stress loads are present. Localized attack can occur while the rest of the material is completely unaffected. The presence of impurities in the boundaries or local enrichment or depletion of one or more alloying elements can be the catalyst for this type of corrosion.

- Stress corrosion cracking, which is a combination of tensile loading and a corrosive medium causing the initiation of cracks and then their growth. Time to failure depends on specific application factors and can vary from just a few minutes to several years. Stress corrosion cracking is a very serious and permanent risk in many industrial applications where materials are often under mechanical loading for sustained periods or indeed permanently. In addition to selecting the correct materials, the risk of this type of corrosion can be avoided by stress relieving or annealing after fabrication of the assembly, avoiding surface machining stresses and controlling the corrosive environment.

8.2.1 Offshore Structures

Offshore structure platforms used for in oil and gas exploration and recovery are continuously exposed to sea water. However, the concentration of dissolved materials in the sea varies greatly with location and time as seawater is diluted by rivers, rain, or melting ice. Thus the most important properties of seawater in relation to corrosion of offshore structure are: (1) high salt concentration, mainly sodium chloride, (2) high electrical conductivity, (3) relatively high pH, (4) soluble gases, of which oxygen and carbon dioxide are particularly important in the context of corrosion, (4) the presence of organic compounds, such as spilled oil, and (5) the existence of biological life, to be further distinguished as micro-fouling—for example, bacteria—and macro-fouling—for example, seaweed, mussels, barnacles, and many kinds of animals or fish).

The main numerical specification of seawater is its salinity, which is the total amount of solid material (in grams) contained in one kilogram of seawater when all halides have been replaced by the equivalent of chloride, when all the carbonate is converted to oxide, and when all organic matter is completely oxidized.

On surface equipment, the simplest solution to the corrosion of offshore structures is to prevent salinity from playing a role, which can be achieved by placement of an insulating barrier over the metal concerned. Offshore installations are often painted with zinc-rich primers to form a barrier against rain, condensation, sea mist and spray. The zinc primer not only forms a physical barrier, but also acts as a sacrificial anode should the harrier be breached.

Offshore structures are also protected in other ways. The zone above the high tide mark, called the splash zone, is constantly in and out of water. The most severe corrosion occurs here. Any protective coating or film is continually eroded by waves and there is an ample supply of oxygen and water. Common methods of controlling corrosion in this zone include further coating and also increasing metal thickness to compensate for higher metal loss.

The part of the structure in the tidal zone is subjected to less severe corrosion than the splash zone and can benefit from CP systems at high tide. CP works by forcing anodic areas to become cathodes. To achieve this, a reverse current is applied to counteract the corrosion current. The current can be generated by an external OC source-impressed CP-c-or by using sacrificial anodes.

The rest of the structure, which is-exposed to less severe seawater corrosion, is protected by CP. However, crustaceans and seaweed attach to the submerged parts adding weight that may increase stress-related corrosion. This mechanism occurs when the combined effects of crevice, or pitting, corrosion and stress propagate cracks leading to structural failure.

However, a covering of life does restrict oxygen reaching metal, and so reduces corrosion.

Other forms of structural stress are also important. Low-frequency cyclic stress—resulting from factors such as waves, tides, and operating loads—can allow' time for corrosion within cracks as they are opened. Modeling and accounting for these stresses are therefore an extremely important part of corrosion prevention.

Corrosion increases with water salinity up to about 5% (w/w) of sodium chloride. Above this, the solubility of oxygen in the water decreases reducing corrosion rates. In fact, when the salt content is above 15%, the rates are lower than with fresh water. When water and acid gases are present, the corrosion rates rise rapidly. Water dissolves carbon dioxide and/or hydrogen sulfide and becomes acidic.

In highly corrosive environments, carbon steel can be protected by corrosion inhibitors during production. Like acid corrosion inhibitors, these adhere to casing and completion strings to form a protective film. Inhibitors can be continuously introduced into a producing well by a capillary tube run on the outside of the tubing as part of the completion design.

Other methods include batch treatment where inhibitor is pumped down the tubing regularly, say every six weeks, or squeeze treatments, where inhibitor is pumped into the formation.

To protect wells and pipelines from external corrosion, CP is used. In remote areas sacrificial ground beds may be used for both wells and pipelines. A single ground bed can protect up to 50 miles (80 km) of pipeline. In the Middle East, solar panels have been used to power impressed current CP systems. Other methods include thermoelectric generators fueled directly from the pipeline. To protect several wells, a central generator may be used and a distribution network.

Under the right conditions, iron sulfide and iron carbonate scales—the corrosion products when H_2S or CO_2 are present—provide protective coatings. The composition of production fluids, however, typically changes during the life of a reservoir so relying on natural protection may not be wise. Corrosion monitoring, in some form, should always be undertaken.

However, many of the materials used in offshore structures are much more expensive than those used in industries which have traditionally handled sea water, such as shipping and power plant. In these industries, carbon steel, cast iron, copper base alloys, and standard grades of stainless steel have been the usual choice for sea water applications. The desalination industry followed traditional practice, and whereas in many cases, this proved satisfactory, in others it did not. Upgrading, therefore, largely involved economic decisions-as the better materials were always available, the main problem was to decide what level of cost and performance was acceptable.

8.2.2 Stainless Steel

There is a wide variety of stainless steels that can be used as condenser tubes in multistage flash (MSF) distillation units.

By way of information, MSF distillation units (Figure 8.3) are used for water desalination process and distill sea water by flashing a portion of the water into steam in multiple stages of what are essentially countercurrent heat exchangers. MSF distillation plants produce about 60% of all desalinated water in the world. The unit is organized into a series of stages, each containing a heat exchanger and a collector for condensate. The sequence has a cold end and a hot end while intermediate stages have intermediate temperatures. The stages have different pressures corresponding to the boiling point of water at the stage temperatures.

When the plant is operating in a steady state condition, feed water at the cold inlet temperature flows, or is pumped, through the heat exchangers in the stages and warms up. When the water reaches the brine heater it already has nearly the maximum temperature. In the heater, an amount of additional heat is added. After the heater, the water flows through valves back into the stages which have ever lower pressure and temperature. As it flows back through the stages the water is now called brine, to distinguish it from the inlet water. In each stage, the temperature is above the boiling point of water at the pressure of the stage, and a small fraction of the brine water boils (*flashes*) to steam thereby reducing the temperature until equilibrium is reached. The resulting steam is a little hotter than the feed water in the heat exchanger. The steam cools and condenses against the heat exchanger tubes, thereby heating the feed water as described earlier.

The total evaporation in all the stages is up to approximately 15% v/v of the water flowing through the system, depending on the range of temperatures used. With increasing temperature there are growing difficulties of scale formation and corrosion. 120°C (250°F) appears to be a maximum, although avoidance of scaling on the internals of the unit may require temperatures below 70°C (160°F).

The feed water carries away the latent heat the condensed steam, maintaining the low temperature of the stage. The pressure in the chamber remains constant as equal amounts of steam is formed when new warm brine enters the stage and steam is removed as it condenses on the tubes of the heat exchanger. The equilibrium is stable, because if at some point more vapor forms, the pressure increases and that reduces evaporation and increases condensation. In the final stage the brine and the condensate has a temperature near the inlet temperature. Then the brine and condensate are pumped out from the low pressure in the stage to the ambient pressure. The brine and condensate still carry a small amount of heat that is lost from the system when they are discharged. The heat that was added in the heater makes up for this loss.

In addition, MSF distillation plants, especially large ones, are often paired with power plants in a cogeneration configuration. Waste heat from the power plant is used to heat the seawater, providing cooling for the power plant at the same time. This reduces the energy needed by one-half to two-thirds, which drastically alters the economics of the plant, since energy is by far the largest operating cost of a MSF distillation plant.

In flowing seawater, the performance of stainless steel tubes is quite satisfactory. They withstand general corrosion and are not affected by the flow velocities operating in a distiller. Normal steels are susceptible to SCC in solutions of high chloride content at high temperatures. As the tubes are not subject to any noticeable stresses, this type of attack is unlikely to develop. In stagnant solutions, on the other hand, normal stainless steels are liable to undergo pitting corrosion. During a long outage, the distiller must be carefully drained and flushed with ample quantities of potable water to prevent chloride attack. Stainless steel tubes are also prone to crevice corrosion, which develops under barnacles, stuck rubber balls and the like. The same procedures described in case of copper-base tubes apply in such cases.

Weld areas and the heat affected zones are weak points where intergranular and pitting attack readily propagate on stainless steel in a desalination unit. Spot and arc welding is to be applied, and annealing should be carried out whenever feasible. Finally, stainless steel tubes mounted on copper-base tube plates will lead to their rapid deterioration by galvanic action. CP using sacrificial mild steel anodes overcomes this problem.

The very high corrosion rates are caused by extended periods of wetness and high concentrations of chlorides that accelerate corrosion. Another factor that needs to be considered in the atmospheric zone is ultraviolet light from the sun as this may have a degrading effect on some types of corrosion protection.

The part of the structure that is alternately above and below the water line due to tide and waves (usually referred to as the *splash zone*) is subject to even higher corrosion rates and may be on the order of eight-to-ten times the corrosion rate of similar structures onshore. Other factors in this area that need consideration include ultraviolet light from the sun, as well as erosion from the water, possible debris and in some places of the world even ice.

The part of the structure that is below the water line at lowest tide (usually referred to as the *immersion zone*) is also subject to high corrosion rates and may be on the order of four-to-five times the corrosion rate of similar structures onshore. Another factor that needs to be considered is fouling.

A special case of immersion is related to the part of the steel structure that is situated below the seabed—typically rammed into the bottom sand. Corrosion rates are typically much lower due to lower concentrations of oxygen. However, pitting corrosion may also be a factor and the corrosion rate will be significantly higher.

8.2.3 Carbon Steel

Sea water is a complex environment consisting of a mixture of inorganic salts, dissolved gases and organic compounds. However, it also supports living matter in the form of both macro-organism (e.g., fish, shellfish, seaweed, etc.) and microorganisms. All of these can have an influence on corrosion processes. Sea water is a well-buffered solution, so the pH remains fairly constant at about 8. This means that most corrosion processes are dependent on the presence of oxygen. For carbon steels immersed in sea water, the rate of corrosion is mainly dependent on the oxygen content and the temperature of the sea water, the composition of the steel has relatively minor influence even when small amounts of alloying elements are added.

8.2.4 Copper Base Alloys

Copper alloys have been used traditionally in marine engineering for heat exchanger tubing and for cast and wrought components in pumps and valves. These alloys form good protective films in sea water, and provided these films are undamaged, corrosion is slight. However, the protective films are susceptible to damage by fast flowing sea water, and this is an important factor in selecting these alloys for applications involving fluid flow.

In deaerated sea water, copper alloys can still form protective films and show higher resistance to corrosion than in natural water. This is because the potential of copper alloys in sea water is much higher than the potential at which hydrogen can evolve; thus, in the absence of oxygen, corrosion is negligible. However, as in the case of carbon steel, low oxygen levels with high flow rates can cause impingement attack but the rate of attack was much lower than in aerated sea water. The data indicate that at low temperatures and oxygen levels, corrosion of the three alloys tested is acceptably low. However, at high temperatures with high oxygen levels, although the cupronickels continue to show low corrosion rates, aluminum brass is attacked. Thus, in MSF plants, where aluminum brass tubing is used, it is confined to the lower temperature recovery stages with the cupronickels being used in the higher temperatures areas.

Any steel structure placed offshore will be subjected to the stresses of a severely corrosive environment. For example, steel structures situated above the water in the oxygen-containing atmosphere (usually referred to as the *atmospheric zone*) are in a high corrosivity category. The corrosion rate of unprotected steel is high and may be three-to-four times the corrosion rate of similar structures onshore.

8.2.5 Corrosion by Polluted Water

Polluted seawater contains ammonia and/or hydrogen sulfide. Both materials are products of the decay of dead animals and organisms. The two pollutants attack copper condenser tubes.

Hydrogen sulfide reacts spontaneously with copper tubes to produce a black, porous copper sulfide film. Being nonprotective, the film allows further attack until the metal is eaten through. Attack is more rapid if the polluted water is loaded with slime and silt.

Under the slime, attack by sulfide is considered a special type of deposit corrosion in which metal deterioration is initiated by the presence of hydrogen sulfide rather than by a deficiency in oxygen.

Attack by sulfide-polluted water is identified by the black coloration of the tube inside. Treatment of the black film with dilute acid sets free the hydrogen sulfide, known by its characteristic bad odor. On the micro-scale range, the presence of sulfide is ascertained through its catalyzing the evolution of nitrogen gas bubbles from a drop of iodine-sodium azide solution. The test is carried out under a magnifying lens or microscope.

Sulfide corrosion can be dealt with in a variety of ways—suitable tube material can be chosen if it is known from the start that only polluted water is available. Extension of the intake facilities away from the source of polluted water is a second alternative. When present in small amounts, hydrogen sulfide is oxidized by chlorination. This must, however, be carefully controlled. The same also applies to rubber ball cleaning if sulfide-polluted sludge is in abundance.

Seawater polluted with ammonia affects the corrosion of copper base condenser tubes in two distinct ways: it induces SCC and the rate of crack propagation increases with pollutant content; and ammonia and ammonium salts enhance the general attack by dissolving and complexing with the copper ions. The tubes lose their protective film and acquire a shiny appearance. Eventually the tubes fail when they become perforated at weak points. Failure can occur anywhere along the tube length and the resulting pore takes any form. Ammonia attack is suspected when daily analysis of the brine shows a constant, high level of copper.

There is no direct solution to the problem of corrosion by ammonia-polluted seawater. If the source of pollution is permanent, relocation of the seawater intake might prove practical.

If pollution is widespread, a change to ammonia-resistant tubes, for example, titanium or stainless steels, might be considered (Mountford, 2002). Chlorination of ammoniated seawater is effective in removing low levels of pollution. The two agents react to produce chloramines which are weaker disinfectants than chlorine itself. The formation of chloramines reduces, but does not completely eliminate the problem of ammonia-corrosion since chloramines readily hydrolyze to the original harmful material. Ferrous sulfate dosing retards the aggressive action of small amounts of ammonia by forming an inhibiting ferric oxide film.

8.3 CORROSION CONTROL

Corrosion is a major issue in offshore environments due to extreme operating environments and the presence of aggressive corrosive elements. Resistance to corrosion can represent the difference between trouble-free long-term operation and costly downtime. With the presence of extreme corrosive elements such as hydrogen sulfide, carbon dioxide, brine and a whole range of hazardous chemicals compounded by extreme temperatures and pressures, there are few more arduous environments than offshore. Thus, there is the need to provide protection for steel and other materials against sharp and abrasive particles, in instances where these particles are in solid form or in suspension. Thus, offshore installations producing oil and gas have a great need for our protective coatings and linings in order that they can operate safely and continuously for the life of the offshore field (typically 20–50 years).

8.3.1 Platforms

The DDCV and SPAR type structures have presented deep water operators with a cost effective production facility for prospects with a moderate
number of subsea wells, and are being widely used in the Gulf of Mexico.
These structures consist of the following main components, each of which
has unique problems: (1) hull components, (2) tanks, (3), riser systems, and
miscellaneous areas.

8.3.1.1 Hull Components

The hull is typically a large cylindrical steel tube with a variety of internal
compartments that provide buoyancy to support the topside drilling and
production equipment with a central annulus (center well) through which
all production is accomplished. Hulls are typically 90–120 ft in diameter
with a center well diameter of 30–50 ft. The overall length of the hull section can be up to or in excess of 600. This construction presents two distinct
sea-water immersed corrosion exposures.

In the *outer hull*, in which normally basic grades of carbon steel exposed
to natural sea-water, corrosion is prevented by using conventional marine
epoxy coatings through the splash zone and topsides and conventional sacrificial anode CP to bare steel in the immersed region. Design criteria are
the same as for conventional platforms. Monitoring of CP system performance is simply accomplished using surface deployed or remotely operated
vehicle (ROV) interfaced monitoring equipment.

In the *center well*, the structure also used carbon steel, which is exposed to
quiescent sea-water. Exclusion of normal sunlight reduces marine growth
accumulation and densely packed riser systems make post installation access difficult to impossible, and can cause temperatures to riser a little above
ambient sea-water. The net result is a slightly lower current requirement for
CP of the steel on the wall of the center well. Coatings are typically used
only in splash zone and emerged areas, sacrificial anodes must be flush-
mounted to avoid obstructions during riser installation—this fact must be
taken into account when computing anode resistance.

There are also a series of guide frame structures within the center well;
these guides restrict lateral movement of the riser systems during vessel
relocation about its operational footprint and offset normal hydrodynamic forces. It is important to locate anodes on these guide frames, as they
could be shielded from anodes on the inside wall. These anodes that will
provide a large proportion of current drained to the risers through fortuitous contact. The limited post installation access to this area makes it a

good candidate for fixed reference electrodes to monitor the CP system performance.

Most recent designs utilize a *truss section* attached to the base of the main hull to support the bottom main ballast tank at the base of the SPAR structure. The truss is typically 300–325 ft (91–100 m) long. Construction is standard tubular steel members with periodic heave plates, which provide guides for the riser systems. The truss section is normally left uncoated and fitted with conventional sacrificial anodes.

The *main hull* will contain a number of internal compartments. Some of these are void tanks that contain only air and are never flooded. Other tanks are variable ballast tanks which will see raw sea-water some of the time and will be largely dry for the majority of the life. Being a part of the hull, these compartments are fabricated from basic grades of carbon steel It is not uncommon for these tanks, particularly the variable ballast tanks, to be heavily baffled and reinforced thus creating a large number of shielded compartments.

8.3.1.2 Tanks

Void tanks are usually vented to atmosphere and thus cannot be fully de-oxygenated. They are often susceptible to condensation on the inner walls. Access is possible through man-ways and periodic visual and nondestructive examination is always a part of the required in-service inspection program. Access to effect coating repairs during operation is possible but economically undesirable.

Coating is the main control against corrosion—possibly supplemented by vapor phase control as a backup method. Where allowed, coal tar epoxies fill the role because of they are generally surface preparation tolerant.

Dehumidification can be a very effective method of void space corrosion control. The general adage, no water = no corrosion hold true and, should the dehumidification system not be 100%, vapor phase control as a secondary method of corrosion control is a reasonable strategy. Coupons for testing should be retrievable (preferably without tank entry) for monitoring corrosion rates.

For *variable ballast tanks* (*hard tanks* coatings and sacrificial anode CP) are required in most cases. It is important to locate anodes in all compartments and distribute them preferentially lower in the tank because of surface wetting. Maintenance current densities for bare steel in these environments should be 3 mA/sq. ft (30 mA/sq M).

Particular attention should be paid to the coating quality in the upper areas of the tank likely to be in wet atmospheric service for much of the time, in these areas it is worthwhile considering the use of metallic based primers or even thermally sprayed aluminum (TSA) as a stand-alone coating or as an epoxy primer.

For *permanent ballast tanks*, construction is typically the basic grades of carbon steel. The external surfaces of these tanks may are simply left bare and cathodically protected. The inside surfaces present a fairly unique problem.

In order to provide maximum negative buoyancy at this location on the structure, it is necessary to fill the tank not only with sea-water but also with additional dense material. The ballast of choice is typically magnetite (Fe_3O_4) in a granular form. It is recommended that anodes be used above *and* below the magnetite and the tank should be coated particularly where the magnetite contacts the steel. Reference electrodes should be installed as well as current density and anode current monitors both above and below the magnetite surface.

8.3.1.3 Riser Systems

The riser systems—as opposed to the riser pipe on a refinery catalytic cracking unit (Speight, 2014a)—on this type of structures are referred to as *top-tensioned risers* (Bai and Bai, 2005). They are actually small structures within a structure, having their own buoyancy systems to support them. The only designed contact to the hull structure is through the topside flow-line connections, which are above water, having said this there is a very high probability of fortuitous contact through mechanical interference with the hull. The riser systems are free to move completely independently of the hull.

The *main riser sections* are of pipe-in-pipe type construction with the outer pipe acting as a conductor. At the base of the riser there is a stress joint (there may also be a similar section near the point of exit at the base of the hull—referred to as the keel). The annulus between the riser and the outer pipe is usually flooded with sea-water. Mechanical spacer centralizers are clamped to the inner riser pipe to control movement within the conductor pipe.

Air buoyancy cans are large tank like structures built around the outer surface of the riser conductor near the top of the riser. Even though there are various designs, they have some common corrosion areas irrespective of specific design.

Outer can surfaces see the same environment as the center well areas of the hull but buoyancy is critical and weight loading must be minimized. Anodes are not a good choice because of the additional weight as well as the possibility of mechanical interference with the risers guide frames. The stroke length on these risers can be as much as 40 feet or more. While mechanical interference is absorbed on wear strips on the outside of the can there is the good possibility of coating damage during installation of the risers.

Thermal sprayed aluminum provides a good solution in this area and, while serving primarily as a barrier coating the TSA can also provide an adequate level of CP to small areas of exposed steel. However, the coating is conductive and will generally be at a potential that is 50 mV or more positive than the anodes in the hull center well, and will drain a small amount of CP from those anodes when contact between the structures exists.

Inner can surfaces are, during normal operation, are mainly void but they may have an open bottom that is exposed to sea water. Depending on the oxygen concentration, it may be necessary to coat the inside surfaces and provide a limited number of anodes at the base of the cans. Typically, the cans are filled with nitrogen in order to exclude the sea-water during installation, periodic refilling is recommended to keep oxygen concentration to a minimum.

Export risers and flow lines from remote wells are either steel catenary design or may utilize flexible sections. Various methods have been used successfully utilizing either conventional fusion bond epoxy coatings or TSA. Anodes can be located on the riser catenary section but some operators prefer to use anodes located at the touch down area and on the hull to provide protection from the ends.

8.3.1.4 Other Areas

The *outer surfaces* are exposed to sea-water (except inside the air cans), and see all resistivity layers in the sea-water as they transit almost the entire water column. The use of conventional coatings with bracelet anodes has several drawbacks: (1) possible shielding and coating damage under clamped stabilizer strakes and buoyancy modules, (2) possible resistive build up through mechanical joints, and (3) weight limitations. For these reasons most systems use sealed TSA as the corrosion control.

The outside of the actual riser pipe should be treated like the outside of the conductor (coated with TSA). The inner wall of the conductor can be left bare if sea-water in the annulus is suitably inhibited.

Stress joints are designed to take the major share of bending moment on the riser and are therefore made from high strength materials with good stress characteristics and are located in the outer conductor pipe. Some titanium grades are particularly suitable from a mechanical standpoint and are therefore a common choice in many systems. The propensity of titanium to suffer hydriding under CP at certain levels provides a dilemma, one made more difficult by the lack of long term field data and the very high consequence of a failure.

In deep water, it is common practice to pre-drill a number of *wells* then temporarily cap them until the production structure can be located on site. Some of these pre-drilled wells have nowhere to attach anodes for corrosion protection. In these cases it is prudent to install a pod of anode material which can be electrically tied back to one or more of these wells using clamps.

Lateral *mooring systems* are required to enable the hull structure to be moved around its operational footprint. These multi-leg systems are subject to very high loading and a corrosion failure could be very costly. CP is a suitable method for corrosion control, but the following precautions are required: (1) ensure that a flexible continuity jumper cable is provided between the fairlead structure and the hull, and (2) ensure that current losses to the chain are calculated when sizing anodes.

8.3.2 Pipelines

Offshore pipelines are protected from seawater corrosion usually by coating, which is supplemented with CP to provide protection at coating defects or *holidays*. In the Gulf of Mexico, the pipeline coatings used until the early to mid-1970s were either asphaltic/aggregate, somastic-type coating (an extrudable mastic consisting of oxidized bitumen with mineral fill materials) or hot-applied coal tar enamels. Since then, the trend has been to use fusion-bonded epoxy powder coatings.

In the earlier days, the trend in CP was to rely on impressed-current systems. In the 1960s and early 1970s, zinc bracelet anodes attached to the pipe were widely used. However, more efficient aluminum alloys have surpassed zinc as the preferred material for offshore galvanic anodes. There are, however, still some operators using impressed current systems and some using zinc anodes.

Virtually all new pipelines installed in the Gulf of Mexico are equipped with aluminum bracelet anodes. There are two basic types: (1) square shouldered and (2) tapered. The square-shouldered anodes are typically used on pipe that has a concrete weight coating. When installed, the anodes are flush with, or slightly recessed inside, the outside diameter of the concrete. The tapered anodes are designed to be installed on pipelines with only a thin film corrosion coating.

When designing a CP system for a pipeline, the corrosion engineer has to consider the following variables, all of which will have an impact on the final anode alloy and size selection: (1) design life required—(minimum is 20 years), (2) pipe diameter length and to-from information, (3) geographic location, (4) type of coating, (5), water depth, (6) product temperature, (7) electrical isolation from platforms or other pipelines, (8) burial method, and (9) pipe-laying/installation method, which will have a direct impact on the amount of coating damage and there is also the risk of having anodes detached during the pipe-laying process.

Higher corrosion rates can be generally expected when the pipe coating has a combination of large damaged areas and adjacent pinhole defects, and when the pipe is exposed to seawater rather than mud. There is also a particular risk of microbiologically influenced corrosion (MIC) on buried lines with bitumen mastic-type coatings and depleted CP. If the CP systems have depleted, corrosion will begin at numerous sites all over the pipeline. Unless detected and retrofitted, the first leak could be the end of the pipeline, as the next several hundred will not be far behind.

Retrofitting the CP system with supplemental anodes would only make sense if the line in question is very old and the required additional life was significant. However, in addition to the cost to perform a pipeline CP inspection will run up to several thousand dollars per mile, retrofitting pipeline CP systems offshore is not always a simple matter, especially when lines are deeply buried.

8.4 CORROSION MONITORING

A major problem for the offshore crude oil and natural gas industry is the effective monitoring of equipment against corrosion in the potentially hazardous working environments (Clark, 2014). Combating or preventing corrosion is typically achieved by a complex system of monitoring, preventative repairs and careful use of materials. Monitoring methods include both off-line checks taken during maintenance and on-line monitoring. Off-line checks measure corrosion after it has occurred, telling the engineer when equipment must be replaced based on the historical information he has collected (*preventative management*).

In the environs of offshore operations, steel pipework and vessels are always at risk of corrosion or erosion, depending on the metallurgy of the equipment, aggressiveness of the process fluids, and operating conditions. Unless monitored and managed carefully, there is a potential risk (in fact, a medium-to-high risk) of equipment failure. Such events could result in a release of crude oil and natural gas, with a possible consequential impact on the environment and/or personnel, apart from the financial costs of equipment repairs and unplanned operational interruptions or damaged reputations. However, there are various well-established, manual techniques for the periodic assessment of equipment. The drivers of corrosion and erosion include: (1) process conditions, (2) feedstock constituents, and (3) the presence of abrasive solids, such as sand. The environmental and reputational costs from leak/release events; and the growing shortage of skilled inspection staff are hurdles confronting traditional plant integrity monitoring processes (Clark, 2014).

However, the widely adopted approach of periodic manual inspections does not capture the often intermittent, and sometimes accelerated and nonlinear nature of corrosion. It is, therefore, very difficult to use this data to correlate directly with either corrosion drivers or inhibitor applications to enable an understanding of the effects of feedstock and process decisions and the use of inhibitors on equipment integrity. Permanent corrosion monitoring can represent progress in enabling effective use of corrosion inhibitors and in providing a real-time update of the detrimental effects of corrosion.

Continuous corrosion monitoring is necessary to evaluate use of feedstock selection and diversification. In fact, processing high-acid crude oils and opportunity crude oils (Chapter 1) are key issue for many refiners. While naphthenic acid corrosion tends to be somewhat localized, it is still necessary to deploy arrays of sensors which enable making a series of point measurements to highlight where a significant increase in corrosion activity is being observed. This approach also enables an understanding of the effectiveness of inhibitor chemical injections in high-risk platform and/or pipeline locations (Speight, 2014b,c).

Thus, the ability to assess corrosion of alloys aids in the management of corrosion in many types of equipment. An important obstacle to assessing and predicting corrosion is the variety of the combinations of alloys and corrosive environments. Many aspects of equipment/process design, process operation, alloy selection, alloy design, and plant maintenance are influenced by the expected lifetimes of equipment in high-temperature, corrosive gases. These lifetimes are greatly affected by the conditions present in process equipment, because process equipment usually has maximum allowable temperatures, or other process conditions, which are limited by the corrosion rates expected for the equipment.

On-line systems are a more modern development, and are revolutionizing the way corrosion is approached. There are several types of on-line corrosion monitoring technologies such as linear polarization resistance (LPR), electrochemical noise, and electrical resistance (ER). On-line monitoring has generally had slow reporting rates in the past (minutes or hours) and been limited by process conditions and sources of error but newer technologies can report rates up to twice per minute with much higher accuracy (referred to as real-time monitoring). This allows process engineers to treat corrosion as another process variable that can be optimized in the system. Immediate responses to process changes allow the control of corrosion mechanisms, so they can be minimized while also maximizing production output (Kane et al., 2005). In an ideal situation having on-line corrosion information that is accurate and real-time will allow conditions that cause high corrosion rates to be identified and reduced (*predicative management*).

Corrosion monitoring is just as important as recognizing the problem and applying controls. Monitoring attempts to assess the useful life of equipment when corrosion conditions change and how effective the controls are. Techniques used for monitoring depend on what the equipment is, what it is used for and where it is located.

Structures—monitoring corrosion on exposed structures is fairly straight forward and is carried out by visual inspection. More rigorous tests are required when a structure is load-bearing. Some form of nondestructive testing is used, such as magnetic particle testing to reveal cracks. Many platforms are inspected on a regular schedule and require underwater divers or ROVs using still or video photography to check the condition of legs and risers. During this inspection, the corrosion rate of sacrificial anodes can be assessed, which are typically designed to last seven or eight years so they will have to be replaced during the typical 20-year life of a rig.

8.4.1 Drill Pipe

To monitor drill pipe corrosion and the effectiveness of mud treatments, coupon rings are installed between joints (*left*). The rate of corrosion can then be assessed by measuring the amount of metal lost from the rings.

During drilling, mud systems are routinely monitored for chemical and physical properties. Tests specifically related to corrosion control include an analysis of oxygen, carbon dioxide (CO_2), hydrogen sulfide (H_2S), and bacteria. Hydrogen sulfide is detected by measuring the total tested further by adding acid to liberate hydrogen sulfide, which can be measured using any standard hydrogen sulfide detector. Bacterial attacks can be recognized by a drop in pH, increase in fluid loss or change in viscosity.

8.4.2 Casing and Tubing

Various corrosion logging tools measure internal corrosion, external corrosion and even evaluate CP of oil wells. One of the most commonly used techniques has been the multi-finger caliper run on either slick line or electric line. This measures the internal radius of casing and tubing using lightly sprung feeler arms. (Heavy springing can cause the fingers to leave tracks through protective scales and chemical inhibitors leading to enhanced corrosion from running the survey itself.) An improvement on contact calipers is the ultrasonic, which uses a rotating ultrasonic transducer to measure the echo time of a high-frequency sonic pulse—the processed signal produces a map of the casing.

Wireline logging provides a good evaluation of downhole corrosion, but disrupts production and may involve pulling completions. Oil companies, therefore, like to use surface monitoring methods to indicate when downhole inspection is required.

8.4.3 Pipelines

Surface monitors include test coupons placed at strategic points in the Flowline and also more sophisticated techniques that attempt to measure corrosion rates directly (resistance devices, polarization devices, galvanic probes, hydrogen probes and iron counts). [1] This approach to monitoring can be hit or miss when trying to relate surface corrosion to downhole corrosion. In the past, the onset of well problems instigated monitoring. While waiting for a failure is not recommended, recovering corroded tubing or casing at least provides valuable after-the-fact information, and every opportunity is taken to find out what caused the corrosion and the failure (Figure 8.4). Some downhole monitoring techniques have been adapted to logging pipelines.

The same surface logging equipment is used, but the logging tools themselves have been made more flexible to pass around sharp bends. Short lengths of pipe may be logged by this method but longer lengths are usually monitored by smart pigs. These are sophisticated instrument packages, which use ultrasonic, flux leakage, and other electromagnetic techniques to check for corrosion. The data are usually stored in the pig itself for later retrieval. The pig is pumped along a pipeline from a launching station to a

Figure 8.4 Corroded bracing. *Source: Figure reproduced from El-Reedy, 2012, Fig. 7.47.*

purpose-built receiving section of the pipeline. Surveys cover tens or even hundreds of miles.

8.5 MONITORING METHODS

There are various established test methods that can be used for the periodic assessment of pipe and vessel integrity. Periodic inspections do not, however, deliver continuous pipework condition data that can be correlated with either corrosion drivers or inhibitor us to understand the impact of process decisions and the inhibitor usage on plant integrity. Manual acquisition of ultrasonic wall thickness data is also frequently associated with repeatability limitations and data-logging errors.

Permanently installed sensor systems, on the other hand, deliver continuous high-quality data. The ultrasonic sensors can be installed on pipes and vessels operating at up to 600°C (1110°F). These sensors have also been certified as intrinsically safe for use in most hazardous environments. The system has been proven in operation over a number of years in various operations. Continuous monitoring through use of appropriate test methods data can validate that, when corrosion is occurring, it is often an intermittent process rather than a continuous event. It is in such cases that it is particularly valuable to be able to correlate thickness data over time with process and/or inhibitor parameters.

8.5.1 Corrosion Coupons

Most data for corrosion of alloys in high-temperature gases are reported in terms of weight change/area for relatively short exposures and inadequately defined exposure conditions. However, metal loss needs to be directly related to the loss of metal thickness used in equipment design and operation decision-making. Corrosion in high-temperature gases is affected by parameters of the corrosive environments such as temperature, alloy composition, time, and gas composition. Online analyzers can be used for continuous monitoring of acidity or basicity (pH), iron (Fe), and chlorine (Cl). Closed-loop automated controllers are also utilized to optimize dosing. Continuous measurement presents a step change in the level of corrosion rates that can be determined and the accuracy of that determination.

In the *coupon method*, the corrosion coupon (*weight loss*) technique is the best known and simplest of all corrosion monitoring techniques. The method involves exposing a specimen of material (the coupon) to a process environment for a given duration, then removing the specimen for analysis. The basic measurement, which is determined from corrosion coupons, is weight loss; the weight loss taking place over the period of exposure being expressed as corrosion rate.

Online corrosion coupons have been widely used in critical process circuits for the assessment of fluid corrosivity. Normally, alloy–steel and CS coupons are used for corrosion monitoring. Coupons can be fixed or retractable—in retractable coupons at high-temperature locations, a proper sealing system should be used. Coupons can also be of different shapes and types—the shape can be rectangular, circular or helical, and may be normal or pre-stressed. Based on the corrosion rate of coupons (for both general and pitting corrosion), the extent of the corrosive can be assessed and dosing with additives can be optimized.

The simplicity of the measurement offered by the corrosion coupon is such that the coupon technique forms the baseline method of measurement in many corrosion monitoring programs. The technique is extremely versatile, since weight loss coupons can be fabricated from any commercially available alloy. Also, using appropriate geometric designs, a wide variety of corrosion phenomena may be studied which includes, but is not limited to: (1) stress–assisted corrosion, (2) bimetallic (galvanic) attack, (3) differential aeration, and (4) heat-affected zones.

Advantages of weight loss coupons are: (1) the technique is applicable to all environments—gases, liquids, solids/particulate flow, (2) visual inspection can be undertaken, (3) corrosion deposits can be observed and analyzed, (4) weight loss can be readily determined and corrosion rate easily calculated, (5) localized corrosion can be identified and measured, and (6) inhibitor performance can be easily assessed.

In a typical monitoring program, coupons are exposed for a 90-day duration before being removed for a laboratory analysis. This gives basic corrosion rate measurements at a frequency of four times per year. The weight loss resulting from any single coupon exposure yields the "average" value of corrosion occurring during that exposure. The disadvantage of the coupon technique is that, if a corrosion upset occurs during the period of exposure, the coupon alone will not be able to identify the time of occurrence of the upset, and depending upon the peak value of the upset and its duration, may not even register a statistically significant increased weight loss.

Therefore, coupon monitoring is most useful in environments where corrosion rates do not significantly change over long time periods. However, they can provide a useful correlation with other techniques such as ER measurements and LPR measurements.

8.5.2 ER Methods

Electrical resistance monitoring involves the use of ER probes, which can be thought of as "electronic" corrosion coupons. Like coupons, ER probes provide a basic measurement of metal loss, but unlike coupons, the value of metal loss can be measured at any time, as frequently as required, while the probe is *in situ* and permanently exposed to the process stream.

The ER technique measures the change in ohmic resistance of a corroding metal element exposed to the process stream. The action of corrosion on the surface of the element produces a decrease in its cross-sectional area with a corresponding increase in its ER. The increase in resistance can be related directly to metal loss and the metal loss as a function of time is by definition the corrosion rate.

Although still a time averaged technique, the response time for ER monitoring is far shorter than that for weight loss coupons. The graph below shows typical response times. ER probes have all the advantages of coupons, plus: (1) direct corrosion rates can be obtained, (2) the probe remains installed in-line until operational life has been exhausted, and (3) they respond quickly to corrosion upsets and can be used to trigger an alarm.

Electrical resistance probes are available in a variety of element geometries, metallurgies, and sensitivities and can be configured for flush mounting such that pigging operations can take place without the necessity to remove probes. The range of sensitivities allows the operator to select the most dynamic response consistent with process requirements.

Online ER probes are the most widely used, intrusive corrosion-monitoring devices that give the corrosivity of the fluid and measure corrosion rate in the stream. The ER probes, which are made of the same metallurgy as that of the main pipe, are inserted through stubs connected to the pipe. The probes detect the process fluid and undergo corrosion in the circuit.

The electric resistance of the probe changes with corrosion and deterioration, and the change in resistance corrosion rate is measured. These probes do not measure the exact corrosion on the pipe; instead, they measure the corrosivity of the fluid in that particular circuit. During high–TAN crude processing, circuits with process fluid temperatures above 180°C (355°F) are most susceptible to corrosion and, therefore, ER probes are inserted into these lines.

Normally, these probes are installed at furnace inlet lines of atmospheric and vacuum units, atmospheric gas oil and vacuum gas oil circuits, reduced crude oils and short residue circuits. They are also used in the feed line to the fluid catalytic cracking unit (preferably in the filter loop) and the hydrocarbon feed line (at approximately 200°C, 390°F) before the charge enters the reactors. Locations are identified based on the flow conditions and line geometry. Ideal locations are pumping discharges and control valve loops downstream of control valves.

Various advanced versions of ER probe-based systems that have high levels of accuracy and sensitivity and that are proven for measuring corrosion in high-temperature lines during naphthenic crude handling are used. These high-sensitivity probes have an expected life of three to 4 years and will almost instantaneously respond to a 10-miles-per-year (mpy) change in corrosion rate. Traditional, flush-mounted ER probes need about a week to detect 10-mpy changes in corrosion rate. The thinner the element, the faster the response; likewise, the thicker the element, the longer the probe life.

Two forms of probe element are available: (1) flush and (2) cylindrical. Also there are several mounting configurations to choose from, the most common of which allows the probes to be inserted and removed under full process operating conditions without shutdown.

Flush probes are used for best thermal performance where flush mounting with the pipe wall is desirable or essential. A typical example is a bottom-of-line location. In these applications, water films commonly collect in the bottom of the line and are the primary cause of corrosion. The flush probe ensures that the whole of the probe element is exposed to the water film.

Cylindrical probes are suited to virtually any aggressive environment, since there is no sealing material other than the parent metal. The measurement area of the element is much greater in this design and is suitable for use in a single-phase flow.

8.5.3 Field Signature Methods

In addition to ER probes, nonintrusive systems like field signature methods (FSMs) are used at certain critical locations, such as furnace outlets, transfer lines, etc., for monitoring the health of equipment and piping at high temperatures and velocities. These are normally used where two-phase flow exists at relatively higher temperatures and inaccessible locations.

In FSMs, corrosion measurements are done directly on the pipe and fittings, whereas ER probes measure the corrosivity of the process fluid. FSMs consist of multiple sensors (metallic pins) that are spot-welded in a rectangular pattern on the pipe/bends at the most critical locations. An electric current is passed from one side of the matrix to the other, and the voltage between the pins and the critical area is measured. This gives a unique electric field signature that depends on the geometry and thickness of the pipe and the electrical conductivity of the metal. Any change in field (or signature) resulting from internal corrosion or erosion is revealed by a change in voltage across the sensor pins.

Online logs are available with permanently installed data-loggers that collect readings on a real-time basis. They have a high level of accuracy, and sensitivity is typically on the order of 0.05% to 0.1% of pipe wall thickness. The system can be set up to communicate with the production system, the pipeline system, or the refinery control system so that data can be reviewed remotely.

A unique advantage of measurement by FSMs is that it gives area coverage rather than point measurement. It detects thickness loss even in the locations between the pins. Advantages are that the corrosion rate are that it can be accurately measured at high temperatures from remote sections of the plant, it provides an operator to the facility to optimize the corrosion control program, and it responds quickly to increased corrosion.

8.5.4 LPR Monitoring

The LPR monitoring technique is based on complex electro-chemical theory. For purposes of industrial measurement applications it is simplified to a very basic concept. In fundamental terms, a small voltage (or polarization potential) is applied to an electrode in solution. The current needed to maintain a specific voltage shift (typically 10 mV) is directly related to the corrosion on the surface of the electrode in the solution.

By measuring the current, a corrosion rate can be derived. The advantage of the LPR monitoring technique is that the measurement of corrosion rate is made instantaneously. This is a more powerful tool than either the coupon method or the ER method where the fundamental measurement is metal loss and where some period of exposure is required to determine corrosion rate. The disadvantage to the LPR monitoring technique is that it can only be successfully performed in relatively clean aqueous electrolytic environments.

The LPR monitoring method will not work in gases or water/oil emulsions where fouling of the electrodes will prevent measurements being made.

8.5.5 Galvanic Monitoring

The galvanic monitoring technique, also known as *zero resistance ammetry* (ZRA) is another electrochemical measuring technique. With zero resistance ammetry probes, two electrodes of dissimilar metals are exposed to the process fluid. When immersed in solution, a natural voltage (potential) difference exits between the electrodes. The current generated due to this potential difference relates to the rate of corrosion which is occurring on the more active of the electrode couple. Galvanic monitoring is applicable to the following electrode couples: (1) bimetallic corrosion, (2) crevice and pitting attack, (3) corrosion assisted cracking, (4) corrosion by highly oxidizing species, and (5) weld decay.

Galvanic current measurement has found its widest applications in water injection systems where dissolved oxygen concentrations are a primary concern. Oxygen leaking into such systems greatly increases galvanic currents and thus the corrosion rate of steel process components. Galvanic monitoring systems are used to provide an indication that oxygen may be invading injection waters through leaking gaskets or deaeration systems.

8.5.6 Other Methods

In addition to FSMs, several other non-intrusive, online monitoring techniques have become popular in recent years. However, most of these techniques are ultrasonic-based and, therefore, are point contact-type devices. This means that they do not cover the entire surface area on which the sensors are positioned. Some of these systems are spot-welded, while others are simply clamped and, therefore, can be moved to other positions. These instruments are generally used at locations such as control valve loops and bends and at pump systems.

These alternative systems are often used to validate ER probe readings over a longer duration. The unique advantage of such systems is wireless communication. Alternatively, some technologies use permanently installed ultrasonic probes on pipes and equipment to directly measure thickness. These instruments can be clamped onto the pipes, and online data can be gathered.

The main disadvantage of these systems is that they measure point thickness and do not capture the area. Also, they are not as accurate or as sensitive as FSMs. However, these systems can be used in place of ER probes, and they are less expensive than FSMs.

8.5.6.1 Hydrogen Penetration Monitoring

In acidic process environments, hydrogen is a by-product of the corrosion reaction. Hydrogen generated in such a reaction can be absorbed by steel particularly when traces of sulfide or cyanide are present. This may lead to hydrogen induced failure by one or more of several mechanisms. The concept of hydrogen probes is to detect the amount of hydrogen permeating through the steel by mechanical or electrochemical measurement and to use this as a qualitative indication of corrosion rate.

Aside from ER probes and FSMs, there are portable instruments that measure hydrogen (H_2) flux on the pipe. The basis of this technology is that certain modes of corrosion result in the generation of atomic hydrogen, and, when atomic hydrogen diffuses through the metal wall and permeates outside the pipe wall, this forms a molecule (Speight, 2014b). By monitoring the changes in hydrogen diffusion rate, the variation of internal corrosion can be inferred. These portable measuring devices can be used whenever required and can operate over a wide temperature range.

Monitoring is done both on-line and off-line. Offline monitoring refers to the monitoring of the effects of corrosion after it has happened and is usually identified while doing maintenance. Of course this is not the most effective way of saving the plant but, this was used in older times. It used to be considered as a step taken that is better late than never. Today on-line system has evolved to make it more effective in solving problem before it gets worse.

8.5.6.2 Radiography

Conventional and digital radiography are other proven, nondestructive techniques for assessing piping and equipment health during handling high-acid crude oils. Conventional film radiography can be used as a scanning tool, but it lacks quantitative capability. Radiography is not a 24/7 online monitoring system, although it has mobility that allows it to be deployed at a much wider area and at many locations.

In digital radiography, the main advantage is a quantitative approach that provides for online, condition-based assessment of risk. Some machines use solid-state X-ray technology in digital radiography, which is fast, portable and shows the percentage of metal loss. It does not require scaffolds or insulation removal.

8.5.6.3 Biological Monitoring

Biological monitoring and analysis generally seeks to identify the presence of *sulfate reducing bacteria*—a class of anaerobic bacteria which consume sulfate from the process stream and generate sulfuric acid, a corrosive which attacks production plant materials.

8.5.6.4 Sand/Erosion Monitoring

These are devices which are designed to measure erosion in a flowing system. They find wide application in oil/gas production systems where particulate matter is present.

8.5.6.5 Real-Time Monitoring

Traditional methods used to monitor and control corrosion may include installation of corrosion monitoring equipment, use of caustic in the crude oil, and a variety of other chemical corrosion control solutions. These traditional approaches, properly applied, provide acceptable corrosion control during operating time—when the unit is functioning normally. However, they may not detect or allow adequate or timely responses to the upsets that occur or the damage they can cause, during 10% of unit operating time. The available methods are often not sufficiently sensitive and the frequency, reliability, and accuracy of measurements are not good enough to facilitate a timely response. However, even the best corrosion-control programs may not detect significant problems before the damage is done.

New solutions are being developed to detect and to capture significant changes in the corrosive environment in real time, to measure the changes accurately, and to address and correct those changes before significant corrosion has occurred. For example, a new analyzer (patent pending) has been developed for the continuous measurement of acidity and basicity (pH), iron (Fe), chlorine (Cl) in water and can provide the accurate, real-time data that are required for effective and timely corrosion control (Hilton and Scattergood, 2010).

There is a variety of wet chemistry tests in conjunction with corrosion monitoring tools to track the corrosive environment in crude oil and natural gas handling equipment. These wet chemistry tests quantify several components, such as acidity and basicity (pH), chloride (Cl), ammonia (NH_3, NH_4^+), sulfide (S^{2-}) and iron (Fe) in the water. However, the effectiveness of this testing is limited by the time it takes to collect and analyze samples. These tests are generally part of the routine service performed by operations personnel or the chemical supplier, and they may only be performed at daily or weekly intervals. Generally, operations staff will run a few of these tests once per shift, typically pH and possibly chloride. The result is the collection of minimal data, most of it during periods of stable operation. Only rarely and by chance is the data collected during a period of unit upset, when 90% of corrosion occurs. When upsets do occur, operations staff are usually busy trying to get the crude unit lined out and back to steady-state, and data collection is a very low priority. Typically, the amount of data collected through a corrosion-control program in the course of one year is just a fraction of the amount of data captured by an operations historian over an equivalent period.

The acidity or salinity (pH value) is the most frequently measured parameter in the crude-unit accumulator boot water, which may be checked from 4 to 10 times a day, possibly more often if a low pH is observed, or if it is a problematic unit. In some case, online pH probes may be required to monitor the water properties but, these probes do require frequent calibration—as a result, manual measurements of the acidity—basicity (pH) are often preferred.

The frequency of sampling and performing the other wet chemistry tests, for chloride, iron and ammonia, is substantially less. The result may be a total of between 52 and 260 data sets per year—the majority of which are collected during periods of stable operation when little or no corrosion occurs. The same limitations apply to corrosion-rate data collected from probes and other monitoring devices. Data loggers may be used to relay these measurements to their central control system, in an attempt to gather more timely information, but this is not a common practice. However, test accuracy and speed of data turnaround are significant concerns when relying on manual wet chemistry testing. Human error, choice of test method, and the temptation to take shortcuts in sampling technique and preparation, can significantly affect the accuracy of the resulting data.

The time lag between sampling and performing the actual test can also make a significant impact on the value of the data for effective corrosion control. Samples are usually collected on a set schedule, at the end of a shift, and 4–6 h may elapse before they are processed in the laboratory and test results communicated to unit personnel. While this may be an adequate response time during periods of stable operation, and when data are used primarily to measure unit performance against a key performance indicator, it is too slow to facilitate a timely response within the corrosion window, the relatively brief period of upset when the most serious corrosion occurs. Without accurate, frequent and timely testing, corrosive incidents may be completely missed or discovered only after significant damage has occurred.

The key to controlling corrosion, without throwing metallurgy at the problem, is the ability to capture accurate data in real time, detecting and closing the corrosion window before significant damage occurs.

Associated water from the separator can be continuously sampled and passed through the analyzer, where specially designed pH electrodes provide a real-time measure of the pH. Simultaneously, the analyzer performs an automated, online analysis of chloride and total iron concentration in the process water. Chloride and iron analyses can be performed at frequencies ranging from one to six times per hour, depending on system conditions. With the online analyzer, the onset of a corrosion window in time can be detected and adjustments made to the corrosion-control chemical program, using closed-loop automated controllers, thus avoiding lag time and under-feed or over-feed of critical chemical components.

Rust never sleeps. To be truly effective, a corrosion-control program must go beyond the industry current practice of periodic sampling and manual sample processing. To this end, the analyzer can provide continuous, accurate, repeatable data, including conditions during the critical 10% of operations when 90% of corrosion occurs. By detecting these "corrosion windows" consistently and in real time, the analyzer provides a continuous view of pH, chloride and iron levels in the system, permitting the application of timely and effective chemical solutions before significant cor-rosion has occurred.

8.6 CORROSION CONTROL AND INHIBITION

Corrosion control is the prevention of this deterioration by mitigat-ing the chemical reactions that cause corrosion in three general ways: (1) change the environment, (2) change the material, or (3) place a barrier be-tween the material and its environment. All methods of corrosion control are variations of these general procedures, and many combine more than one of them. The material does not have to be metal but *is* a metal or an al-loy of metals in most cases. The metal does not have to be steel, but, because of the strength and cheapness of this material, it usually *is* steel or an alloy of steel. Again, the environment is, in most cases, the atmosphere, water, or the earth, that is, the constituents of the soil. There are, however, enough excep-tions to make corrosion control more complex.

It is essential to conclude this chapter by a mention of corrosion control in specific instances. For example, the potential for corrosion by naphthenic acids is assessed by the TAN number, the temperature, the sulfur content, the velocity (shear stress), and the flow regime. During naphthenic crude processing, corrosion at high temperature is mitigated by injecting either phosphate-based ester additives or sulfur-based additives, which provide an adherent layer that does not corrode or erode due to the effect of naphthenic acids. It has been suggested and partially proven that corrosion during pro-cessing of high-acid crude oils is a lower risk of the sulfur content is high—the relationship between the acid number and amount if sulfur is not fully understood but it does appear that the presence of sulfur-containing con-stituents has an inhibitive effect (Piehl, 1960; Mottram and Hathaway, 1971; Slavcheva et al., 1998, 1999).

To mitigate high-temperature corrosion, a high-temperature corrosion inhibitor is dosed into the process streams, normally at a concentration of 3 ppm (w/w) to 15 ppm (w/w). The additive is mixed with the suit-able process stream, typically at a ratio of 1:30 and the mix is injected into the cooled stream (<100°C, <212°F) with specially designed corrosion-mitigation quills, since the inhibitor is highly corrosive to alloy steel and even stainless steel. The dosage limit of the phosphorus-based inhibitor is calculated from the allowable phosphorus level in the products—products such as the vacuum gas oil that is used as the feedstock for a hydrocracking unit or as the feedstock for a fluid catalytic cracking unit.

The pH stabilization technique can be used for corrosion control in wet gas pipelines when no or very little formation water is transported in the pipeline. This technique is based on precipitation of protective corrosion product films on the steel surface by adding pH-stabilizing agents to increase the pH of the water phase in the pipeline. This technique is very well suited for use in pipelines where glycol is used as hydrate preventer, as the pH stabilizer will be regenerated together with the glycol—thus, there is very little need for replenishment of the pH stabilizer.

Equipment reliability during the handling of high-acid crude oils is paramount (Speight, 2014c). Hardware changes—such as upgrading materials construction from carbon steel (CS) and alloy steel to stainless steel (SS) 316/317, which contains molybdenum and is significantly resistant to naphthenic acid corrosion—are complicated tasks and require large capital investment as well as a long turnaround for execution. Alternatives to hardware changes are corrosion mitigation with additives and corrosion monitoring with the application of inspection technologies and analytical tests.

Some of the ways adapted today to overcome the effects of corrosion are:

— Use of certain types of metal in certain areas of the plant to facilitate the plant to be longstanding in spite of the effects of corrosion.
— For example, carbon steel is used for 80% of plant requirements as it is cost efficient and withstands most forms of corrosion due to hydrocarbon impurities below a temperature of 205°C (400°F). But as it is not able to resist other chemicals and environments, it is not for general use. Other kinds of metals used are, low alloys of steel containing chromium and molybdenum, and stainless steel containing high concentrations of chromium for excessively corrosive environments.
— Tougher metals like nickel, titanium, and copper alloys are used for the most corrosive areas of the plant which are mostly exposed to the highest of temperatures and the most corrosive of chemicals (Burlov et al., 2013).

Since, of late, with the increase in knowledge in the field gained over the years, it has become a lot easier and more effective in fighting the occurrence or corrosion. Some effective ways in preventing corrosion would be monitoring and preventing the occurrence of these effects even before they occur.

Many problems of correct use of corrosion control measures (for example, injection of chemicals such as inhibitors, neutralizers, biocides and others) may be solved by means of corrosion monitoring methods

(Groysman, 1995, 1996, 1997). For example, hydrocarbons containing water vapors, hydrogen chloride and hydrogen sulfide, leave the atmospheric distillation column at 130°C (265°F). This mixture becomes very corrosive when cooled below the dew point temperature of 100°C (212°F). In order to prevent high acidic corrosion, neutralizers and corrosion inhibitor can be injected at appropriate points and, in addition, corrosion monitoring equipment such as weight loss coupons and ER probes should also be installed. The ER probes show the corrosion situation continuously. The weight loss should be changed every several months, in order to compare with the results of the ER probes and to examine the danger of chloride attack (pitting corrosion). The more points in the unit that are included for corrosion monitoring, the better and more efficient is the corrosion coverage.

Also, problems exist in every cooling water system in the oil refining industry: (1) corrosion, (2) inorganic deposits containing carbonate scale, and (3) corrosion products of iron, phosphates, silicates and some others, and biofouling (microbial contamination). On-line corrosion and deposit monitoring systems used in the cooling water system allow monitoring the general corrosion of carbon steel (or any other alloy) at ambient temperature (non-heated steel surface) and at the drop temperature in the heat exchanger (heated steel surface), pitting tendency also for heated and nonheated surface, and heat transfer resistance—the quantitative value of inorganic and organic deposits (fouling).

Corrosion costs the crude oil and natural gas industry billions of dollars annually. Although one of the major contributors to corrosion is the acidity of associated water, pH measurements have been less-than-satisfactory because of the less-than-satisfactory ability to measure pH in the aggressive environment they have to contend with. When the correct equipment is chosen however, in-line pH measurement and control facilities have proven to be of great value in reducing plant-wide corrosion, and the consumption of chemicals such as pH control reagents and corrosion inhibitors. This not only results in significant cost savings but also in increased earnings through increased on-stream process time.

Advances in sensor technology and intelligent automation of the measurement point, enable pH measurement in the most challenging marine and acid water environments. From that perspective an investment in pH measurement is highly recommended.

In summary, control of corrosion requires: (1) evaluation of the potential corrosion risks, (2) consideration of control options—principally inhibition as well as materials selection, (3) monitoring whole life cycle suitability, (4) life cycle costing to demonstrate economic choice, and (5) diligent quality assurance (QA) at all stages.

High-temperature crude corrosion is a complex problem. There are at least three corrosion mechanisms: (1) furnace tubes and transfer lines where corrosion is dependent on velocity and vaporization, and is accelerated by naphthenic acid, (2) vacuum column where corrosion occurs at the condensing temperature, is independent of velocity, and increases with naphthenic acid concentration, and (30 side-cut piping where corrosion is dependent on naphthenic acid content and is inhibited somewhat by sulfur compounds (Chapter 1).

Mitigation of process corrosion includes blending, inhibition, materials upgrading, and process control. Blending may be used to reduce the naphthenic acid content of the feed, thereby reducing corrosion to an acceptable level. Blending heavy crude oil and light crude oil can change shear stress parameters and might also help reduce corrosion. Blending is also used to decrease the level of sulfur content in the feed and inhibit, to some degree, naphthenic acid corrosion (Chapter 1).

In terms of high acid crude oils, corrosion is predominant at temperatures higher than 180°C (355°F), where shear stress on pipe walls is significant. Such corrosion problems in the high-temperature section of the atmospheric distillation column are normally mitigated by dosing phosphate ester-based or sulfur (S)-based inhibitors at certain critical locations inside the process units. The inhibitors are dosed in process streams with the help of injection quills.

In addition to inhibitor dosing, intensive corrosion monitoring and analytical tests play a major role in equipment and piping health monitoring and in analyzing stream properties online. Corrosion-monitoring equipment includes customized inspection technologies with both intrusive and nonintrusive systems.

The first phase of an engineered solution is to perform a comprehensive high TAN impact assessment of a crude unit processing a target high TAN blend under defined operating conditions. An important part of the any solution system is the design and implementation of a comprehensive corrosion monitoring program. Effective corrosion monitoring helps confirm which areas of the unit require a corrosion mitigation strategy, and provides essential feedback on the impact of any mitigation steps taken.

With a complete understanding of the unit operating conditions, crude oil and distillate properties, unit metallurgies and equipment performance history, a probability of failure analysis can be performed for those areas which would be susceptible to naphthenic acid corrosion. Each process circuit is assigned a relative failure probability rating based on the survey data and industry experience.

Corrosion inhibitors are often the most economical choice for mitigation of naphthenic acid corrosion. Effective inhibition programs can allow operators to defer or avoid capital intensive alloy upgrades, especially where high acid crude oils are not processed on a full time basis.

Injection of corrosion inhibitors may provide protection for specific fractions that are known to be particularly severe. Monitoring needs to be adequate in this case to check on the effectiveness of the treatment. Process control changes may provide adequate corrosion control if there is the possibility of reducing charge rate and temperature. For long-term reliability, upgrading the construction materials is the best solution. Above 288°C (550°F), with very low naphthenic acid content, cladding with chromium (Cr) steels (5%–12%, w/w Cr) is recommended for crudes of greater than 1% (w/w) sulfur when no operating experience is available. When hydrogen sulfide is evolved, an alloy containing a minimum of 9% (w/w) chromium is preferred. In contrast to high-temperature sulfidic corrosion, low-alloy steels containing up to 12% (w/w) Cr do not seem to provide benefits over carbon steel in naphthenic acid service. Type 316 stainless steel (>2.5%, w/w molybdenum) or Type 317 stainless steel (>3.5%, w/w molybdenum) is often recommended for cladding of vacuum and atmospheric distillation columns.

Mild steel has been the most widely used alloy for structural and industrial applications since the beginning of industrial revolution. The use of acid media in the study of corrosion of mild steel has become important because of its industrial applications such as acid pickling, industrial cleaning, acid descaling, oil-well acid in oil recovery, and petrochemical processes. The refining of crude oil were carried out in a variety of corrosive conditions and in such, the corrosion of equipment is generally caused by a strong acid through attacking on equipment surface. And in many of structural and industrial applications of mild steel, they are also exposed to corrosive environments and they are susceptible to different types of corrosion. Therefore, the use of corrosion inhibitors to prevent metal dissolution will be inevitable and the use of inhibitors is found to be one of the most practical methods for protection against corrosion, especially in acidic media.

Control of the corrosion rate can be affected by reducing the tendency of the metal to oxidize, by reducing the aggressiveness of the medium, or by isolating the metal from the fluid. The latter can be done by coating the metal with a millimeter thick impervious noncorroding coating. These have wide spread use, but their effect may not be permanent because of breaks in the coating over time. Also in some systems coating might interfere with the process for which the equipment is used because they might change the heat transfer properties for example.

In cases where a fairly thick coating is not acceptable, the use of corrosion inhibitors comes into play. These chemicals are continually fed into the fluid with the objective of having them move to the metal–fluid interface. There the intact inhibitor molecularly attaches to the metal, or it reacts with the surface to form a thin adherent compound. In the first case they act by adsorption. In either case, the films are only one to a few molecules thick (i.e., nanometer thickness).

Some inorganic (non–carbon) chemicals like the chromates function in both ways and do so very well. However, they are hazardous to human and animal health and are not normally used now. Other inorganics like phosphates, borates, nitrites, and silicates function by reaction to form micrometer thick films. They are used for some metals in near neutral (pH $\sim$ 7) aqueous solutions, for instance, in water treatment plants.

For many aqueous systems, for elevated temperature equipment, for crude oil production, and so on, organic (carbon based) chemicals find more use. These materials are likely to function by adsorption. Here, the organic molecule, which orients itself suitably, becomes attached to the solid surface often via a less than total reaction between the inhibitor molecule and the solid surface. The attachment does not require a total electron transfer in either direction. A columbic force, for example ion-dipole attraction, suffices to attach the inhibitor molecule to the solid surface which, in turn, interferes with access of the corrosive entity to the surface.

The adsorbed layer can be formed all over the surface either in a single layer or as a multilayer or a mixture of both. The more complete the coverage the better. This process has the advantage of being molecularly thin and thus not too intrusive in heat conduction for example.

But there are problems. The amount of a given material adsorbed from a mixture depends on its concentration, temperature, fluid flow rate, as well as on the nature of the adsorbent that is, the solid surface. The film has to be kept intact by continually adding inhibitor to the medium to maintain a predetermined concentration of inhibitor. The continuing concentration is generally lower than that used initially, but both are at the milli-molar level.

That is not all. If temperature or flow rate of the system changes, the amount adsorbed is apt to change. For temperature, the change is energetic, a basic change in the amount adsorbed. For fluid flow, the change is dynamic, that is a change of fluid movement at the interface. This may affect the amount left at the interface. Faster flow generally causes removal of some of the physically adsorbed material from the solid surface.

To this point the object was to get you to visualize the system. The point now is to get some insight into how the efficiency of inhibition is determined. Corrosion rates can be measured in a variety of ways. In fact, corrosion rates can be determined from any measurable change available. Examples are metal weight loss, rate of gaseous production, and changes in solution composition. The basic approach is to expose small pieces of the metal in question to the fluid environment, preferably under flow conditions. Measurement of the corrosion rate can be carried out electrochemically or chemically. It is best to acquire not only the initial rate but also the steady rate. The latter is more useful for practical purposes.

To repeat, the rates can be determined from any quantity that changes regularly as a function of time. Thus, change in solution concentration of iron (for steels), pH of the solution, or weight of metal coupon can be used.

There are some caveats. Laboratory experiments are useful in screening candidate inhibitors. Following that, they should be subjected to a lab system that emulates conditions in the field, composition of the fluid, temperature and its changes, flow rates and their changes, time, and so on. Then there should be tests in the real system itself, at least long enough to cover any condition cycles the system may have. However, continuous exposure with ongoing data output is best.

To repeat, inhibitors are useless if they do not reach the metal surface intact. They can be lost by reaction with chemicals in the stream or to "thieves" in the system, that is any other solid surface exposed to the fluid stream like sand. Also, the inhibitor should not be detrimental to either process or product. Further, since the system may change with time, so must the corrosion control. Thus the control system must be monitored consistently.

The question may be asked, why go to all of the trouble that is entailed in using corrosion inhibitors? There are two important reasons to do so.

First, there is the matter of safety. Industrial corrosion can lead to process breakdowns that culminate in explosions or the venting of chemicals dangerous to health. In either case the ounce of prevention looms large over the pound of cure.

Second, there is the matter of cost. An article in 1995 (Battelle) estimated the cost of corrosion in the United States per year was $300 billion, an appreciable portion of the Gross Domestic Product. It also estimated that 30% of that amount could be saved by corrosion control, which of course includes corrosion inhibition.

The key to effective corrosion management is information since it is on the basis of that information that on-going adjustments to corrosion control are made. Information is valid data. Thus, to make effective corrosion management decisions on a day-to-day basis, the monitoring data must be valid. This is not simply a requirement for the probes to be operating correctly. It requires that they be placed in the most appropriate places, that is, at those points where the corrosion controlling activity might be expected to work, but where it might equally be expected to be least effective, for example, remote from the inhibitor injection point.

In many cases, specially designed traps are introduced into a plant so that corrosion probes may be inserted. These often produce their own microenvironment, atypical of the plant itself, and with little hope of effective entry for an inhibitor. Data from a probe in such a location are unlikely to be relevant to corrosion management elsewhere in the system. Invalid data leads to ineffective corrosion management.

From time to time a corrosion management program should be reviewed at both the strategic and tactical level. In human affairs, things change. The management of a facility will always be alive to current market trends, competitor's activities, interest rate movements and so on. Inevitably, it may be necessary to revise the management objectives from time to time. Since the corrosion management program was constructed to meet the objectives of an earlier plant management plan, it will be necessary to review the program and possibly to alter it. Likewise, the pace of technological change is rapid compared to the anticipated lifetime of most facilities. Thus, newer, more effective, cheaper means of achieving the same ends may emerge, and indeed, it may be possible to adopt them in place of existing tactics within the corrosion management program. Thus, the program is not a fixed blueprint, but a means to an end that must be reviewed and revised to meet the current management objective.

Corrosion cannot be ignored for it will not go away. However, there is little merit in controlling corrosion simply because it occurs, and none in ignoring it completely. The consequences of corrosion must always be considered. If the consequence of corrosion can be tolerated, it is entirely proper to take no action to control it. If the consequences are unacceptable, steps must be taken to manage it throughout the life of the facility at a level that is acceptable.

Good corrosion management aims to maintain, at a minimum life cycle cost, the levels of corrosion within predetermined acceptable limits. This requires that, where appropriate, corrosion control measures be introduced and their effectiveness ensured by judicious, and not excessive, corrosion monitoring and inspection. Good corrosion management serves to support the general management plan for a facility. Since the latter changes as market conditions, for example, change, the corrosion management plan must be responsive to that change. The perceptions of the consequences and risk of a given corrosion failure may change as the management plan changes. Equally, some aspects of the corrosion management strategy may become irrelevant. Changes in the corrosion management plan must, inevitably, follow.

Crude blending is the most common solution to high TAN crude processing. Blending can be effective if proper care is taken to control crude oil and distillate acid numbers to proper threshold levels.

8.7 ASSESSMENT OF DAMAGE REMAINING LIFE

Assessment of the extent of damage due to corrosion depends on (1) inspection, or (2) on an estimation of the accumulation of damage based on a model for damage accumulation, or (3) both. Sound planning of inspections is critical so that the areas inspected are those where damage is expected to accumulate and the inspection techniques used are such as will provide reliable estimates of the extent of damage. If the extent of the damage is known or can be estimated, a reduced strength can be ascribed to the component and its adequacy to perform safely can be calculated.

The general philosophy for estimating carrion and the ability of a metal or alloy to continue in service is involved several steps (API, 2000): (1) identification of flaws and damage mechanisms, (2) identification of the applicability of the assessment procedures applicable to the particular damage mechanism, (3) identification of the requirements for data for the assessment, (4) evaluation of the acceptance of the component in accordance with the appropriate assessment techniques and procedures, (5) remaining life evaluation, which may include the evaluation of appropriate inspection intervals to monitor the growth of damage or defects, (6) remediation if required, (7) in-service monitoring where a remaining life or inspection interval cannot be established, and (8) documentation, providing appropriate records of the evaluation made.

Life estimates can also be made based on the predicted life at the temperature and stress that are involved, by subtracting the calculated life used up, and making an allowance for loss of thickness by oxidation or other damage. Recently there has been increased use of the procedures of continuum damage mechanics for creep damage and remaining life assessment (Kachanov, 1986; Webster and Ainsworth, 1994; Penny and Marriott, 1995; Prager, 1995).

REFERENCES

API, 2000. Recommended Practice No. 579 – Fitness-for-Service. American Petroleum Institute (API), Washington, DC.

Ashworth, V., 1994. Corrosion. Shreir, L.L., Jarman, R.A., Burstein, G.T. (Eds.), Corrosion Control, 2, 3rd Edition Butterworth-Heinemann, Oxford, United Kingdm.

Bai, Y., Bai, Q., 2005. Subsea Pipelines and Risers. Elsevier, Oxford, United Kingdom.

Braestrup, J.B., Andersen, J.B., Andersen, L.W., Bryndum, L.B., Christensen, J.C., Nielsen, N.J.R., 2005. Design and Installation of Marine Pipelines. Blackwell Science Ltd., Oxford, United Kingdom.

Burlov, V.V., Altsybeeva, A.I., Kuzinova, T.M., 2013. A new approach to resolve problems in the corrosion protection of metals. International Journal of Corrosion and Scale Inhibition 2 (2), 92–101.

Clark, K., 2014. Continuous corrosion monitoring enhances operational decisions. Hydrocarbon Processing, March 1.

El-Reedy, M.A., 2012. Offshore structures: Design, Constructions, and Maintenance. Gulf Professional Publishers, Elsevier, Boston, Massachusetts.

Groysman, A., 1995. Corrosion Monitoring in the Oil Refinery. Paper No. 07. Proceedings. International Conference on Corrosion in natural and industrial environments: problems and solutions, Italy, May 23–25.

Groysman, A., 1996. Corrosion Monitoring in the Oil Refinery. Proceedings. 13th International Corrosion Congress, Melbourne, Australia, November 25–29.

Groysman, A., 1997. Corrosion Monitoring and Control in Refinery Process Unit. Paper No. 512. Proceedings. CORROSION/97, New Orleans, Louisiana.

Guo, B., Song, S., Chacko, J., Ghalambor, A., 2005. Offshore Pipelines. Gulf Professional Publishing, Elsevier, Burlington, Massachusetts.

Hilton, N.P., Scattergood, G.L., 2010. Mitigate corrosion in your crude unit. Hydrocarbon Processing 92 (9), 75–79.

Kachanov, L.M., 1986. Introduction to Continuum Damage Mechanics. Martinus Nijhoff, Dordrecht, Netherlands.

Kane, R.D., Eden, D.C., Eden, D.A., 2005. Innovative solutions integrate corrosion monitoring with process control. Materials Performance, 36–41, February.

Knag, M., 2005. Fundamental behavior of model corrosion inhibitors. Journal of Dispersion Science and Technology 27, 587–597.

Mottram, R.A., Hathaway, J.T., 1971. Some Experience in the Corrosion of a Crude Oil Distillation Unit Operating with Low Sulfur North African Crudes. Paper No. 39. Proceedings. CORROSION/71. NACE International, Houston, Texas.

Mountford Jr., J.A., 2002. Titanium—Properties, Advantages and Applications Solving the Corrosion Problems in Marine Service. Paper 02170. Proceedings. Corrosion 2001. NACE International, Houston, Texas.

Palmer, A.C., King, R.A., 2008. Subsea Pipeline Engineering, 2nd Edition Pennwell Corporation, Tulsa, Oklahoma.

Penny, R.K., Marriott, D.L., 1995. Design for Creep, 2nd Edition Chapman & Hall, London, United Kingdom.

Piehl, R.L., 1960. Correlation of corrosion in a crude distillation unit with chemistry of the crudes. Corrosion 16, 6.

Prager, M., 1995. Development of the MPC omega method for life assessment in the creep range. Journal of Pressure Vessel Technology 117, 95–103.

Rowlands, J.C., 1994. Sea Water Corrosion: Metal/Environment, 3rd Edition Butterworth-Heinemann, Oxford, United Kingdom.

Sastri, V., 1998. Corrosion Inhibitors: Principles and Applications. John Wiley & Sons Inc., Hoboken, New Jersey.

Slavcheva, E., Shone, B., Turnbull, A., 1998. Factors Controlling Naphthenic Acid Corrosion. Paper No. 98579. Proceedings. CORROSION/98. NACE International, Houston, Texas.

Slavcheva, E., Shone, B., Turnbull, A., 1999. Review of naphthenic acid corrosion in oil refining. British Corrosion Journal 34 (2), 125–131.

Speight, J.G., 2014a. The Chemistry and Technology pf Petroleum, 5th Edition CRC Press, Taylor & Francis Group, Boca Raton, Florida.

Speight, J.G., 2014b. Oil and Gas Corrosion Prevention. Gulf Professional Publishing, Elsevier, Oxford, United Kingdom.

Speight, J.G., 2014c. High Acid Crudes. Gulf Professional Publishing, Elsevier, Oxford, United Kingdom.

Webster, G.A., Ainsworth, R.A., 1994. High Temperature Component Life Assessment. Chapman & Hall, London, United Kingdom.

Environmental Impact

Outline

Handbook of Offshore Oil and Gas Operations. http://dx.doi.org/10.1016/B978-1-85617-558-6.00009-X

9.1 INTRODUCTION

Offshore activities arising from crude oil and natural gas exploration and production take place in waters of more than half the nations on the Earth and require careful monitoring. Gone are the times when primitive, shore-bound wooden wharves confine offshore operators—instead, wells are drilled and production actives involve from manufactured steel and/or concrete structures—which are, in many cases, movable (Chapter 3). Furthermore, offshore vessels and platform rigs are regularly used to drill for (and produce) crude oil and natural gas in waters over 7,500 ft deep and as far as 200 miles from the shore.

The exploration and production activities that are involved with crude oil and natural gas require the constant handling of flammable gases and liquids at high temperature and pressure—many of the liquids and gases are also highly toxic to flora and fauna (including humans) in addition to being highly corrosive (Speight, 2014a,b,c). The main risks are typically associated with the uncontrolled escape of hydrocarbons (gases and liquids) and other hazardous chemicals, which can cause fire and explosions from which environmental contamination arises. The environmental effects can be amplified by the working environment, particularly in offshore structures where working in a limited space in a remote location is the norm. In addition, exploration and production activities also induce serious effects on the environment—air, water, and soil (especially when the crude oil or crude oil derivative is washed ashore) and, accordingly, planned efforts are necessary to limit any such effects.

Thus, the complexity and flammability of petroleum and natural gas (Speight, 2014a) and offshore drilling for oil and gas has the potential to critically impact pristine marine ecosystems and can also lead to industrialization of our coastlines. The offshore oil development impacts on the environment in many different ways (Moore et al., 1987; Kingston, 1992; Windom, 1992; Hyland et al., 1994; Olsgård and Gray, 1995; Peterson et al., 1996; Patin, 1999; Hartley et al., 2003). Initially, exploration activities, which usually start with one or more seismic surveys, are followed by exploratory drilling. These activities generate significant amounts of waste products and attract further adverse effects on the environment arising from the support vessel traffic and transportation traffic—notably from crude oil tankers and undersea pipelines. Furthermore, all of the exploration and drilling activates may occur consecutively or simultaneously. In fact, crude oil and natural gas production in the oceans (as well as lakes and other bodies of water) is more challenging than for the analogous onshore activities and much of the efforts applied to offshore activities involves overcoming these challenges (Chapter 2, Chapter 3). This includes the following provisos: (1) manned facilities need to be kept above sea-level, and (2) the need for offshore manned facilities and both of these issues present challenges.

Manned facilities above sea level can be accomplished with enormous platforms with the feet placed on the bottom of the sea (Chapter 3). An example is the Troll A platform operated by Statoil Hydro in the North Sea, which is standing on a depth of approximately 1,000 ft. Other platforms may be floating, only anchored to the bottom of the sea. Construction costs may be reduced by using some units but additional caution is needed along with mechanisms for mitigating the effects of the movement due to waves and working at such depths adds several hundred feet to the fluid column in the drill-string. This, in turn, increases bottom-hole pressure and also increases the energy needed to lift sand and cuttings for crude oil-sand separation operation on the platform. As a result, the current trend is to attempt to accommodate more of the production in subsea equipment, such as subsea facilities to separate the crude oil and the sand followed by reinjection of the into the well—subsea production systems can range in complexity from a single satellite well with a flow line linked to a fixed platform or an onshore installation, to several wells on a template or clustered around a manifold.

Thus, the use of subsea equipment can save the work of pumping the sand (with the crude oil) up to the platform followed by separation on the platform or even pumping the sand to the shore using an undersea installations—such as a pipeline direct from the undersea wellhead to the shore. In fact, subsea installations assist in the goal of developing resources covered by progressively deeper waters and also negates many (but not all) of the challenges related to sea ice, such as in the northern part of the Barents Sea, the southern half of which (including the ports of Vardø (Norway) and Murmansk (Russia) remain ice-free year round (other than the occurrence of the occasional errant iceberg) due to the warm North Atlantic drift.

Offshore oil and gas operations produce a number of waste streams that can initially contaminate the sea and bottom sediments in the vicinity of offshore platforms. These waste streams include produced water, ballast water, displacement water, deck drainage, drilling muds, drill cuttings, produced sand, cement residue, blow-out preventer fluid, sanitary and domestic wastes, gas and oil processing wastes, slop oil, cooling water, and desalination brine. Of the list of possible contaminants, drilling muds, drilling mud cutting, and produced water pose the greatest potential threats to aquatic environments and are the subject of considerable legislation (Chapter 10) (Neff et al., 1987; Wills, 2000).

Produced water and formation water can (but generally do) contain crude oil, inorganic salts, technological chemicals, and trace metals (Neff et al., 1987). In addition, produced water discharge and drilling mud cutting can include heavy metals (such as mercury, cadmium, zinc, chromium, and copper), biocides, corrosion inhibitors, scale inhibitors, detergents, emulsifiers, and oxygen adsorbents, which all have a measure of toxicity to aquatic organisms. Indeed, there is a great deal of diversity of the impact of offshore drilling and production of oil and gas (Neff et al., 1987; Patin, 1999) from the changes temperature and pressure in underground formations as well as the discharge of toxic effluents, water consumption, damage of organisms during seismic surveys, and destruction of the shoreline (Neff et al., 1987; Patin, 1999).

9.2 POLLUTION

Subsea crude oil and natural gas reservoirs reside in shallow water and deep water and shallow water worldwide. The production systems used to recover such resources can be used to develop reservoirs, or parts of reservoirs, which require drilling wells from more than one location and, in fact, deep water conditions and ultra-deep water conditions dictate resource development (Chapter 6).

The development of subsea crude oil and natural gas resources requires equipment that must be sufficiently reliable enough to protect the environment. The deployment of such equipment requires specialized vessels (or platforms) which need equipment suitable for work at relatively shallow depths and equipment for deep water development. Generally, the various operations for offshore crude oil and natural gas production can be accompanied by undesirable (accidental or deliberate) discharges of wastes—gases, liquids, and solid. These waste streams include produced water, ballast water, displacement water, deck drainage, drilling muds, drill cuttings, produced sand, cement residue, blow-out preventer fluid, sanitary and domestic wastes, gas and oil processing wastes, slop oil, cooling water, and desalination brine. Amongst these, drilling muds and produced water are considered to pose the highest potential threats to the environment (Patin, 1999; Wills, 2000).

9.2.1 Causes of Pollution

Pollution at sea by crude oil production operations is generally attributed to the offshore installations and during transportation of the crude oil top an onshore depot. However, it must also be recognized that offshore installations are not the only source of pollution by crude oil. A wide range of oil-based pollutants also enters the marine environment by way of coastal discharges of sewage and industrial waste waters, land-based crude oil and natural gas operations, dumping of dredged materials, and pollutants that enter the ocean via the rivers.

Pollutants from crude oil and natural gas operations have complex and diverse compositions and produce physical damage and physicochemical damage to organisms as well as a variety of carcinogenic effects. Some of these pollutants—especially chlorinated hydrocarbons which are not constituents of crude oil or natural gas but are often used as solvents to clear up for grease and lubricating oil spills on an offshore drillship or platform—cause toxic and mutagenic effects. Other effects that can be caused by offshore drilling for crude oil and natural gas include: (1) mechanical effects on marine life which damage the natural functions of an organism, (2) toxic effects, which damage feeding, reproduction, and respiration, and (3) mutagenic effects, which cause cancer. In fact, the effects from spills of crude oil (from a ship or from a platform) can quickly devastate the marine environment. While each of these effects can be devastating to the marine environment and, in order the estimate the effects of such pollutants, not only the hazardous properties of the pollutants but also (1) the volume of the pollutant, (2) the means of distribution in water, (3) the behavior of the pollutant in water, (4) bio-accumulation in living organisms, (5) the composition, and (6) the stability of the pollutant in water should also be taken into account and not omitted from any assessment of effect and/or damage to the environment (Atlas and Cerniglia, 1995; Speight and Arjoon, 2012).

9.2.2 Entry into the Environment

There are several possible disposal methods for pollutants from arising from offshore oil and gas exploration and production to enter the environment: (1) direct discharge of effluents and solid wastes into the seas and oceans—overboard discharge, (2) ship-to-shore transport, (3) reinjection, and (4) disposal in especially drilled underground structures, (5) land runoff into the coastal zone, mainly with rivers, and (6) atmospheric fallout of pollutants transferred by the air mass onto the surface of the waterways oceans) (Speight and Arjoon, 2012).

Overboard discharge is the easiest and cheapest method of disposal but also the most environmentally damaging method. Overboard disposal of oil-related waste will generally result in limited and short-term environmental impact—it is believed (but not always the case) that crude oil will be degraded fairly rapidly and lose any toxic properties. However, such beliefs are often misguided and are not based on any scientific logic whatsoever the beliefs do not provide any proof that will lead to exclusion the possibility of long-term and cumulative ecological impacts (Patin, 1999; Speight and Arjoon, 2012; Speight, 2014a). For some pollutants (metals, crude oil-based constituents, and other hydrocarbons which are indigenous to the oceans because they are produced by natural bio-geochemical cycles), the task of determining the manner in which the pollutants arise from offshore drilling activities is complicated. In fact, there are many examples when extremely high concentrations of crude-oil based and natural gas-based constituents, as well as heavy metals, are not connected with offshore drilling activities (Neff et al., 1987).

Natural processes such as (1) volcanic activity, (2) oil and gas seepage on the bottom of the ocean or waterway, (3) mud flows, and (4) river flooding are all capable of introducing such pollutants into the ocean. Indeed, such occurrences should be taken into consideration (they are often not in the emotion of the moment) in order to produce an objective assessment of the impact of offshore oil and gas drilling and production activities. In addition, natural seeps are purely natural phenomena that occur when crude oil seeps from the geologic strata beneath the seafloor to the overlying water column and have been occurring for millennia (Etkin, 2009; Speight, 2014a). Although theses natural seeps release substantial amounts of crude oil (or crude oil derivatives) annually, the oil is usually released at a rate low enough for the surrounding ecosystem to adapt. On the other hand, crude oil and natural gas production can result in variable releases of both crude oil and refined products that can, when excessive, immediately overwhelm the environment. Thus, new lines of thinking need to be recognized and developed—including precise analytical methods to determine trace amounts of contaminants to obtain more reliable estimates of the contribution of the various *modes of pollution* of the waterways, particularly pollution of the marine environment (Speight, 2005; Speight and Arjoon, 2012; Speight, 2014a,b).

The combination of these factors under specific conditions ultimately defines the ecological situation in a given area—different marine regions are subjected to various and specific impact factors.

9.2.3 Types of Pollutants

The proportions and amounts of discharged wastes from offshore crude oil and natural gas exploration and production can change considerably during the life-time of the operation. For example, the amount of solid drill cuttings usually decreases as the well increases in depth and the borehole diameter becomes correspondingly smaller. The volume of produced water increase as the hydrocarbon resources are being depleted and production moves from the first stages toward its completion. In addition, the need may arise for a discharge of short duration—such situations include: chemical discharges during construction, hydrostatic testing, commissioning, pigging, and maintenance of the pipeline systems. The pipeline discharges usually contain corrosion and scale inhibitors, biocides, oxygen scavengers, and various anti-corrosion agents (Speight, 2005; Speight and Arjoon, 2012; Speight, 2014c) and the volume of such wastes can be rather considerable.

9.2.3.1 Ballast Water

Pollution arising from maritime traffic is usually associated with tank cleaning of large oil carriers. When oil tankers have discharged their cargo at the onshore facility, they return to the producing platform. In the early days of maritime crude oil transport, the oily residues in empty tanks were cleaned with water which was subsequently discharged into the sea. In addition, ballast water was directly loaded into the empty cargo tanks and crude oil-polluted ballast water was discharged on a large scale during this era. Most modern tankers have segregated ballast water tanks and the oil/water mixture arising from tank cleaning is separated onboard. Another contemporary method for cargo tank washing is *clean oil washing* in which empty tanks are washed with pressurized oil instead of water.

9.2.3.2 Offshore Installations

Offshore development of crude oil and natural gas resources usually starts with seismic surveys and is followed by exploratory drilling accompanied by increased support vessel traffic and oil tanker traffic. The general impacts of exploration and exploitation include noise and vibration, solid and liquid production wastes, increased water column turbidity from dredging, disturbance of the sea bed areas, avoidance of the area by marine wildlife due to construction noise, vibration and the presence of erected facilities, and possible invasions by nonindigenous bio-species carried in ballast water of support vessels and oil tankers (Wills, 2000; Patin, 1999). Furthermore, the environmental stress caused by offshore oil development may cause different biological responses including complex transformations at all levels of the biological hierarchy.

For example, mobile drilling rigs are towed or move under their own power to sites for exploratory drilling and are anchored at multiple mooring points, or they may be dynamically positioned in deeper waters. Initial drilling into the seabed to place risers to the surface results in direct discharge of sediment, drill cuttings, and drilling fluids at the seafloor. After the drilling rig is secured in place, drilling fluids necessary to cool, lubricate, and transport solids from the drill bit to the surface are separated from cuttings (pulverized rock from the formation) on the rig.

During well drilling and production of oil and gas offshore, a variety of liquid, solid, and gaseous wastes are produced on the platform, some of which are discharged into the ocean. They include crude oil, natural gas, petroleum condensates, nonhydrocarbon waste gases, and water produced from sedimentary rock formations. Submerged parts of the platform must be protected against corrosion and growth of biological organisms with antifouling paints and use of electrodes that may release small amounts of toxic metals, such as aluminum, copper, mercury, indium, tin, and zinc (Dicks 1982; Neff et al., 1987).

The most significant discharges associated with exploratory drilling are drill cuttings and drilling fluids. Cuttings are particles of crushed sedimentary rock, ranging in size from clay to coarse gravel, produced by the drill bit as it penetrates into the rock strata. They have an angular configuration, as distinguished from the rounded shape of most weathered natural sediments, and represent a potential input of trace metals, hydrocarbons, drilling fluids, and suspended sediments to receiving waters (Neff et al., 1987).

Drilling fluids are mixtures of natural clays and polymers, weighting agents, and other materials suspended in water or a petroleum material. Discharge of oil-based drilling fluids to the waters of the United States is prohibited, but they are used widely in other parts of the world. During drilling, a mud engineer continually tests the drilling fluid and adjusts its composition to accommodate downhole conditions so as to provide adequate cooling and lubrication of the drill bit and to float cuttings up the borehole for removal at the platform. Consequently, no two drilling fluids are identical in composition (Neff et al., 1987).

Cuttings are usually discharged overboard continuously, while drilling fluids are reused and disposed of later, generally overboard at the drill site. Water drainage from the drill platform may contain drilling fluids, oil, and small quantities of industrial chemicals. Exploration wells are sparsely distributed and generally produce brief operational discharges compared with production facilities, in the absence of a blowout (Neff et al., 1987). Furthermore, during exploratory drilling, several drilling fluid and cuttings-related effluents are discharged to the ocean. Used drilling fluids may be discharged in bulk quantities of substantial quantities (thousands of barrels) several times during a drilling operation and over the life of an exploratory well and 50%–80% (v/v) of this material may be discharged to the ocean (Neff et al., 1987).

Discharges from offshore installations occur mainly through produced water and to a lesser degree from spills and cuttings contaminated with drilling muds (Payne et al., 1989; Clark, 1999; Hartley et al., 2003). These discharges can cause surface contamination and the chemical components of crude oil can cause acute toxic effects as well as a long-term impact. In addition, disposals of cuttings contaminated with oil and chemicals in the immediate vicinity of the installations affect the benthic communities (Chapter 1) that are near to the offshore installations by imposing anoxia and toxic contamination (Maurer et al., 1980, 1981, 1982; Menzie, 1984; Maurer et al., 1986; Olsgård and Gray, 1995). Changes to benthic communities (reduction in species diversity near platforms, with opportunistic species dominating the biomass) are possible in areas surrounding offshore installations. Moreover, as the offshore industry expands into previously unexploited areas (deep-waters, ultra deep-water domains, and into environments seasonally covered by ice) there is the requirement for planned environmental management so that any environmental impacts can be reduced.

Oil-based muds were developed for situations where water-based drilling muds could not provide enough lubrication or other desired characteristics. Usually, this would be when a job required directional, or deviated, drilling (Chapter 5) in which the drill bit can be *steered downhole* so that the well deviates from the vertical by a controlled angle. Such deviated drilling has revolutionized the economics of offshore oil and gas drilling and has become standard procedure on many fields allowing small, discrete reservoirs to be penetrated with a single well. Although the radii of such curved wells may be large, deviated drilling still requires drilling mud with higher lubrication qualities than the ordinary, water-based mud traditionally used for spudding in and drilling vertical wells—particularly when cutting through layers of very hard rock or when drilling smaller radius holes a long way down.

Produced water also contains a range of other natural organic compounds including monocyclic aromatic hydrocarbons (i.e., BTEX—benzene, toluene, ethylbenzene, xylene derivatives) (Figure 9.1), two-ring and three-ring polynuclear aromatic hydrocarbons (PNAs) (Figure 9.2), phenol derivatives (Figure 9.3), and organic acids. Another possible source of contamination is from the leaching of oil and chemicals (particularly toxic sulfide derivatives) from drill cuttings in the immediate (horizontal or vertical) vicinity of installations.

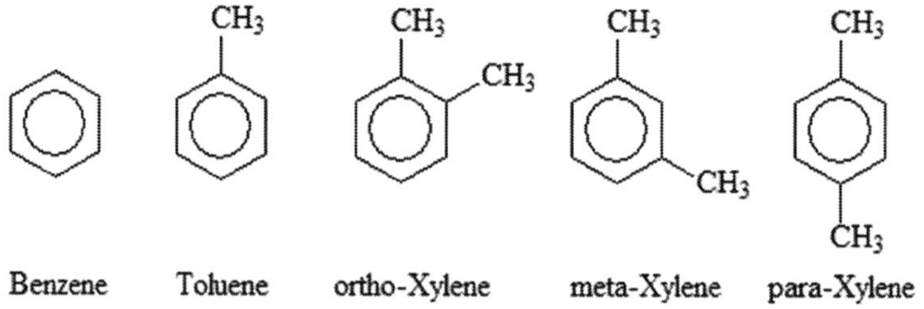

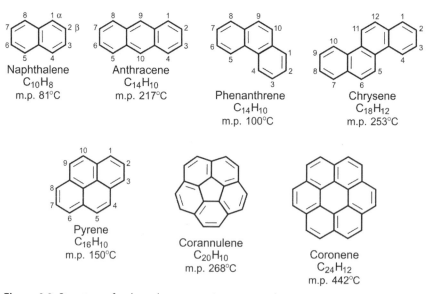

Benzene   Toluene   ortho-Xylene   meta-Xylene   para-Xylene

Figure 9.1 Structure of benzene, toluene, and the isomeric xylenes—ethylbenzene $(C_6H_5C_2H_5)$ is a homolog of toluene.

Naphthalene
$C_{10}H_8$
m.p. 81°C

Anthracene
$C_{14}H_{10}$
m.p. 217°C

Phenanthrene
$C_{14}H_{10}$
m.p. 100°C

Chrysene
$C_{18}H_{12}$
m.p. 253°C

Pyrene
$C_{16}H_{10}$
m.p. 150°C

Corannulene
$C_{20}H_{10}$
m.p. 268°C

Coronene
$C_{24}H_{12}$
m.p. 442°C

Figure 9.2 Structure of polynuclear aromatic compounds.

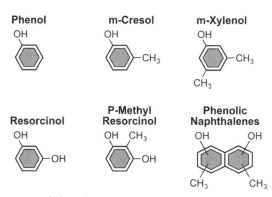

Figure 9.3 Derivatives of phenol.

9.2.3.3 Seismic Surveys

The first development stage of offshore oil development, seismic surveys, involves generating loud and mostly low frequency sound waves. Their reflection off the seafloor and sub seafloor strata provides data on the oil and gas potential of the area (Woodside, 2003). Industry and some scientists argue that seismic surveys have only limited and temporary effects; sound produced is comparable in magnitude to many naturally occurring and other man-made sounds.

Legislation that will help offshore crude oil and natural gas companies choose how, where, and when to carry out these surveys must continually evolve to assist in the significantly minimization of any negative impacts.

9.2.3.4 Drilling Mud and Drill Cuttings

When seismic surveys reveal an area where the potential for oil discovery exists, exploratory drilling starts. Drilling operations can introduce oil and a wide range of other complex chemical compounds into the environment via drilling fluids and muds (Engelhard et al., 1989; Dulfer, 1999; Hartley et al., 2003.). There are two major classes of drilling fluids: (1) oil-based and (2) water-based which are circulated in the borehole to control temperatures and pressures, to cool and lubricate the drill bit, and to remove drill cuttings from the borehole. The cuttings are small fragments of subsurface rock that break and are incorporated into the drilling fluid, the drilling mud (Wills, 2000; Kloff and Wicks, 2004).

No precise, standard formulation exists for drilling fluids—the composition of the fluid depends on the needs of the particular situations, which differ considerably in different regions and may even radically change during each drilling process while drilling rocks of very different structure (from solid granite formations to salt and slate strata). At present, two main types of drilling fluids are used in offshore drilling. They are based either on crude oil, oil products, and other mixtures of organic substances (diesel, paraffin oils, and so on) or on water (freshwater or seawater with bentonite, barite, and other components added) (Neff et al., 1987). During the last 10 years, the preference is given to using the less-toxic water-based drilling fluids. However, in some cases, for example during the drilling of deviated wells through hard rock, the use of oil-based fluids may be inevitable. The oil-based fluids, in contrast with the water-based ones, are usually not discharged overboard after a single application. Instead, they are regenerated and reintroduced into the drilling cycle.

Originally, the oil-based drilling muds included diesel fuel as their base component due to its availability and low cost. However, starting in the 1980s, especially after many countries prohibited the use of diesel in drilling muds, new formulations were developed that replaced diesel oil with less hazardous substances. More recently, a new generation of drilling fluids based on the products of chemical synthesis with ethers, esters, olefins, and poly-alpha-olefins has been developed. From the environmental perspective, the most important fact is that these drilling fluids have low toxicity as compared with other drilling formulations.

Each component of a drilling fluid has one or more chemical and technological functions. For example, (1) barite—$BaSO_4$—is used to control and regulate hydrostatic pressure in the well, (2) emulsifiers—such as alkylacrylate sulfonate derivatives—form and maintain emulsions, (3) sodium chloride and calcium chloride create conditions for maintaining an isotonic osmotic balance between the water phase of the emulsion and surrounding formation water, (4) organophilic clays—such as amine treated bentonite clay—as well as organic polymers and polyacrylate derivatives ensure the optimal fluid viscosity necessary for drilling under different geological conditions, (5) sodium sulfite, ammonium bisulfite, zinc carbonate, and other oxygen scavengers are pumped into the well to prevent the corrosion of drilling equipment in the oxidizing environment, and (6) lime is added to increase the basicity (increase the pH) of drilling fluids, which helps to reduce corrosion and stabilize the emulsions in the muds. Drilling discharges also contain many heavy metals (mercury, lead, cadmium, zinc, chromium, copper, and others) that come from components of both drilling fluids and drill cuttings (Neff et al., 1987; Gerard, 1996; Gerrard et al., 1999).

Various kinds of wastes are associated with drilling, including drilling muds and cuttings. Drill cuttings are removed from drilling muds and cleaned in separators and the separated drilling muds and cleaning fluids used to treat cuttings are partially returned to the circulating system. Cuttings piles present some level of environmental problem but the extent of the area affected varies markedly among platforms (Gerrard et al., 1999). After separation of the cuttings from the drilling mud and in the absence of a viable and safe alternative were historically discharged into the sea underneath or adjacent to platforms. As much mud as possible is recovered for re-use, however some adheres to the cuttings and is also discharged. These discharges often accumulate into a *drill cuttings pile* on the seabed in the area of the platform (Gerrard et al., 1999).

Drill cuttings material can be dispersed by tides and currents before reaching the seabed (as in the southern North Sea) or some piles may also be unstable after formation as they have a soft surface which allows material to be redistributed by bottom currents—this means the *edge* of the pile cannot be easily be delineated. In the oceans or seas with significant current strength, cuttings piles did not form where current strengths are weak and the water is deep cuttings piles have a greater potential to form. For platforms using oil-based drilling muds, the area adversely affected may extend up to two-to-three miles from the central platform (Gray et al., 1990). On the other hand, platforms that use water-based drilling muds may show a much-reduced adverse effect (Daan and Mulder, 1993, 1996).

9.2.3.4.1 Drilling Muds

Drilling muds fulfil several functions, and because they may be used in a very wide variety of rock formations and circumstances, their specific composition varies from site to site, and with different depths in the same well. Since offshore oil and gas exploration began, many different types of substances have been, and are being, used in the formulation and modification of drilling muds and included emulsifiers, lubricants, wetting agents, corrosion inhibitors, surfactants, detergent, caustic soda, salts, and organic polymers (Bell et al., 1998). In fact, it is more correct to consider drilling mud as a *mud system*—a combination of different substances that can used and modified to suit particular conditions encountered during drilling. In addition to the chemicals mentioned above, drilling muds also consist of gelling and deflocculating agents (bentonite clays) filtration control agents, pH and ion-control substances, barites, biocides, corrosion inhibitors, lubricants, defoaming agents, and trace elements of heavy metals (such as arsenic, barium, chromium, cadmium, lead, and mercury) which can be retained in the marine environment (Neff et al., 1987; Annis, 1997; Patin, 1999; Wills, 2000; CAPP, 2001).

The function of drilling muds conduct cuttings away from the drill face, provide a hydrostatic head that counters the pressure of gas or oil, and provide a physical and chemical means of stabilizing and protecting the rock formation that is being drilled. Three main types of mud have been used and any or all of them may be present in the drill cuttings piles: (1) water-based muds, (2) oil-based muds, and (3) synthetic muds (SMs).

In the case of *water-based muds*, the base fluid in these muds is water, and they are considered to be the most environmentally favorable mud systems. Water based muds are typically used when drilling the top sections of the well. On the other hand, in the *oil-based muds* the base fluid in these muds is oil and is subject to the release of hydrocarbons to the environment (Hird and Tibbetts, 1995). Diesel was used until 1984 when it was banned as a base fluid and replaced more refined oils, the *low toxicity oils*. The discharge of oil-based muds has been phased out in the 1990s, and they have been replaced by SM systems (pseudo-oil based muds (POBM)). Finally, in the in the SMs (also called POBM), the base fluids for these muds are synthesized from a variety of non-petroleum-related products. However, these muds are also being phased out.

No precise, standard formulation exists for drilling fluids. Their composition depends on the needs of the particular situations. These differ considerably in different regions and may even radically change during each drilling process while drilling rocks of very different structure (from solid granite formations to salt and slate strata). At present, two main types of drilling fluids are used in offshore drilling. They are based either on crude oil, oil products, and other mixtures of organic substances (diesel, paraffin oils, and so on) or on water (freshwater or seawater with bentonite, barite, and other components added). During the last 10 years, the preference is given to using the less-toxic water-based drilling fluids. However, in some cases, for example during drilling of deviated wells through hard rock, using oil-based fluids is still inevitable. The oil-based fluids, in contrast with the water-based ones, are usually not discharged overboard after a single application. Instead, they are regenerated and included in the technological circle again.

The environmental hazard of drilling muds is connected, in particular, with the presence of lubricating materials in their composition. These lubricating substances usually have a hydrocarbon base. They are needed for effective drilling, especially in case of slant holes or drilling through solid rock. The lubricants are added into the drilling fluids either from the very beginning as a part of the original formulations or in the process of drilling when the operational need emerges. In both cases, the discharges of spent drilling muds and cuttings coated by these muds contain considerable amounts of relatively stable and toxic hydrocarbon compounds and a wide spectrum of many other substances.

9.2.3.4.2 Drill Cuttings

Drill cuttings are produced when wells are drilled—the drill bit cuts into the rock sediments and produces fragments which range in diameter from microns-size pieces to two-centimeter pieces, depending largely on the nature of the rock. Many cuttings are composed of shale or sandstone (from the drilled sediments) but the overall morphology of the cuttings is influenced by the drilling mud, which can affect the size of the particles and their tendency to agglomerate (McFarlane and Nguyen, 1991; Bell et al., 1998). Cuttings are carried away from the drill face by the drilling mud (drilling fluid) which is pumped down the inside of the drill string and then circulated back to the surface inside the drill casing, carrying the cuttings with it.

Drill cuttings separated from drilling muds have a complex and variable composition depending upon factors such as the type of rock, drilling regime, and formulation of the drilling fluid. However, in all cases, drilling muds play the leading role in forming the composition of drill cuttings. Thus, the types of cuttings vary depending on the point of release, which may also have a bearing on the size: (1) cuttings discharged directly onto the seabed, (2) cuttings discharged from a drilling rig, and (3) cuttings discharged from a fixed platform.

In the first case (*cuttings discharged directly onto the seabed*), the upper part of the well may be drilled without any drill casing in place and the circulated cuttings cannot be returned to the surface – the cuttings exit the well at the level of the seabed. When *cuttings are discharged from a drilling rig* after drilling the upper section of the hole, a drill casing is installed and the cuttings can be returned to the drilling rig. After appropriate cleaning and the recovery of the drilling mud, the cuttings are discharged to sea via a rigid or flexible pipe, which carries the cuttings to a point below sea level, where they are discharged. The point of discharge is usually high in the water column, and the cuttings will therefore fall a considerable distance, which exposes them to the effects of tides and currents for a longer period of time, and may lead to the creation of a more widely dispersed accumulation.

Typically, *cuttings discharged from a fixed platform* after the installation of the drill casing are returned to the platform for clean-up and/or mud recovery, and then discharged to sea through a caisson which is attached to the platform by secure attachment points—on a fixed platform the discharge caisson can be longer than on a semisubmersible drilling rigs. Consequently, since the platform is fixed and the cuttings from every well are discharged through the same caisson, the cuttings may also fall onto the same area of the seabed, more or less directly below the platform. In addition, the cuttings may be released at a greater depth, away from the more energetic surface zone and closer to the seabed, allowing less opportunity for lateral spreading but any influence are site-specific (Bell et al., 1998).

9.2.4 Produced Water

Produced water, including injection water and solutions of chemicals used to intensify crude oil and natural gas production, is one of the main sources of oil pollution in the areas of offshore oil and gas production (EEA, 2008). It is significant that, as a crude oil reservoir is being depleted, the ratio between the water and crude oil in the extracted product increases and water becomes the prevailing phase with a concomitant increase in the volume of discharged water and the difficulty associated with of the water treatment. Produced water usually includes dissolved salts and organic compounds, oil hydrocarbons, trace metals, suspensions, and many other substances that are components of formation water from the reservoir (including *connate water*—sometimes referred to as *fossil water*—is water entrapped in the interstices of the sedimentary rocks at the time of deposition) or components used during drilling and other production operations—which give rise to a variety of toxicological effects (Connell and Miller, 1981; Neff et al., 1987; Speight and Arjoon, 2012).

The discharge of produced water considerably dominates the discharge of other wastes. Depending on its quality, the produced water is either discharged into the sea or injected into the disposal well. In any case, the composition of the discharged produced water is complex and changeable—there is no so-called average or typical parameters that describe the composition of produced water and each sample should be subject to reliable and complete analysis to describe the composition. Very generally, produced water contains oil, inorganic salts, and trace metals along with heavy metals (such as mercury, cadmium, zinc, chromium, copper), biocides, corrosion and scale inhibitors, detergents, emulsifiers, and oxygen adsorbents.

Produced waters, including injection waters and solutions of chemicals used to intensify hydrocarbon extraction and the separation of the oil–water mixtures, are one of the main sources of oil pollution in the areas of offshore oil and gas production (Table 9.1). It is significant that, as a hydrocarbon

Table 9.1 Sources and Scale of Oil Pollution Input into the Marine Environment

Types and source of input	Environment		Scale of distribution and impact		
	Hydrosphere	Atmosphere	Local	Regional	Global
Natural:					
Natural seeps and erosion of bottom sediments	+	−	+	?	−
Biosynthesis by marine organisms	+	−	+	+	+
Anthropogenic:					
Marine oil transportation (accidents, operational discharges from tankers, etc.)	+	−	+	+	?
Marine nontanker shipping (operational, accidental, and illegal discharges)	+	−	+	?	−
Offshore oil production (drilling discharges, accidents, etc.)	+	+	+	?	−
Onland sources: sewage waters	+	−	+	+	?
Onland sources: oil terminals	+	−	+	−	−
Onland sources: rivers, land runoff	+	−	+	+	?
Incomplete fuel combustion	−	+	+	+	?

Note: +, −, and ? mean, respectively, presence, absence, and uncertainty of corresponding parameters.

reservoir is being depleted, the ratio between the water and oil fraction in the extracted product increases, and water becomes the prevailing phase. At the same time, both the volumes of discharged waters and the difficulties of their treatment increase.

9.2.5 Other Wastes

Large quantities of produced water, drilling muds, and drill cuttings, discussed above, as well as discharges of storage displacement water and ballast water are the source of regular and long-term impacts of the offshore industry on the marine environment. Besides these discharges, sometimes the need arises to conduct a one-time discharge of short duration. Such situations include, in particular, chemical discharges during construction, hydrostatic testing, commissioning, pigging, and maintenance of the pipeline systems. The pipeline discharges usually contain corrosion and scale inhibitors, biocides, oxygen scavengers, and other agents.

Similar situations emerge during other technological and maintenance activities. Examples include cleaning and anticorrosion procedures, discharging the ballast waters from the hydrocarbon storage tanks, well repairing, well workover operations, replacing the equipment, and others. These aqueous discharges often contain surface-active substances, such as lignosulfonate derivatives, lignite, sulfo-methylated tannins, and many other chemicals.

9.3 OIL SPILLS

Accidents inevitably accompany offshore development and such accidents are the sources of environmental pollution at all stages of crude oil and natural gas production. The causes, scale, and severity of the accidents' consequences of such accidents are variable and, to a certain extent, each accidental situation develops in accordance with a site specific scenario. Nevertheless, the environmental consequences of accidental episodes are especially severe, sometimes dramatic, when they happen near the shore, in shallow waters, or in areas with slow water circulation.

Oil spills impose serious damage on the environment and there is a need to comply the environmental rules and regulations to improve oil spill response and cleanup performance. Current oil spill control technologies have failed to ensure environmental safety and ecosystem integrity. Cleanup efforts often lead to more environmental hazards in one way or the other causing public outcry for alternate energy sources. Since hydrocarbon driven energy demand is expected to rise in the foreseeable future, it becomes imperative to work towards sustainable oil operations (Al-Majed et al., 2011).

9.3.1 Drilling and Production Accidents

Drilling accidents are usually associated with unexpected blowouts of liquid and gaseous hydrocarbons from the well as a result of encountering zones with abnormally high pressure.

There are two major categories of drilling accidents (1) catastrophic events which involve intense and prolonged crude oil blow out and occurs when the pressure in the drilling zone is sufficiently high that usual technological methods of well control are not operative—such abnormally high pressure is most often encountered during exploratory drilling in new fields, (2) regular, routine episodes of crude oil spill and blowout during drilling operations—these accidents can be controlled rather effectively by shutting in the well with the help of the blowout preventers and by changing the density of the drilling fluid.

Accidents that result in large oil spills involving offshore oil installations can be caused by many different factors. *Blowouts* of wells or pipeline ruptures are the best known-examples. A blowout (loss of well control) can take place if a drilling rig encounters a pocket of subsea crude oil and/or natural gas under excessive geological pressure or when errors are made or from technical failures.

9.3.2 Platform and Terminal Accidents

Accidental oil spills also arise during routine operations when crude oil is loaded and discharged. This normally occurs in ports or at oil terminals such as offshore production platforms. The magnitude of the problem is quite serious. The amount of oil spilled during terminal operations is three times of an order greater than the total amount of oil spilled after accidents with oil tankers. However, there are several examples of global best practice in port management and tanker traffic control systems, where the problem has been reduced to very small proportions using existing technology and careful management.

The consequences of a large spill can reach the lethal limits of sea-life and are especially disastrous when the oil washes ashore and accumulates in sediments of shallow coastal zones.

9.3.3 Tanker Accidents

Traditional shipping and oil transportation routes are often exposed to the impacts of oil-polluted discharges from tankers and other vessels—in fact risks are involved in all forms of crude oil and natural gas transport (Neff et al., 1987). Large spills may arise from maritime traffic after the grounding of a crude oil tanker, collision with another vessel, and due to cargo fires and explosions. Significantly, both large drilling accidents and large tanker catastrophes occur relatively rarely and the frequency of such incidents as well as the crude oil volumes released in large spills differs from year to year. In addition, the probabilistic nature of accidental situations and highly variable volumes of spilled oil do not allow accurate predictions of future accidents to be made and although the level of crude oil spillage from tankers has tended to decrease, a major accident resulting in the release of a large volume of crude oil could change this situation and revise the outlook.

9.3.4 Other Accidents

Underwater reservoirs for storing liquid hydrocarbons (crude oil, crude oil–water mixtures, and gas condensate) are a necessary element of many crude oil and natural gas development projects. They are often used when tankers instead of pipelines are the main means of hydrocarbon transportation. Underwater *storage tanks* are either built near the platform foundations or are anchored in the semi-submerged position in the area of developments and near the onshore terminals. Sometimes, the anchored tankers are used for this purpose. As a result, a risk exists of damaging the underwater storage tanks and releasing their content, especially during tanker loading operations and under severe weather conditions.

Complex and extensive systems of underwater *pipelines* have a total length of thousands of miles and transport oil, gas, condensate, and their mixtures. These pipelines are among the main factors of environmental risk during offshore oil developments, along with tanker transportation and drilling operations. The causes of pipeline damage can differ greatly and range from material defects and pipe corrosion to ground erosion, tectonic movements on the bottom, and encountering ship anchors and bottom trawls. The main causes of these accidents are material and welding defects that can be overcome by quality control programs. Depending on the cause and nature of the damage (cracks, ruptures, and others), a pipeline can become either a source of small and long-term leakage or an abrupt (even explosive) blowout of crude oil at the wellhead near the bottom of the sea.

9.4 IMPACT OF OIL ON THE OCEAN

When crude oil is spilled into the ocean it undergoes a number of physical and chemical changes, some of which lead to its removal from the sea surface while others cause it to persist. The removal or persistence of crude oil in the environment depends upon factors such as the amount of oil spilled, the physical and chemical characteristics of the oil as well as the prevailing climatic and sea conditions and whether the oil remains at sea or is washed ashore. An understanding of the processes involved and how they alter the nature, composition, and behavior of crude oil over time is fundamental to all aspects of the response to a crude oil spill response.

However, it must be recognized that no single process may be operative—any two or more of the processes described below may be operative as soon as crude oil is spilled, although their relative importance varies with time. Spreading, evaporation, dispersion, emulsification and dissolution are most important during the early stages of a spill while oxidation, sedimentation and biodegradation are longer term processes which determine the ultimate fate of oil. An understanding of the way in which weathering processes interact is important when attempting to forecast the changing characteristics of oil during the lifetime of a slick at sea.

It must also be recognized that the movement of an oil slick on the sea surface is due to winds and surface currents, and may be influenced by the combined weathering processes. The actual mechanisms governing spill movement are complex, but experience shows that oil drift can be predicted from a simple vector calculation of wind and surface current direction. Thus, reliable prediction of slick movement is clearly dependent upon the availability of good wind and current data. Accurate current data are sometimes difficult to obtain. For some areas it is presented on charts or tidal stream atlases but often only general information is available. In shallow waters near the coast or among islands, currents may be complex and are often poorly understood, rendering accurate prediction of slick movement particularly difficult.

9.4.1 Chemical and Physical Changes

A series of complex chemical and physical processes leading to the transformation of crude oil in the marine environment start developing almost immediately after the crude oil comes into contact with seawater. Moreover, the progression, duration, and result of the transformations depend on the properties and composition of the crude oil, the extent of the spill, and environmental conditions. In more general terms, the physical and chemical changes that spilled oil undergoes are collectively known as *weathering* (sometimes chemically referred to as *oxidation*) and although the individual processes causing these changes may act simultaneously, the relative importance of each process varies with time. Collectively, the processes affect the behavior of crude oil and determine its ultimate fate.

Crude oils of different origin vary widely in their physical and chemical properties, whereas many refined products tend to have well-defined properties irrespective of the crude oil from which they are derived. Some crude contain varying proportions of high molecular weight nonvolatile constituents (such as resin and asphaltene constituents) that tend to linger in the environment (Speight and Arjoon, 2012).

The properties of crude oil that are of most interest after a spill are: (1) API gravity or specific gravity, (2) distillation profile, (3) viscosity, and (4) pour point. All of these properties are dependent on chemical composition (e.g., the amount of asphaltene constituents, resins and waxes which the oil contains) (Hsu and Robinson, 2006; Gary et al., 2007; Speight, 2005; Speight and Arjoon, 2012; Speight 2014a).

The *specific gravity* (*relative density*) of crude oil is its density in relation to pure water. Conventional crude oil (Chapter 1) has a specific gravity less than 1.00 and is lighter than sea water which has a specific gravity of approximately 1.025. The American Petroleum Institute gravity scale (°API) is commonly used to describe the specific gravity of crude—in this scale conventional crude oil has a value on the order of 20 to 50 (Speight, 2005, 2009, 2014a,b). In addition to determining whether or not the oil will float, the specific gravity can also give a general indication of other properties of the oil. For example, oils with a low specific gravity (high °API) tend to contain a high proportion of volatile components and to be of low viscosity (Speight, 2014a).

As a general rule, the lower the specific gravity of the oil the less persistent it will be. The concept of a half-life is helpful in defining removal rates of less persistent oils. This is the time taken for the removal of 50% (v/v) of the oil from the sea surface so that after six half-lives, little more than 1% (v/v) of the oil will remain. Half-life calculations are less useful for heavier oils and water-in-oil emulsions. However, it is important to appreciate that some apparently light oils behave more like heavy ones due to the presence of waxes. Crude oil with a wax content greater than approximately 10% (w/w) tends to have a high pour points and if the ambient temperature is low, the crude oil will be either a solid or a highly viscous liquid, and any natural breakdown processes will be slow.

The *distillation characteristics* of crude oil describe its volatility and the distillation characteristics are expressed as the proportions of the parent oil which distil within given temperature ranges. Some oils contain bituminous, waxy or asphaltenic residues which do not readily distil, even at high temperatures. These are likely to persist for extended periods in the environment.

The *viscosity* of crude oil is an indication of the resistance of the oil to flow and is usually (in the United States) is expressed in centistokes. High viscosity oils do not flow as easily as those with lower viscosity but all crude oils become more viscous (i.e., flow less readily) as their temperature falls, some more than others depending on their composition. Since sea temperatures are often lower than cargo or bunker temperatures on board a vessel, viscosity-dependent clean-up operations such as skimming and pumping generally become more difficult as the spilled crude oil cools.

The *pour point* is the temperature below which crude oil will not flow—the test was originally developed as a function of the wax and asphaltene content of the oil (Speight, 2001, 2014a). As crude oil cools, it will reach a temperature, the *cloud point*, at which the wax components begin to form crystalline structures. This increasingly hinders flow of the oil until it eventually changes from liquid to semi-solid at the pour point.

The processes described below come into play as soon as oil is spilled, although their relative importance varies with time. Spreading, evaporation, dispersion, emulsification and dissolution are most important during the early stages of a spill while oxidation, sedimentation and biodegradation are longer term processes which determine the ultimate fate of oil (Burns et al., 1993). An understanding of the way in which weathering processes interact is important when attempting to forecast the changing characteristics of crude oil during the lifetime of a slick at sea.

For fixed installations such as oil terminals and offshore oil fields, where a limited number of oil types are involved and prevailing conditions are well known, fairly accurate predictions can be made, which simplifies the development of an effective plan. Plans for areas where a wide range of oil types are handled or where tankers pass in transit cannot cover all eventualities. It is therefore even more important that the type of oil spilled is established at the earliest opportunity so that when a response is required, the most appropriate techniques may be used.

9.4.1.1 Aggregation

Oil aggregates in the form of petroleum lumps, tar balls, or pelagic tar can be presently found both in the open and coastal waters as well as on the beaches. They derive from crude oil after the evaporation and dissolution of its relatively light fractions, emulsification of oil residuals, and chemical and microbial transformation. The chemical composition of oil aggregates is rather changeable. However, most often, its base includes the high-molecular-weight constituents of the oil.

Oil aggregates look like light gray, brown, dark brown, or black sticky lumps. They have an uneven shape and vary from 1 mm to 10 cm in size (sometimes reaching up to 50 cm). Their surface serves as a substrate for developing bacteria, unicellular algae, and other microorganisms. Oil aggregates can exist from a month to a year in the enclosed seas and up to several years in the open ocean—the life-cycle is completed when slow degradation occurs in the water, on the shore (if they are washed there by currents), or on the sea bottom (if they lose their floating ability).

9.4.1.2 Biodegradation

Biodegradation is the transformation of a substance into new compounds through biochemical reactions or the actions of microorganisms such as bacteria or, alternatively, biodegradation is the process by which microbial organisms transform or alter (through metabolic or enzymatic action) the structure of chemicals introduced into the environment (Speight and Arjoon, 2012).

Sea water contains a range of marine microorganisms capable of metabolizing crude oil constituents—the organisms include bacteria, molds, yeasts, fungi, unicellular algae, and protozoa which can utilize oil as a source of carbon and energy. Such organisms are distributed widely throughout the oceans although they tend to be more abundant in chronically polluted coastal waters, such as those with regular vessel traffic or which receive industrial discharges and untreated sewage. The main factors affecting the rate and extent of biodegradation are the characteristics of the oil, the availability of oxygen and nutrients (principally compounds of nitrogen and phosphorus) and temperature. Each type of microorganism involved in the process tends to degrade a specific group of crude oil constituents (such as hydrocarbon constituents) and thus a wide range of microorganisms, acting together or in succession, are needed for degradation to occur.

As degradation proceeds, a complex community of microorganisms develops. Although the necessary microorganisms are present in relatively small numbers in the open sea, they multiply rapidly when crude oil is available and degradation will continue until the process is limited by nutrient or oxygen deficiency. While microorganisms are capable of degrading most of the wide variety of compounds in crude oil, some large and complex molecules are resistant to attack.

Because the microorganisms live in the water, from which they obtain oxygen and essential nutrients, biodegradation can only take place at an oil/water interface. At sea, the creation of oil droplets, either through natural or chemical dispersion, increases the interfacial area available for biological activity and so may enhance degradation. On the other hand, crude oil stranded in thick layers on shorelines or above the high water mark will have a limited surface area and will be subject to drier conditions which will render degradation extremely slow, resulting in the oil persisting for many years. Similarly, once oils become incorporated into sediments on the shoreline or sea bed, degradation is very much reduced or may stop due to a lack of oxygen and/or nutrients.

The variety of factors influencing biodegradation makes it difficult to predict the rate at which oil may be removed. Although biodegradation is clearly not able to remove bulk oil accumulations, it is one of the main mechanisms by which dispersed oil or the final traces of a spill on shorelines are eventually removed.

9.4.1.3 Dispersion

Waves and turbulence at the sea surface can cause all or part of a slick to break up into droplets of varying sizes which become mixed into the upper layers of the water column. While some of the smaller droplets may remain in suspension, the larger ones rise back to the surface, where they either coalesce with other droplets to reform a slick or spread out in a very thin film, often referred to as *sheen*. Droplets which are small enough are kept in suspension by the turbulent motion of the sea, which mixes the oil into ever greater volumes of sea water, so reducing its concentration. The increased surface area presented by dispersed oil can promote processes such as biodegradation, dissolution and sedimentation.

The rate of dispersion is largely dependent upon the nature of the oil and the sea state, proceeding most rapidly with low viscosity oils in the presence of breaking waves. Oil that remains fluid and spread unhindered by other weathering processes may disperse completely in moderate sea conditions within a few days. The application of dispersant chemicals can speed up this natural process. However, crude oil properties and the manner in which way the properties change with time on the sea are extremely important when assessing the likelihood of successful chemical dispersion. The viscosity and pour point of a crude oil provide an indication of ability of crude oil to disperse when spoiled on to the sea.

The choice of dispersant and the dosage will affect the amount of oil actually dispersed. In many circumstances it is preferable to use undiluted concentrate dispersants in open waters. Dispersants are manufactured to slightly different formulations, and their effectiveness varies to a greater or lesser degree with the type of oil treated. Some dispersants have been formulated specifically with the aim of treating viscous oils. Laboratory tests may be carried out to compare the effectiveness of one dispersant relative to another for a particular oil and some countries require operators of oil terminals or rigs to undertake such studies to identify the most effective dispersant for the oil involved. As in many cases, with the variation in crude oil, the dispersant must be selected accordingly.

Conversely, viscous oils and oils at temperatures below their pour point, or oils that form stable water-in-oil emulsions, tend to form thick lenses on the water surface that show little tendency to disperse, even with the addition of dispersant chemicals. Such oils can persist for weeks and on reaching the shore may eventually form hard asphalt if not removed.

The use of dispersants tends to be controversial since it is often viewed not only as a method of minimizing potential impacts on sensitive resources by preventing or reducing shoreline contamination, but it is also seen as adding another pollutant to the environment. Despite improvements in dispersant formulations, toxicity of the dispersant/oil mixture to marine fauna and flora is often the major environmental concern. Approval processes for dispersant use are normally in place which are designed to take both effectiveness and toxicity into account. For example, the environmental regulations some countries require that the dispersant/oil mixture is no greater in toxicity than the crude oil.

In summary, when used appropriately, the use of a chemical dispersants is an effective method of response to an oil spill. A dispersant is capable of rapidly removing large amounts of certain types of crude oil from the sea surface and transferring it into the water column. Following dispersant application, wave energy will cause the oil slick to break up into small droplets of crude oil that are rapidly diluted and subsequently biodegraded by microorganisms occurring naturally in the marine environment. A dispersant can also delay the formation of persistent water-in-oil emulsions. In common with other response techniques, the decision to use dispersants must be given careful consideration and take into account oil characteristics, sea and weather conditions, and environmental sensitivities. Significant environmental and economic benefits can be achieved, particularly when other at-sea response techniques are limited by weather conditions or the availability of resources. In certain situations, the use of a dispersant may provide the only means of removing significant quantities of surface crude oil quickly, therefore minimizing or preventing environmental damage.

9.4.1.4 Dissolution

Some components of crude oil are water-soluble to a certain degree, especially low-molecular-weight aliphatic and aromatic hydrocarbons. Polar compounds formed as a result of oxidation of some oil fractions in the marine environment also dissolve in seawater. Typically, because of the relatively nonpolar nature of crude oil, dissolution takes more time than evaporation, dissolution takes more time and the physicochemical conditions at the ocean surface strongly affect the rate of the process—the rate and extent to which crude oil dissolves depends upon oil composition, spreading, water temperature, water turbulence, and degree of dispersion.

The higher molecular weight constituents of crude oil crude oil are virtually insoluble in sea water whereas lower molecular weight constituents, particularly aromatic hydrocarbons such as benzene and toluene, are slightly soluble (Speight, 2001, 2014a). However, the lower molecular weight constituents are also the most volatile and are lost very rapidly by evaporation. The concentration of dissolved hydrocarbons in sea water thus rarely exceeds 1 ppm and dissolution does not make a significant contribution to the removal of oil from the sea surface.

9.4.1.5 Emulsification

Emulsification of crude oil (constituents) in the marine environment depends on the crude oil composition and the turbulent regime of the water mass. Water-in-oil emulsions usually appear after strong storms in the spill zone and can exist in the marine environment for more than three months—the stability of these emulsions usually increases with decreasing temperature. The reverse emulsions (oil-in-water emulsions—droplets of oil suspended in water) are much less stable because surface-tension forces quickly decrease the dispersion of oil. This process can be slowed with the help of emulsifiers—surface-active substances with strong hydrophilic properties used to eliminate crude oil spills. Emulsifiers assist in the stabilization of crude oil emulsions and promote dispersal of the crude oil to form microscopic (invisible) droplets, which accelerates the decomposition of oil products in the water.

The formation of water-in-oil emulsion reduces the rate of other weathering processes and is the main reason for the persistence of high-API gravity and medium-API gravity crude oil on the sea surface.

9.4.1.6 Evaporation

Once crude oil has been spilled, the viscosity rapidly increases from its initial value due to the loss of volatile components through evaporation and through emulsification. The more volatile components of crude oil evaporate and the rate of evaporation is dependent on ambient temperatures and wind speed. In general, those oil components with a boiling point below 200°C (390°F) will evaporate within a period of 24 h in temperate conditions. In addition, some crude oils are particularly prone to forming water-in-oil emulsions (especially those that have a relatively high asphaltene content (>0.5%) and a combined nickel/vanadium concentration greater than 15 parts per million). Emulsification causes an increase in both viscosity and volume.

The increase in viscosity caused by evaporation and by emulsion formation restricts the ability of the dispersant to reach the oil/water interface and makes it difficult to overcome the mechanical resistance to mixing. This prevents the formation of small oil droplets. However, if the emulsion is unstable, concentrate dispersants may be able to break it back to its parent oil, releasing the water and allowing the relatively fresh oil to be dispersed by a second application of dispersant. As a result of these changes in oil properties over time, the opportunity for the successful application of dispersants is limited. The time available usually ranges from a few hours to a few days depending on the type of oil involved and the environmental conditions.

The initial spreading rate of the oil affects evaporation since the larger the surface area, the faster light components will evaporate. Rough seas, high wind speeds and warm temperatures will also increase the rate of evaporation. Any residue of crude oil remaining after evaporation will have an increased density and viscosity, which affects subsequent weathering processes and also has an adverse effect on the effectiveness of clean-up techniques.

9.4.1.7 Microbial Degradation

The fate of most crude oil in the marine environment is ultimately defined by transformation and degradation by microbial activity. The degree and rates of hydrocarbon biodegradation depend, first of all, upon molecular structure (Burns et al., 1993; Speight and Arjoon, 2012)—for example, paraffin compounds (alkanes) biodegrade faster than aromatic and naphthenic substances. With increasing complexity of molecular structure (increasing the number of carbon atoms and degree of chain branching) as well as with increasing molecular weight, the rate of microbial decomposition usually decreases. Besides, this rate depends on the physical state of the oil, including the degree of its dispersion.

The main factors affecting the rate and extent of biodegradation are the characteristics of the oil, the availability of oxygen and nutrients (principally compounds of nitrogen and phosphorus) and temperature (Speight and Arjoon, 2012). Each type of microorganism involved in the process tends to degrade a specific group of hydrocarbons and thus a wide range of micro-organisms, acting together or in succession, are needed for degradation to occur. As degradation proceeds, a complex community of microorganisms develops. Although the necessary microorganisms are present in relatively small numbers in the open sea, they multiply rapidly when oil is available and degradation will continue until the process is limited by nutrient or oxygen deficiency. While microorganisms are capable of degrading most of the wide variety of compounds in crude oil, some large and complex molecules are resistant to attack. However, the variety of factors influencing biodegradation makes it difficult to predict the rate at which crude oil may be removed.

9.4.1.8 Oxidation and Weathering

Chemical transformation of crude oil (typically oxidation, often referred to as *weathering* but other processes such as the evaporation of lower boiling constituents are also involved) on the water surface commences shortly (usually one day) after the crude oil spill. The oxidation processes are catalyzed by some trace elements (e.g., vanadium) and inhibited (slowed) by compounds of sulfur. The final products of oxidation (hydroperoxides, phenols, carboxylic acids, ketones, aldehydes, and others) usually have increased water solubility. Oxidation is promoted by sunlight and the reactions—photo-oxidation—initiate the polymerization and decomposition of the most complex molecular species in crude oil, which increases the viscosity of the crude oil and promotes the formation of solid oil aggregates.

Thick layers of viscous crude oils or water-in-oil emulsions tend to oxidize to persistent residues rather than degrade because higher molecular weight compounds are formed (by the photo-oxidation process) and create a protective surface layer. This phenomenon is reflected in the formation of *tar balls* which deposit on shorelines and which usually consist of a solid outer crust of oxidized oil and sediment particles, surrounding a softer, less weathered interior.

9.4.1.9 Physical Transport

The distribution of oil spilled on the sea surface occurs under the influence of gravitation forces. It is controlled by oil viscosity and the surface tension of water—which also influence the potential for emulsion formation.

As soon as crude oil is spilled on to water, it starts to spread over the surface and the speed at which spreading occurs is dependent to a great extent on the viscosity of the crude oil and the volume spilled. Conventional crude oil spreads more readily than high-viscosity heavy—at temperatures below the pour point, crude oil rapidly solidifies and spreading is minimized. The rate at which crude oil spreads on the ocean is also affected by tidal streams and currents—the stronger the combined forces, the faster the spreading process.

An oil slick usually drifts in the same direction as the wind and, while the slick thins, especially after the critical thickness of about 0.1 mm has been reached, the slick disintegrates into separate fragments that spread over larger and more distant areas. A considerable part of oil disperses in the water as fine droplets that can be transported over large distances away from the place of the spill.

9.4.1.10 Sedimentation

After a spill, some of the oil (up to 30%) will be adsorbed on the suspended material and deposited on the sea bottom (Choi and Cloud, 1992; Ahmad et al., 2005; Speight, 2005; Islam et al., 2010; Cojocaru et al., 2011; Speight and Arjoon, 2012). This mainly happens in the narrow coastal zone and shallow waters where particulates are abundant and water is subjected to intense mixing. In deeper areas remote from the shore, sedimentation of oil (except for the heavy fractions) is an extremely slow process.

Simultaneously, the process of bio-sedimentation occurs—in this process plankton and other organisms absorb the emulsified oil and cause it to sediment on the ocean floor with the plankton metabolites and remainders. The suspended forms of oil and its components undergo intense chemical and biological (microbial in particular) decomposition in the water but this situation radically changes when the suspended oil reaches the sea bottom—the rate of decomposition of the sea-bottom oil ceases and the rate of oxidation also drops under the anaerobic conditions in the bottom environment. In addition, shallow coastal areas and the waters of river mouths and estuaries are often laden with suspended solids that can bind with dispersed oil droplets, thereby providing favorable conditions for sedimentation of oily particles to the sea bed. The deposited oil that accumulates inside the sediments can be preserved for many months and even years. After a spill, any claims of recovery or accounting for all of the spilled oil may be completely unjustified (and incorrect) in the light of the sedimentation phenomenon (Speight and Arjoon, 2012).

9.4.1.11 Self-purification

As a result of the processes previously discussed, crude oil in the marine environment rapidly loses its original properties and disintegrates into hydrocarbon fractions which have different chemical composition and structure and exist in different migrational forms. The intermediate products of the process undergo transformations that slow after reaching thermodynamic equilibrium with the environmental parameters. Eventually, the original and intermediate compounds formed in the transformation disappear resulting in the formation of carbon dioxide and water. Such self-purification of the marine environment inevitably happens in water ecosystems if, of course, the toxic load does not exceed tolerable or acceptable limits.

9.4.1.12 Spreading

Spreading, as has already been mentioned (above) in conjunction with other phenomena (above) occurs immediately upon (or very shortly after) spillage of crude oil on the sea surface. The speed at which spreading takes place depends on the viscosity of the oil, the volume of crude oil spilled, and the temperature of the water. The more fluid, low viscosity crude oils spread more quickly than those with a high viscosity and fluid crude oil initially spreads as a coherent oil slick, which begins to break up rapidly. On the other hand, heavy oil (highly viscous crude oil (Chapter 1) fragments and, at temperatures below the pour point, crude oil (conventional or heavy) rapidly solidifies (Speight, 2001, 2014a) and spreading is minimized—the oil layer may be inches (many centimeters) thick. Climatic effects, such as wind, wave action, and water turbulence tend to cause oil to form narrow bands or *windrows* parallel to the wind direction. At this stage the properties of the oil become less important in determining slick movement. The rate at which oil spreads or fragments is also affected by tidal streams and currents—the stronger the combined forces, the faster the process.

9.4.2 Persistent Oil and Nonpersistent Oil

Oil chemicals entering the ocean have many fates—volatile constituents are lost by evaporation to the atmosphere while other constituents are degraded broken up by photochemical reactions (photodegradation) and bacteria can also cause degradation (biodegradation). The combination of biological, physical, and chemical processes is usually referred to as weathering and these weathering reactions have different rates depending on the chemical structure of the oil, habitat conditions (such as water temperature or oxygen and nutrient supply), and mixing of the water by wind, waves, and currents. In some spills, oil does not last much beyond weeks to months.

But when oil pours into shallow waters with muddy sediments—such as marshes or lagoons—and conditions allow the oil to become mixed into the mud, it will generally persist for a long time. This is a result of the fundamental chemistry of oil compounds. Since they don't dissolve in water, oil compounds tend to adhere to particles in the water or get incorporated into biological debris, such as fecal matter or dead organisms. These oiled particles and debris settle from the water column and become part of the sediments on the bottom. Once mired in the sediment, some oil chemicals can persist for years or decades, depending on the environment. In areas swept by high-energy currents, the material may be dispersed. In areas where sediments accumulate (such as ship channels through urban harbors), the contaminated sediments become an environmental concern—both when simply lying on the bottom and when channels are dredged and the mud must be disposed.

Heavy oil (or a fractions of the oil) which has a specific gravity greater than sea water at the prevailing temperature (>1.025), causing them to sink once spilled or after evaporation of the more volatile constituents. Most crude oil has a sufficiently low specific gravity to remain afloat unless the constituents interact with and attach to more dense sediment or organic particles. Dispersed oil droplets can interact with sediment particles suspended in the water thus becoming heavier and sinking. However, adhesion to heavier particles mast often takes place when oils strand or become buried on beaches.

On exposed, high energy beaches, large amounts of sediment can be incorporated and the oil can form dense tar mats. Seasonal cycles of sediment build-up may cause oil layers to be successively buried and uncovered. Even on less exposed sandy beaches, stranded oil can become covered by wind-blown sand. Once crude oil has been mixed with beach sediment, it will sink if washed back out to sea by storms, tides or currents. On sheltered shorelines, where wave action and currents are weak, muddy sediments and marshes are common. If crude oil becomes incorporated into such fine grained sediments, it is likely to remain there for a considerable time (persistent oil).

Shallow coastal areas and the waters of river mouths and estuaries are often laden with suspended solids that can bind with dispersed oil droplets, thereby providing favorable conditions for sedimentation of oily particles to the sea bed. Like some heavy crude oil, most heavy fuel oils and water-in-oil emulsions have specific gravities close to that of sea water, and even minimal interaction with sediment can be sufficient to cause sinking. Fresh water from rivers also lowers the salinity of sea water, and therefore its specific gravity—this may be common in deltaic areas—and can encourage neutrally buoyant droplets to sink. Oil may also be ingested by planktonic organisms and incorporated into fecal pellets, subsequently falling to the seabed.

When oil droplets in the water column adhere to very fine sediment particles or particles of organic matter they can form flocculates, which may be widely dispersed by currents or turbulence. Small quantities of oil in sea bed sediments or on beaches may also become attached to such particles and become suspended in the water as flocculates as a result of storms, turbulence or tidal rise and fall. This process (clay-oil flocculation) can result over a period of time in the removal of oil from beaches.

9.5 IMPACT OF NATURAL GAS ON THE OCEAN

Natural gas is closely related to crude oil since both resources were believed to have been formed substances are thought to have formed in the crust of the Earth as a result of transformation of organic matter due to the heat and pressure of the overlying rock (Chapter 1, Chapter 2). For the most part (there are always exceptions), crude reservoirs also contain natural gas, although natural gas is often found as the sole occupant of some reservoirs.

9.5.1 Composition

Like petroleum but with a lesser number of variations because of the lower number of constituents, the composition of natural gas varies widely and depends on the origin, types of precursor, genesis, and location of the deposit, geological structure of the region, and other factors. Natural gas chiefly consists of saturated aliphatic hydrocarbons, that is, methane and its homologues often up to the C_8 hydrocarbons—the deeper the location of gas deposit, the higher the number of methane homologues. Other components commonly found in natural gas are carbon dioxide, hydrogen sulfide, nitrogen, and helium which constitute an insignificant proportion of natural gas composition but in some reservoirs the nonhydrocarbon constituents are present in considerably higher concentrations (Mokhatab et al., 2006; Speight, 2007, 2014a). Furthermore, in gas condensate fields, the concentration of methane homologues is usually considerably higher than the concentration of methane. On the other hand, in natural gas that is associated with crude oil, the concentration of the methane homologues is comparable with the concentration of methane. In addition, substantial amounts of hydrocarbon gases associated with crude oil are dissolved in this oil. During oil extraction, as the pressure goes down, gas comes to the surface of the oil and is released into the environment.

Besides the previously mentioned sources of natural gas (transformations of organic matter in the crust of the Earth, microbial decomposition of organic substances, and reduction of mineral salts), *gas hydrates* are another extremely promising source of gas hydrocarbons on the sea bottom (Mokhatab et al., 2006; Speight, 2014a). From the physicochemical point of view, gas hydrates can be considered as a modification of ice that has a high content of gas—the hydrates are solid crystalline species that look like compressed snow. Hydrates form during the interaction of many components of natural gas (methane, ethane, propane, isobutane, carbon dioxide, and hydrogen sulfide) with water under certain combinations of high pressure and relatively low temperature—the sea bed approximately one mile below the surface has ideal temperature and pressure régimes for the formation of gas hydrates. While gas hydrates are a valuable resource, hydrate formation usually accompanies and complicates gas and crude oil extraction and transportation because hydrates can accumulate on the sides of wells and pipelines, thus causing plugging. The methods used to overcome these difficulties include pumping inhibitors (methanol, glycol, and solutions of potassium chloride) into the wells and pipelines, dehydrating the gas, and heating it up to temperatures higher than the temperature of hydrate formation (Mokhatab et al., 2006; Speight, 2014a).

Similar to oil, gas enters the environment due to both natural and anthropogenic processes. Among the major mechanisms of methane natural production in the biosphere, the decomposition of organic matter by methane-producing bacteria (such as *Methanococcus*, *Methanosarica*) deserves a special mention. These bacteria receive energy from the reaction which causes the reduction of carbon dioxide to methane and water:

$$CO_2 + 4H_2 \rightarrow CH_4 + 2H_2O$$

This type of reaction is typical of the reactions occurring in the silt deposits of lakes and marshes and for marine sediments that are lacking in oxygen but rich in organic matter. Microbial methane formation in the oceans is usually accompanied by sulfur reduction and the release of hydrogen sulfide:

$$S_{organic} + 2[H] \rightarrow H_2S$$

The sulfur-reducing reactions occur in the upper part of sediments at and below the seafloor. In regions with a cold and moderate climate at depths of over 1,600 ft, methane can accumulate in a form of gas hydrates. In areas with a warmer climate, some methane from shallow formations is often released from the sediments into the water column and then into the atmosphere.

Methane can appear in the marine environment not only due to microbial and biochemical decomposition of the organic substance in bottom sediments. This gas can also occur as a result of the natural bottom seepage of combustible gases from shallow crude oil-bearing and natural gas-bearing structures—for example, in the Gulf of Mexico, the North Sea, the Black Sea, and the Sea of Okhotsk as well as other marine areas. This process can lead to intensive vertical flows of hydrocarbon gases from the bottom to the sea surface. Sometimes it is accompanied by gas hydrate decomposition that can be explosive in nature.

Hydrogen sulfide (H_2S)—another component of natural gas—is water soluble in contrast to the water-insolubility of methane. Hydrogen sulfide can cause hazardous pollution situations in both the atmosphere and the water environment and this gas belongs to the group of poisons with acute effects. The appearance of hydrogen sulfide in the atmosphere and hydrosphere can cause serious economic damage and medical problems among local population.

An important anthropogenic source of gaseous hydrocarbons in the water environment is the offshore drilling accidents and the environmental consequences can be extremely hazardous. Another potential source of gaseous hydrocarbons in the hydrosphere is damaged gas pipelines, both on the seafloor and on land where they cross over rivers and other water bodies. The causes of such damage can vary from corrosion processes (Chapter 8) to natural disasters (severe ice conditions, seismic activity, and earthquakes). Possible pipeline damage can lead to hazardous environmental impacts on water ecosystems.

9.5.2 Toxicity

The toxicity data on different gaseous poisons can help to reveal some general features of interaction between gaseous traces and marine organisms (Patin, 1999). Furthermore, during toxicological studies of different gases, including methane and its derivatives, consideration must be given to the influence of other factors (especially temperature and oxygen regime) that can radically change the direction and symptoms of the effect. In particular, increasing temperature usually intensifies the toxic effect of practically all substances on faunal species (including humans) because of the direct correlation between the level of metabolism and temperature. Thus, toxic substances may have a lesser effect at lower temperatures (not proven for all faunal species) but will become lethal with increasing temperature. This circumstance should be taken into consideration during environmental assessment of the potential impact of natural gas and other toxic gases, especially when studies are conducted in cooler regions. In such regions, methane hydrates may be accumulated during the winter and dissociate during the increased temperatures in the summer which may be followed by the releasing of free methane with corresponding environmental consequences.

Another critical environmental factor that directly influences the impact of natural gas (or natural gas constituents) on aqueous organisms is the concentration of dissolved oxygen. This decrease sometimes depends more on the species characteristics and the rate of their gas metabolism rather than on the nature of the poison. Clearly, such effects are of special interest when interpreting the data on faunal response to natural gas in situations of significant change in the oxygen regime (e.g., during eutrophication of water bodies or seasonal and weather variations of the oxygen content.

9.6 ATMOSPHERIC EMISSIONS

Atmospheric emissions accompany most of the crude oil and natural gas operations and according to some estimates up to 30% of the hydrocarbons emitted into the atmosphere during well testing precipitate onto the sea surface and create distinctive and relatively unstable slicks around the offshore installations. In addition, the constant or periodical burning of associated gas and excessive amounts of hydrocarbons during well testing and development as well as flaring to eliminate gas from the storage tanks and pressure-controlling systems (Kingston, 1992). In addition, the combustion of gaseous fuels and liquid fuels in the energy-generating units (such as diesel-powered generators and pumps, gas turbines, internal combustion engines) on the platforms, ships, and onshore facilities also produces emissions. Finally, there are also effects from the evaporation or venting of hydrocarbons during different operations of their production, treatment, transportation, and storage.

In spite of the fact that some countries now prohibit flaring of crude oil-associated gases, it remains one of the major sources of atmospheric emissions in the world. Components of atmospheric pollution caused by

oil and gas development include gaseous products of hydrocarbon evaporation and burning as well as aerosol particles of the unburned fuel. From
the ecological perspective, the most hazardous components are nitrogen
and sulfur oxides, carbon monoxide, and the products of the incomplete
burning of hydrocarbons. The oxides of nitrogen, oxygen and carbon (primary pollutants) interact with atmospheric moisture, transform under the
influence of solar radiation, and precipitate onto the land and sea surfaces
to form acid rain (secondary pollutants—also known as acid deposition)
(Speight, 2014a):

$$SO_2 + H_2O \rightarrow H_2SO_3 \text{ (sulfurous acid)}$$
$$SO_3 + H_2O \rightarrow H_2SO_4 \text{ (sulfuric acid)}$$
$$NO + H_2O \rightarrow HNO_2 \text{ (nitrous acid)}$$
$$3NO_2 + 2H_2O \rightarrow HNO_3 \text{ (nitric acid)}$$
$$CO_2 + H_2O \rightarrow H_2CO_3 \text{ (carbonic acid)}$$

Technical means to rectify and prevent atmospheric pollution during
offshore oil and gas production are practically identical to the analogous
methods that are widely and often effectively used on land and in other
industries.

9.7 ENVIRONMENTAL MANAGEMENT

Offshore oil production involves environmental risks—most notably oil spills, both from oil tankers transporting oil from the platform to
onshore facilities, pipelines doing the same and leaks and accidents on the
platform as well as emission of gases to the atmosphere—and assiduous environmental management of such risks is necessary (Kloff and Wicks, 2004).
There is also the impact of produced water, which is excess water from well
drilling or production and which contain varying amounts of oil, drilling
fluid or other chemicals used in or resulting from oil production.

9.7.1 Hurricanes

Every year, hurricanes (so-called in the Atlantic Ocean) and cyclones or
typhoons (so-called in the Pacific Ocean) of varying strength lay siege to
various coastal regions of the United States and other countries. For example, the impact of the 2005 storms on the vital energy infrastructure of
the Outer Continental Shelf of the United States was severe, resulting in
the damage and/or destruction of more than 100 offshore platforms, which

does not include serious damage to another 52 platforms and causing serious disruption in the production of crude oil and f the nation's oil and natural gas production.

The offshore industry continues to work to improve the safety and environmental records, as well as improve the time needed to bring shut-in production back to consumers (Sutton, 2012). Also, the offshore industry contends with hurricane forces—the structural design innovations, the safeguards for employees and the environment, and the improvements being made in advance of future storms.

9.7.2 Decommissioning

When the in-service life and efficiency of an offshore oil and gas production facility is at an end and the facility has produced as much oil and gas as is technically and economically feasible on the basis of the best available techniques, oil field practices typically require that the environment should be restored to approximately its pre-exploration condition.

Decommissioning (also called *abandonment*) is the process by which the owner-operator of an offshore oil or gas installation will plan, gain approval for, and implement the removal, disposal, or reuse of an installation when it is no longer needed for its current purpose (Jahn et al., 1998; Ekins et al., 2006). This applies not only to entire field developments at the end of their economic life, but also to individual installations such as wells that will no longer be used while other parts of the field are still producing.

Well abandonment requires the following actions: (1) isolation of all oil-producing and gas-producing intervals, (2) containment of all over-pressurized zones, (3) protection of overlying aquifers, which is particularly important for onshore well-abandonment, and (4) removal of wellhead equipment. The process begins with a process (sometimes called the *well-killing* process) in which the produced fluids are circulated out of the well, pushed back into the formation, and replaced by fluids that are sufficiently heavy to contain any open formation pressure. Once the well is *dead*, the (Christmas) tree is removed and replaced by a blow-out preventer—any production tubing may be removed through the blow-out preventer.

The perforations are then cemented to seal off all production zones. It is normal to set a series of cement plugs in both the production liner and the production casing. The production casing is cut and removed above the top cement, and another cement plug over the casing stub will isolate the annulus and any formation that may still be open below. On the other hand, an offshore platform or facility requires different procedures.

A typical large steel offshore deep-water structure consists of: (1) *topside*, which is the actual platform above the surface of the sea on which offshore activities take place, (2) *jacket*, which supports the topside and is a structure largely of tubular steel, which may be 450 to 500 ft high and weigh 15,000 to 35,000 tons, (3) *footings*, which is the lowest and heaviest section of the jacket, which are considered separately for decommissioning purposes, and (4) a pile of drill cuttings on the seabed beneath the platform, which consists of drilled rock particles and drilling fluids arising from drilling the wells. In addition, there are likely to be pipelines for the export of oil and gas, which may also need to be decommissioned.

The three decommissioning possibilities are as follows: (1) leave *in situ*, (2) shallow disposal, and (3) recovery. *Leaving the structure in situ* after the cleaning of all hydrocarbons—the individual component parts of the structure (such as the topside and jacket) can be considered separately, obviously lower components have to be left *in situ* for this to be considered for a higher component. In the case of *shallow disposal*, this involves dismantling the structure and depositing it onto the seabed around the site of the operational structure. *Recovery* involves the removal and transport to shore, and dismantling and reprocessing or landfilling, of all the components of the structure.

In addition, there are also a number of other decommissioning possibilities, which may be legal in some jurisdictions: (1) monitoring, (2) toppling, and (3) deep-sea disposal. *Monitoring* involves leaving the structure *in situ* with a program of ongoing monitoring of the fate of the abandoned structure and associated materials (such as pipelines and drill cuttings piles). Furthermore, appropriate monitoring will need to be carried out for any materials left *in situ*. It has therefore been considered in the discussion of the options that envisage this, rather than as a separate option. *Toppling* involves performing the minimum tasks required to topple the structure so that it simply lies on its side at the site and may be an option for fixed steel structures, toppling would have many similarities to the shallow disposal option but with moderately lower material, energy and financial requirements. On the other, *deep-sea disposal* involves removing the structure for disposal in the deep ocean, where it would be effectively impossible for there to be any further human interaction with the material comprising it.

Important issues also arise when assessing the decommissioning scenarios for oil and gas pipelines are whether the pipeline is presently trenched or covered, and if so, the status of the coverings and the prospects of the pipelines being uncovered in the future. This is important as exposed pipelines (in addition to other objects on the seabed) can pose a risk to trawling operations. If recovery of pipelines is desired, for smaller flexible pipelines this can be achieved by reeling them in. For less flexible pipelines, they may be cut and made buoyant and towed to shore, or cut and lifted onto a vessel.

The material, energy and value assessment of decommissioning pipelines is similar in form to the assessment of the decommissioning of main structures. The key difference is that the *in situ* scenario may have significant material demands, if the pipeline is to be covered with rocks. A further consideration is the relatively dispersed nature of pipelines, and whether the material, energy and value benefits of recovering such dispersed material justifies the material, energy and financial cost associated with such recovery.

The extremely high cost of decommissioning and removal off offshore installations led to the need to revise some of the national and international regulations that were adopted approximately 40 years ago. Such a revision covered, in particular, the requirement set by the Convention on the Continental Shelf (Treves, 2013) and the United Nations Convention on the Law of the Sea (Chapter 10) to remove abandoned offshore installations totally. At present, a more flexible and phased approach is used. It suggests immediate and total removal of offshore structures (mainly platforms) weighing up to 4,000 tons in the areas with depths less than 250 ft and after 1998—at depths less than 330 ft. In deeper waters, removing only the upper parts from above the sea surface to 180 ft deep and leaving the remaining structure in place is allowed. The removed fragments can be either transported to the shore or buried in the sea. This approach considers the possibility of secondary use of abandoned offshore platforms for other purposes.

From the technical-economic perspective, the larger the structures are and the deeper they are located, the more appropriate it is to leave them totally or partially intact. In shallow waters, in contrast, total or partial structure removal makes more sense—the fragments can be taken to the shore, buried, or reused for some other purposes. However, any options when the structures or their fragments are left on the bottom may cause physical interference with fishing activities. In these cases, the possibility of vessel and gear damages and corresponding losses does not disappear with termination of production activities in the area. Instead, abandoned structures pose the threat to fishing for many decades after the oil and gas operators leave the site. The obsolete pipelines left on the bottom are especially dangerous in this respect—the degradation and uncontrolled dissipation of these pipelines over wide areas may lead to the most unexpected situations occurring during bottom trawling in the most unexpected places. At the same time, national and international agreements about the decommissioning and abandonment of offshore installations refer mostly to large, fixed structures such as the drilling and production platforms.

From many perspectives, the most interesting projects are the ones aimed at converting the fixed marine structures into artificial reefs (Baine, 2001; Ponti et al., 2002). Artificial reefs are usually created to increase fish production and yield, to support recreational activities (diving and angling), and to promote nature conservation and coastal protection (Bohnsack and Sutherland, 1985; Baine, 2001; Svane and Petersen, 2001). These reefs should increase the surface available for settling of sessile organisms (Relini et al., 1994), offer shelter and refuge to a variety of plants and animals, and provide habitats for reproduction and egg deposition (Frazer and Lindberg, 1994; Lozano Alvarez et al., 1994; Bombace et al., 1995). Moreover, many represent an important tourist attraction for both divers and anglers because they host an interesting benthic flora and fauna and act as fish aggregating devices (Bombace et al., 1995).

Thus, artificial reefs are known to be an effective means of increasing the bio productivity of coastal waters by providing additional habitats for marine life and such reefs are not a new idea (Wolfson et al., 1979; Bombace, 1989; Aabel et al., 1996; Ponti et al., 2002). In fact, the concept of creating an artificial reef has become a popular coastal management concept that has been reduced to practice in many countries as they benefit sessile species, benthic fish, and algal growth (Baine, 2001; Ponti et al., 2002). To establish a reef, different shapes and structures are made with concrete, rocks, vessels, plastics, wood, and steel.

Complete or partial removal of steel or concrete fixed platforms that weigh several thousand tons is practically impossible without using explosive materials. Bulk explosive charges have been used in many cases, which are short-term impacts on the marine environment and biota and should not be neglected. However, it is not an easy task to obtain data for reliable estimates of possible mortality of marine organisms, especially fish, during an explosive activity even if the initial data, such as the type of explosive, depth of the water, bottom relief, and others, are known.

When partial removal takes place, portions of the platforms, such as jackets of bases, remain on the seafloor. Even after the complete removal of major structures, many small parts are left, including unused pipes and stabilizer mattresses, which are typically made of concrete. It is reported that these materials left on the seafloor affect the normal benthic habitat. The presence of larger pieces left behind may cause mechanical damage to fishing gear and can create obstacles in shipping navigation. Abandoned offshore oil and gas facilities on the seafloor also create environmental issues as the pieces may contain hazardous wastes such as heavy metal sludge, polychlorobiphenyl fluids, and hydrocarbon compounds. Some of these compounds have been found to remain for a period as long as ten years in the marine environment. The effects of these compounds on marine organisms can be chronic or lethal, depending on their concentrations (Khan and Islam 2005, 2007, 2009; Islam et al., 2010).

9.8 PRESENT TRENDS

Catastrophic spills make the headlines, but it is the chronic dribble, dribble, dribble of seemingly small inputs that supplies most of the oil polluting the world's oceans. In recent decades scientists have made substantial progress in understanding how oil enters the oceans, what happens to it, and how it affects marine organisms and ecosystems. This knowledge has led to regulations, practices, and decisions that have helped us reduce sources of pollution, prevent and respond to spills, clean up contaminated environments, wisely dredge harbors, and locate new petroleum handling facilities. But tracking the sources, fates, and effects of oil in the marine environment remains a challenge for a number of reasons. For starters, oil is a complicated mixture of hundreds, sometimes thousands, of chemicals. Every source of oil, and even the same general types of oil (crude oils or fuel oils, for example) can have distinctive compositions depending on which oil field or well they came out of and how they were refined. This varying, complex mixture of chemicals gets spilled or seeps into an already complex chemical chowder of sea water, mud, and marine organisms in the ocean. There, the oil is stirred by currents, tides, and waves, altered by other physical processes, and changed further by chemical reactions and interactions with organisms in the sea (Farrington and McDowell, 2004).

Between one-third and one-half of the oil in the ocean comes from naturally occurring seeps. These are seafloor springs where oil and natural gas leak and rise buoyantly from oil-laden, sub-seafloor sediments that have been lifted close to the surface of the Earth by natural processes. In fact, since crude oil is a natural product and if oil is natural to the oceans and if it is the biggest source of input, some observers wonder about the question raised about crude oil as a pollutant. The obvious response to such wonderment lies in the locations where spills occur and the crude oil is a nonindigenous material as well as the rate of crude oil input into the ecosystem. Oil seeps are generally old, sometimes ancient, so the marine plants and animals in these ecosystems have had hundreds to thousands of years to adjust and acclimate to the exposure to crude oil constituents. On the other hand, the production, transportation, and consumption of crude oil often results in the input of crude oil as a nonindigenous and hazardous visitor that overstays its welcome to environments and ecosystems that have not experienced significant direct inputs and have not become acclimatized to the crude oil.

In the midst of this dynamic situation, scientists seek to pinpoint the impacts of oil on myriad individual species, as well as on entire ecosystems. Here, the challenge comes full circle because the impact of oil can vary greatly, depending on its distinctive chemical composition. But let's start at the beginning.

Progress in prevention—through more stringent laws, rules, and guidelines, and increased vigilance by industry and regulations (Chapter 10)—has reduced accidental spills, at least in developed countries. For instance, studies of tanker spills have prompted regulations for the steady, ongoing replacement of single-hulled tankers in the world fleet with double-hulled tankers.

Finally, the cost of decommissioning an offshore structure is high. Offshore oil and gas fields (like onshore fields) tend to reach a plateau level of production, which can be maintained for a number of years, after which production gradually decreases. Declining oil and gas production, increased water production, and increasing maintenance costs of aging facilities have a major (detrimental) effect on the economics of production. The decision becomes whether or not production from a field should be discontinued and the facilities decommissioned when production has declined to a point when the economics of operating the field is no longer viable and all options for improving production have been exhausted. This is a time when enhanced recovery operations might be considered.

Enhanced recovery processes such as enhanced oil recovery and improved oil recovery should be considered as a means of deferring decommissioning by increasing the hydrocarbon production (Speight, 2009, 2014a). Such methods are often very sensitive to the oil price, and while common for onshore developments, they are harder to justify offshore. In some cases, there may be incentives to produce oil and gas beyond the point when production revenues no longer meet expenses. However, decommissioning is a substantial expense for any offshore project and revenues for enhanced recovery operations might provide some financial relief.

REFERENCES

Aabel, J.P., Cripps, S., Kjeilen, G., 1996. Oil and Gas Production Structures as Artificial Reefs. In: Jensen, A.C. (Ed.), Proceedings. European Artificial Reef Research. Proceedings. 1st EARRN Conference, Ancona, Italy, 26–30 March 1996, National Oceanography Center, University of Southampton Waterfront Campus, Southampton, United Kingdom, pp. 391–404.

Ahmad, A.L., Sumathi, S., Hameed, B.H., 2005. Residual oil and suspended solid removal using natural adsorbents chitosan, bentonite and activated carbon: A comparative study. Chemical Engineering Journal 108, 179–185.

Al-Majed, A.A., Adebayo, A.R., Hossain, M.E., 2011. A novel sustainable oil spill control technology. Environmental Engineering and Management Journal 10, 333–353.

Annis, M.R., 1997. Retention of Synthetic-Based Drilling Material on Cuttings Discharged to the Gulf of Mexico. Report for the American Petroleum Institute (API). Ad hoc Retention on Cuttings Work Group under the API Production Effluent Guidelines Task Force. American Petroleum Institute, Washington, DC, August 29.

Atlas, R.M., Cerniglia, C.E., 1995. Bioremediation of petroleum pollutants. Bioscience 45, 332–339.

Baine, M., 2001. Artificial reefs: A review of their design, application, management and performance. Ocean & Coastal Management 44, 241–259.

Bell, N., Cripps, S.J., Jacobsen, T., Kjeilan, G., Picken, G.B., 1998. Review of Drill Cuttings Piles in the North Sea. A Report for the Offshore Decommissioning Communications Project. Report No. Cordah/ODCP.004/1998 (Final). Cordah Environmental Consultants, Aberdeen, United Kingdom.

Bohnsack, J.A., Sutherland, D.L., 1985. Artificial reef research: A review with recommendations for future priorities. Bulletin of Marine Science 37, 11–39.

Bombace, G., 1989. Artificial reefs in the Mediterranean Sea. Bulletin of Marine Science 44, 1023–1032.

Bombace, G., Castriota, L., Spagnolo, A., 1995. Benthic communities on concrete and coal-ash blocks submerged in an artificial reef in the central Adriatic Sea. Proceedings. 30th European Marine Biological Symposium, Southampton, United Kingdom, September, pp. 281–290.

Burns, K., Dahlman, G., Gunkel, W., 1993. Distribution and activity of petroleum hydrocarbon degrading bacteria in the North and Baltic Seas. Deutsche Hydrographisches Zeitschrift 6, 359–369.

CAPP, 2001. Technical Report. Offshore Drilling Waste Management Review. Report 2001-0007. Canadian Association of Petroleum Producers, Halifax, Nova Scotia, Canada.

Choi, H.M., Cloud, R.M., 1992. Natural sorbents in oil spill cleanup. Environmental Science and Technology 26, 772–776.

Clark, R.B., 1999. Marine Pollution, 4th Edition Oxford University Press, Oxford, United Kingdom.

Cojocaru, C., Macoveanu, M., Cretescu, I., 2011. Peat-based sorbents for the removal of oil spills from water surface: Application of artificial neural network modeling. Colloids and Surfaces A: Physicochemical and Engineering Aspects 384, 675–684.

Connell, D.W., Miller, G.J., 1981. Petroleum hydrocarbons in aquatic ecosystems behavior and effects of sub-lethal concentrations. Critical Reviews in Environmental Control 11, 105–162.

Daan, R., Mulder, M., 1993. A Study on Possible Environmental Effects of a WBM Cutting Discharge in the North Sea, One Year After Termination Of Drilling. NIOZ Report No. 1993-16. Royal Netherlands Institute for Sea Research (NIOZ), Amsterdam, Netherlands.

Daan, R., Mulder, M., 1996. On the short-term and long-term impact of drilling activities in the Dutch Sector of the North Sea. ICES Journal of Marine Science 53, 1036–1044.

Dicks, B.M., 1982. Monitoring the biological effects of North Sea platforms. Marine Pollution Bulletin 13, 221–227.

Dulfer, J.W., 1999. OBM Drill Cuttings Discharges: Assessment Criteria. Report No. RIKZ-99.018. Ministry of Transport, Public Works and Water Management Directorate-General of Public Works and Water Management National Institute for Coastal and Marine Management/RIKZ, Amsterdam, Netherlands.

EEA, 2008. EN14 Discharge of Oil from Refineries and Offshore Installations. European Environment Agency, Copenhagen, Denmark.

Ekins, P., Vanner, R., Firebrace, J., 2006. Decommissioning of offshore oil and gas facilities: A comparative assessment of different scenarios. Journal of Environmental Management 79 (4), 420–438.

Engelhard, F.R., Ray, J.P., Gillam, A.H. (Eds.), 1989. Drilling Wastes. Elsevier, Amsterdam, Netherlands.

Etkin, D.S., 2009. Analysis of U.S. Oil Spillage. Publication No. 356. American Petroleum Institute, Washington, DC, August.

Farrington, W., McDowell, J.E., 2004. Tracking the Sources and Impacts of Oil Pollution in the Marine Environment43Oceanus Magazine, Woods Hole Oceanographic Institution, No. 1.

Frazer, T.K., Lindberg, W.J., 1994. Refuge spacing similarly affects reef-associated species from three phyla. Bulletin of Marine Science 55, 388–400.

Gary, J.G., Handwerk, G.E., Kaiser, M.J., 2007. Petroleum Refining: Technology and Economics, 5th Edition CRC Press, Taylor & Francis Group, Boca Raton, Florida.

Gerard, A.L.D., 1996. Laboratory Investigations on the Fate and Physicochemical Properties of Drilling Cuttings after Discharged into the Sea. Physical and Biological Effects of Processed Oily Drill Cuttings, E&P Forum Report, Paper 3, 16–24.

Gerrard, S., Grant, A., Marsh, R., London, C., 1999. Drill Cuttings Piles in the North Sea: Management Options During Platform Decommissioning. Research Report No 31. Centre for Environmental Risk, University of East Anglia, Norwich, United Kingdom.

Gray, J.S., Clarke, K.R., Warwick, R.M., Hobbs, G., 1990. Detection of initial effects of pollution on marine benthos: an example from the Ekofisk and Eldfisk oilfields, North Sea. Marine Ecology Progress Series 66, 285–299.

Hartley, J., Neff, J., Fucik, K., Dando, P., 2003. Drill Cuttings Initiative. Food Chain Effects: Literature Review. United Kingdom Offshore Operators Association, Aberdeen, Scotland.

Hird, S.J., Tibbetts, P.J.C., 1995. An Examination of Biodegradation of Aliphatic Hydrocarbons from Oil-Based Drilling Muds. In: The Physical, Biological Effects of Processed Oily Drill, Cuttings, MetOcean Report No. 2.61/202 to the E&P Forum, 104-114, MetOcean Consultants, New Plymouth, New, Zealand.

Hsu, C.S., Robinson, P.R. (Eds.), 2006. Practical Advances in Petroleum Processing, Volume 1 and Volume 2, Springer Science, New York.

Hyland, J., Hardin, D., Steinhauer, M., Coats, D., Green, R., Neff, J., 1994. Environmental impact of offshore oil development on the outer continental-shelf and slope off Point Arguello, California. Marine Environmental Research 37, 195–229.

Islam, M.R., Chhetri, A.B., Khan, M.M., 2010. The Greening of Petroleum Operations: The Science of Sustainable Energy Production. Scrivener Publishing, Salem, Massachusetts.

Jahn, F., Cook, M., Graham, M., 1998. Hydrocarbon Exploration and Production. Elsevier, Amsterdam, Netherlands.

Khan, M.I., Islam, M.R., 2005. Sustainable wealth generation: community based offshore oil and gas operations. Proceedings. CARICOSTA 2005. International Conference on Integrated Coastal Zone Management, University of Oriente, Santiago de Cuba, May 11–13.

Khan, M.I., Islam, M.R., 2007. Handbook of Sustainable Oil and Gas Engineering Operations Management. Gulf Publishing Company, Austin, Texas, Page 18–29.

Khan, M.I., Islam, M.R., 2009. Technological analysis and quantitative assessment of oil and gas development on the Scotian Shelf, Canada. Advances in Sustainable Petroleum Engineering Science 1 (3), 247–266.

Kingston, P.F., 1992. Impact of offshore oil production installations on the benthos of the North Sea. Science 49, 49–53.

Kloff, S., Wicks, C., 2004. Environmental Management of Offshore Oil Development and Maritime Oil Transport: A Background Document for Stakeholders of the West African Marine Eco Region. CEESP (IUCN Commission on Environmental, Economic and Social Policy), International Union for the Conservation of Nature, Gland, Switzerland.

Lozano Alvarez, E., Briones Fourzan, P., Negrete Soto, F., 1994. An Evaluation of Concrete Block Structures as Shelter for Juvenile Caribbean Spiny Lobsters, *Palinurus argus*. Bulletin of Marine Science 55, 351–362.

Maurer, D., Keck, R.T., Tinsman, J.C., Leathem, W.A., 1980. Vertical Migration and Mortality of Benthos in Dredged Material. Part 1: Mollusca. Marine Environmental Research 4, 299–319.

Maurer, D., Keck, R.T., Tinsman, J.C., Leathem, W.A., 1981. Vertical Migration and Mortality of Benthos in Dredged Material. Part II - Crustacea. Marine Environmental Research 5, 301–317.

Maurer, D., Keck, R.T., Tinsman, J.C., Leathem, W.A., 1982. Vertical Migration and Mortality of Benthos in Dredged Material: Part III - Polychaeta. Marine Environmental Research 6, 49–68.

Maurer, D., Keck, R.T., Tinsman, J.C., Leathem, W.A., Wethe, C., Lord, C., Curch, T.M., 1986. Vertical migration and mortality of marine benthos in dredged material: A synthesis. Internationale Revue der Gesamten Hydrobiologie 71, 49–63.

McFarlane, K., Nguyen, V.T., 1991. The Deposition of Drill Cuttings on the Seabed. Paper SPE 23372. Proceedings. 1st International Conference on Health, Safety and Environment. Society of Petroleum Engineers, The Hague, Netherlands.

Menzie, C.A., 1984. Diminishment of Recruitment: A Hypothesis Concerning Impacts on Benthic Communities. Marine Pollution Bulletin 15, 127–128.

Mokhatab, S., Poe, W.A., Speight, J.G., 2006. Handbook of Natural Gas Transmission and Processing. Elsevier, Amsterdam, Netherlands.

Moore, C.G., Murison, D.J., Long, S., Mills, D.J.C., 1987. The impact of oily discharges on the Meiobenthos of the North Sea. Philosophical Transaction of the Royal Society of London B 316, 525–544.

Neff, J.M., Rabalais, N.N., Boesch, D.F., 1987. Offshore oil and gas development activities potentially causing long-term environmental effects. In: Boesch, D.F., Rabalais, N.N. (Eds.), Long-Term Environmenal Effects of Offshore Oil and gas Development. Elsevier Applied Science Publishers, New York.

Olsgård, F., Gray, J.S., 1995. A comprehensive analysis of the effects of offshore oil and gas exploration and production on the benthic communities of the Norwegian continental shelf. Marine Ecology Progress Series 122, 277–306.

Patin, S., 1999. Environmental Impact of the Offshore Oil and Gas Industry. EcoMonitor Publishing, East Northport, New York.

Payne, J.F., Kiceniuk, U., Williams, L.F., Melvin, W., 1989. Bioaccumulation of polycyclic aromatic hydrocarbons in flounder (*Pseudopleuronectes americanus*) exposed to oil-based drill cuttings. In: Engelhard, F.R., Ray, J.P., Gillam, A.H. (Eds.), Drilling Wastes. Elsevier, Amsterdam, Netherlands, pp. 427–438.

Peterson, C.H., Kennecott, II, M.C., Green, R.H., Montagna, P., Harper, Jr., D.E., Powell, E.N., Roscigno, P.F., 1996. Ecological consequences of environmental perturbations associated with offshore hydrocarbon production: a perspective on long-term exposures in the Gulf of Mexico. Canadian Journal of Fisheries and Aquatic Sciences 53, 2367–2654.

Ponti, M., Abbiati, M., Ceccherelli, V.U., 2002. Drilling platforms as artificial reefs: Distribution of macrobenthic assemblages of the Paguro Wreck (Northern Adriatic Sea). ICES Journal of Marine Science 59, S316–S323.

Relini, G., Zamboni, N., Tixi, F., Torchia, G., 1994. Patterns of sessile macrobenthos community development on an artificial reef in the Gulf of Genoa (Northwestern Mediterranean). Bulletin of Marine Science 55, 745–771.

Speight, J.G., 2001. Handbook of Petroleum Analysis. John Wiley & Sons Inc., Hoboken, New Jersey.

Speight, J.G., 2005. Environmental Analysis and Technology for the Refining Industry. John Wiley & Sons Inc., Hoboken, New Jersey.

Speight, J.G., 2007. Natural Gas: A Basic Handbook. Gulf Publishing Company, Houston, Texas.

Speight, J.G., 2009. Enhanced Recovery Methods for Heavy Oil and Tar Sands. Gulf Publishing Company, Houston, Texas.

Speight, J.G., Arjoon, K.K., 2012. Bioremediation of Petroleum and Petroleum Products. Scrivener Publishing, Salem, Massachusetts.

Speight, J.G., 2014a. The Chemistry and Technology of Petroleum, 5th Edition CRC Press, Taylor & Francis Group, Boca Raton, Florida.

Speight, J.G., 2014b. Handbook of Petroleum Product Analysis, 2nd Edition John Wiley & Sons Inc., Hoboken, New Jersey.

Speight, J.G., 2014c. Oil and Gas Corrosion Prevention. Gulf Professional Publishing, Elsevier, Oxford, United Kingdom.

Sutton, I., 2012. Offshore Safety Management. Elsevier, Amsterdam, Netherlands.

Svane, I., Petersen, J.K., 2001. On the problems of epibioses, fouling and artificial reefs: A review. Marine Ecology – Pubblicazioni delia Stazione Zoologica di Napoli I 22, 169–188.

Treves, T., 2013. 1958 Geneva Conventions on the Law of the Sea, Geneva, 29 April 1958. http://legal.un.org/avl/ha/gclos/gclos.html (accessed June 6, 2014).

Wills, J.W.G., 2000. Muddied Waters – A Survey of Offshore Oilfield Drilling Wastes and Disposal Techniques to Reduce the Ecological Impact of Sea Dumping. Ekologicheskaya Vahkta Sakhalina, Yuzhno-Sakhalinsk, Russia.

Windom, H.L., 1992. Contamination of the marine environment from land-based sources. Marine Pollution Bulletin 25, 1–4.

Wolfson, A., Van Blaricom, G., Davis, N., Lewbe, G.S., 1979. The marine life of an offshore oil platform. Marine Ecology Progress Series 1, 81–89.

CHAPTER TEN

Legislation and The Future

Outline

10.1 INTRODUCTION

In order for crude oil and natural gas to be recovered from onshore reservoirs, and beneath-water reservoirs in a rational manner, the oceans (or waterways) and submerged lands need to be defined and managed in terms of the various boundaries (Hedberg, 1976). From the start of the exploration and development of subsea crude oil and natural gas resources (suing the mid-1930s to the mid-1950s period as the example of the start of modern work), there was a series of legal claims by the Federal Government of the United States and by various state governments regarding the control of submerged lands adjacent to the states (within the United States) that occupy shorelines which were predominant issues in defining the offshore resources. This back-and-forth debate (the so-called *Tidelands Controversy*) was eventually settled by a series of decisions by the Supreme Court of the United States during the period 1947 to 1960 which, in summary, granted control of the resources that lay further than three miles from the coastline.

Thus, legislation and regulations regarding crude oil and natural gas exploration, development, and production of the resources from US offshore

Handbook of Offshore Oil and Gas Operations. http://dx.doi.org/10.1016/B978-1-85617-558-6.00010-6

lands developed over several decades (and not by a series of overnight decisions) in response to the concerns of governments and the affected populations. In addition, and as the time period progressed with more and more development taking place, there were (and still remain) various environmental issues that required federal governmental (Congressional) action as well as legislative action by relevant state governments leading to passage of laws relating to offshore development—especially on items such as ecological damage (Dudley and Stolton, 2002).

Within national jurisdictions of the shoreline countries—especially those shores within the jurisdiction of the United States—the offshore crude oil and natural gas industry has been growing at an unprecedented remarkable pace. Depletion of known onshore resources and instability in oil prices as well as some instability in the price of natural gas—until the discovery of the resources known as shale gas (Speight, 2013)—has led to the need to discover and develop new sources of crude oil-based and natural gas-based energy. Worldwide, offshore crude oil and natural production has grown significantly from the 1980s to the present and current production of offshore oil accounts for approximately 30% (v/v) of the total world crude oil production, while the share of the offshore gas industry in world natural gas production is a little higher.

In recent decades, due to increasing world demand for crude oil and natural gas, offshore exploration and development have shifted to new frontiers: the Gulf of Mexico, the North Sea, and offshore West Africa, offshore South-East Asia, offshore South America, and offshore of a variety of Arctic coastline countries—have become the focus of exploration and development activity. Indeed, the traditional onshore sources of crude oil and natural gas, the unseeded reservoirs and exhibited the reality and further potential of new sources of energy, such as Arctic regions.

However, Arctic regions pose a unique operating environment characterized by remoteness, the lack of ancillary supporting infrastructure, the presence of sea ice, extended periods of darkness and cold, and hurricane-strength storms. In addition, a diverse natural ecosystem and the presence of indigenous communities call for the highest standard of environmental protection and responsible development. These factors, along with regulatory uncertainties, add considerable risk and thus cost to exploiting offshore oil and gas. Although this reality recently has tempered the enthusiasm of some oil and gas companies and even cast some development plans in doubt, there is broad agreement that there will be increased offshore hydrocarbon activity in the future. The key issue is whether or not the US government will be prepared to meet the challenges posed by this activity (Ebinger et al., 2014).

This chapter presents an overview of the evolution of offshore developments and the major legislation and regulations that have affected offshore crude oil and natural gas development in the past half century. The most common early disputes revolved around ownership of coastal waters and as the development of offshore activities became more numerous and prominent other issues arose over the need to ensure that offshore operations include the protection of marine and coastal environments as well as worker safety (Rundmo, 1994, 1997; Mearns et al., 1998).

10.2 ENVIRONMENTAL REGULATIONS

From the 1880s, when offshore oil production first began (Chapter 2) until the mid-1900s, there was little or no regulation of crude oil and natural gas exploration and development activities in offshore regions by the US Federal government. During this time, technological advances and increasing demand for crude oil and natural gas provided incentives for offshore exploration and the development of an infrastructure related to offshore crude oil and natural gas production. By 1949, 11 offshore fields had been found and 49 production wells were operating in the Gulf of Mexico and it was in the 1950s that the US government began to respond to increased concerns regarding issues such as offshore jurisdiction, economic factors, safety, and especially (Chapter 9) the environmental impact of offshore activities.

10.2.1 Environmental Legislation

There is a variety of environmental risks that are associated with offshore crude oil and natural gas and oil exploration and production, among them such as incidents relating to discharges or spills of toxic materials whether intentional or accidental, interference with marine life, damage to coastal habitats owing to construction and operations of producing infrastructure, and effects on the economic base of coastal communities (Chapter 9) (Hopkins, 2012; Mannan et al., 2014a,b).

During the 1960s, increasing environmental awareness set the stage for the development of numerous environmental laws, regulations, and executive orders that have affected crude oil and natural gas in offshore reservoirs. In the United States, as in the European Union and many other countries (Von Sydow, 2014), all activities related to crude oil and natural gas exploration and development must now pass through a number of environmental reviews by the respective government agencies. In the United States, federal agencies that play a role in regulating and coordinating environmental laws include the Department of the Interior (DOI), the Environmental Protection Agency (EPA), the National Oceanic and Atmospheric Administration – Department of Commerce, and the Fish and Wildlife Service. Several major pieces of environmental legislation have been enacted by the US Congress in the past several decades to safeguard the environment and protect coastal and marine communities. Furthermore, as technological advances have increasingly allowed the crude oil and natural gas industry to explore and produce in deeper water and from farther beneath the ocean floor over the past half-century (Chapter 3, Chapter 4), developments in the management of these submerged lands have aimed to balance the conflicting interests and needs associated with these activities. Environmental laws and regulations have also had a large impact on the crude oil and natural gas industry. These are notably: (1) the National Environmental Policy Act, (2) the Clean Air Act (CAA), (3) the Coastal Zone Management Act, (4) the Endangered Species Act (ESA), and (5) the Clean Water Act (CWA).

The *National Environmental Policy Act* requires that the Federal Government to consider the environmental impacts of any proposed actions as well as reasonable alternatives to those actions. Through management techniques such as Environmental Assessments, Environmental Impact Statements (EIS), and Categorical Exclusion Reviews, companies who propose an offshore project can better understand and make decisions on how to manage for environmental consequences. An EIS is prepared for every lease sale.

In terms of the *CAA*, all air pollutants resulting from industrial activities were first regulated at the Federal level by the CAA. Proposed and existing crude oil and natural gas facilities must prepare, as part of their development plans and reporting procedures, detailed emissions data to prove compliance with the CAA. The amendments added in 1977 and 1990 set new attainment goals for ambient air quality and updated the Act to account for issues such as acid rain and ozone. Furthermore, the amendments of 1990 established jurisdiction of offshore regions regarding regulation of air quality.

The Coastal Zone Management Act is based on the perceived need to preserve, protect, develop, and restore or enhance the resources of US coastal zones. This Act encourages coastal States to complete an individual Coastal Zone Management Plan for their coastal areas and requires state review of federal actions that affect land and water use in these coastal areas. The consistency clause of this Act gives individual states the power to object to any Federal action that they deem not consistent with their approved Coastal Zone Management Plan.

The ESA protects and promotes the conservation of all species listed as endangered by restricting Federal actions that are likely to harm, harass, or pursue them. Under this Act, plant and animal species can be listed as facing potential extinction after a detailed legal process. The list includes marine and coastal species that could be affected by natural gas and oil operations in the offshore. In 1995, the Supreme Court of the United States ruled that *significant habitat modification* was a reasonable interpretation of the term *harm*. The Act can therefore affect crude oil and natural operations in all areas near or where habitats exist that are considered critical to the designated marine species.

The CWA is the primary law governing the discharge of pollutants into all US surface waters. Under this law, the EPA requires that a National Pollutant Discharge Elimination System (NPDES) permit be obtained before any pollutant is released. The CWA holds certain industries, including crude oil and natural production, to strict standards regarding direct pollution discharges into waterways. These standards are outlined in the NPDES permits and may be based on the age of a facility. For example, new facilities may be subject to more strict standards than existing facilities and, since the permits are issued on a 5-year basis, crude oil and natural gas companies must renew their NPDES permits every 5 years or face penalties handed down by the EPA.

Offshore natural gas and oil exploration, drilling, production, and transportation have all been affected and legislative action has ranged from imposition of a wide range of requirements on operations related to the development of offshore resources.

10.2.2 Maritime Oil Transport

Virtually all aspects of maritime traffic are covered by international conventions—vessels navigate around the globe and an accident could impact on the environment at any point and at any time. Environmental regulation of this sector of offshore resource development on an international level is therefore highly necessary. In fact, in addition to international legislation some countries have written extra stringent regulations for ships that trade in their *exclusive economic zone* (EEZ)—the 200 nautical mile (230 miles) zone as defined by the United Nations Convention on Law of the Sea (IMO, 2007). For example, the USA and countries of the European Union will no longer accept any single-hulled oil tankers in their ports and will not allow such oil tankers to load oil from their offshore facilities. However, the governments of coastal countries have no jurisdiction over international vessels that are on *innocent passage* through the EEZ—vessels that do not trade in that zone and constitute no acute environmental hazard. However, Article 25, paragraph 3, of the United Nations Convention on the Law of the Sea of 10 December 1982 stipulates that a coastal country may, without discrimination in form or in fact among foreign ships, suspend temporarily, in specified areas of its territorial sea the innocent passage of foreign ships if such suspension is essential for the protection of its security, including weapons exercises. Such suspension takes effect, according to the same article, only after having been duly published (UNOLS, 2012).

Vessels with a double-hull configuration provide a significant degree of reduction in risk of oil spills in the event of relatively low impact collision or grounding. There are opinions that double-hull tankers are not necessarily the answer to safer shipping. The concerns relate to increased corrosion and more work to inspect larger surface areas during regular maintenance. It is also thoughts that they argue that poorly designed, constructed, maintained and operated double-hull tankers have as much if nor more potential for disaster than their single-hulled predecessors. Well maintained, well operated, high quality tankers, whatever hull configuration, are to be preferred. Another concern is that a double hull is not a guarantee of non-spillage of the cargo and a high-impact collision will also perforate cause destructive hull perforation of a double-hulled tanker.

10.2.2.1 Historical Perspective

Maritime traffic has been operative for centuries but there was little, if any, regulation of the behavior of the traffic and protocols for behavior were not in place. However, with the rapidly developing and increasing ship transport in the 19th century and in the 20th century (Table 10.1), it became evident that internationally agreed rules had to be put in place. The most important international treaty concerning maritime safety–in response to the sinking of the Titanic, a British passenger liner that sank in the North Atlantic Ocean in the early morning of April 15 1912 after colliding with an iceberg during her maiden voyage from Southampton to New York— was the International Convention for the Safety of Life at Sea (SOLAS) first signed in 1914. The next stage, in 1948, was the setup of an international agency under the auspices of the United Nations that was devoted solely to maritime traffic. Furthermore, maritime safety was one of the main tasks that fell under the wing of the International Maritime Organization (IMO), which started with the adoption of the SOLAS convention. During the

Table 10.1 Timeline for Regulating Maritime Transport of Oil and Offshore
Oil Development.

Year	Action
1958	The First United Nations Conference on the Law of the Sea convened.
1960	The Second United Nations Conference on the Law of the Sea convened.
1968	The Committee on the Peaceful Uses of the Seabed and the Ocean Floor beyond the Limits of National Jurisdiction established.
1970	The General Assembly adopts the Committee's Declaration of Principles and convenes the Third United Nations Conference on the Law of the Sea.
1973	The Third United Nations Conference on the Law of the Sea opens.
1975	The first draft of the Convention on the Law of the Sea submitted to delegations for negotiation.
1982	The Convention adopted by the Third United Nations Conference on the Law of the Sea.
1982	The Convention opened for signature at Montego Bay, Jamaica.
1994	The Convention enters into force.
1995	The International Seabed Authority becomes operational.
1996	The International Tribunal for the Law of the Sea becomes operational.
1997	The Commission on the Limits of the Continental Shelf holds first session.

Source: www.un.org/Depts/los/convention_agreements/convention_20years/oceanssourceoflife.pdf

1950s, ships increased in size and cargo vessels had moved from sail power to motorized power and, it was also in this decade that the maritime transport of crude oil and natural gas increased frequency and importance.

At that time—the decade of the 1950s—maritime oil pollution became an obvious problem in the 1950s. Crude oil-carrying tankers discharged significant quantities of oily wastewater and oil pollution of the oceans and waterways became an important issue for the IMO to tackle. In 1954, a treaty was adopted to deal with the problem—the International Convention for the Prevention of Pollution of the Sea by Oil and in 1959, the IMO took over responsibility for this treaty and the potential disaster of crude oil shipment came to the fore when the tanker *Torrey Canyon* ran aground off the coast of the United Kingdom and spilled more than 750,000 barrels of crude oil into the sea. Until that time, it has been assumed (without any justification or evidence to support the assumption) that the oceans were sufficiently large to cope with any pollution caused by human activity. Since that time, the IMO has developed numerous measures to combat marine pollution, including measures to prevent dumping into the seas of wastes generated by land-based activities.

10.2.2.2 Discharge of Oily Wastes

The International Convention for the Prevention of Pollution from Ships (MARPOL) is the main international convention covering prevention of pollution of the marine environment by ships from operational or accidental causes. The MARPOL Convention was adopted on 2 November 1973 at the IMO. The Protocol of 1978 was adopted in response to tanker accidents that occurred in the 1976–1977 time period. As the 1973, MARPOL Convention had not yet entered into force, the 1978 MARPOL Protocol absorbed the parent Convention. The combined instrument entered into force on 2 October 1983. In 1997, a Protocol was adopted to amend the Convention and a new Annex VI was added which entered into force on 19 May 2005. The Convention includes regulations aimed at preventing and minimizing pollution from ships—both accidental pollution and that from routine operations—and currently includes several technical Annexes.

The main objective of MARPOL is to reduce the discharge of oil products by maritime traffic—during usual operations certain tankers are allowed to discharge a limited amount of oil contained in ballast water and tank washings into the sea. Regulation 9 of MARPOL 73/78 limits the amount of discharge of oil into the sea to 1/30,000 of the total volume of the crude oil cargo. The extra requirement is that the oil content of the discharged effluent cannot exceed 15 ppm (1 mg/L is approximately 1 ppm) and, in effect, limits the operational discharge to amounts much less than the specified maximum value. In addition, the discharge of oil-bearing wastewater within 50 nautical miles (approximately 50.7 onshore miles) from the shoreline is strictly prohibited.

Under regulation 13 of MARPOL 73/78, oil tankers of 20,000 tonnes (22,046 US short tons) deadweight and above are required to have segregated ballast tanks (SBTs), dedicated clean ballast tanks (CBTs), and/or clean oil washing systems (COW), depending on the vessels type, when they were built, and their size. For crude oil carriers of 20,000 and product tankers of more than 30,000 tonnes (33,069 US short tons) deadweight delivered since 1983, it is mandatory to have SBTs. These ballast tanks are completely separated from the cargo oil and fuel oil system and are exclusively allocated to carry ballast water. This system greatly reduces the likelihood of oil-containing ballast water discharge and tankers with a CBT system have a pipe system that may be connected with the crude oil cargo pump and piping system. Discharge of fuel oil sludge from machinery room is strictly forbidden anywhere in the world by MARPOL and the sludge oil should be discharged at reception facilities in ports.

10.2.2.3 Prevention of Accidental Pollution

Without due caution large quantities of oil can end up in the sea after tanker accidents—safer vessels will obviously reduce the risks of accidents. International legislation for making shipping safer is contained in several IMO conventions.

Human failure is another important factor causing maritime accidents—collisions, technical failure and shipboard fires and explosions are all factors that could be caused by human error. It is therefore important that a crew of a tanker should have a thorough technical knowledge and possess the necessary qualifications. To this end, the International Maritime Organization International Convention on Standards of Training, Certification and Watch keeping for Seafarers, 1978 was the first internationally agreed convention to address the issue of minimum standards of competence for the crews of crude oil (and other) cargo ships. This Convention has been revised and updated to clarify the standards of competence required and provide effective mechanisms for enforcement of its provisions.

10.3 THE LAW OF THE SEA

The development of offshore oil, gas, and other mineral resources in the United States is impacted by a number of interrelated legal regimes, including international, federal, and state laws. International law – The Law of the Sea (UN, 1994) provides a framework for establishing national ownership or control of offshore areas, and domestic federal law mirrors and supplements these standards. Governance of offshore minerals and regulation of development activities are bifurcated between state and federal law. Generally, states have primary authority in the three-geographical-mile area extending from their coasts. The federal government and its comprehensive regulatory regime govern those minerals located under federal waters, which extend from the states' offshore boundaries out to at least 200 nautical miles (230 miles) from the shore (Figure 10.1). The basis for most federal regulation is the Outer Continental Shelf Lands Act (OCSLA), which provides a system for offshore oil and gas exploration, leasing, and ultimate development. Regulations run the gamut from health, safety, resource

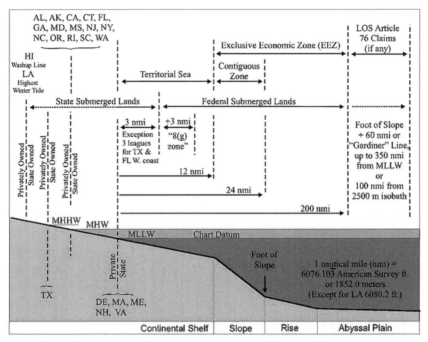

Figure 10.1 Offshore ownership boundaries *(Source: USMinerals Management Service).*

conservation, and environmental standards to requirements for production based royalties and, in some cases, royalty relief and other development incentives (Vann, 2010). At present, over 70 international conventions and agreements are directly concerned with protecting the marine environment (Patin, 1999).

The health of the oceans is vital for the world's economic and ecological well-being (Burden et al., 2011). Urgent action is needed to protect the marine environment and its resources from all sources of pollution, in particular from land-based activities. The Convention on the Law of the Sea entered into force on 16 November 1994, one year after it had reached the 60 ratifications necessary. The United Nations Convention on the Law of the Sea is perhaps one of the most significant but less recognized 20th century accomplishments in the arena of international law. It established for the first time one set of rules for the oceans, bringing order to a system fraught with potential conflict. The scope of the Law covers (1) all ocean space, with all its uses, including navigation and overflight, (2) all uses of all its resources, living and non-living, on the high seas, on the ocean floor and beneath, on the continental shelf and in the territorial seas, (3) the protection of the marine environment, and (4) and basic law and order. The Convention, often referred to as the *constitution of the sea*, is based on the all-important idea that the problems of the oceans are closely interrelated and must be addressed as a whole. Early on in the negotiating process, and possibly key to its success, it was agreed that the treaty must be taken as a whole, not bartered and argued piece by piece. And thus it was adopted on 30 April 1982. Today, it is one of few international agreements that almost all countries abide by in practice, even those that are not States parties. The Convention was opened for signature on 10 December 1982 in Montego Bay, Jamaica, when a record number of States, 119, signed on. Today, there are 137 States Parties plus the European Community. Several States that had previously found some provisions problematic are now taking steps for future ratification or accession. The United States of America has publicly stated its intention of becoming a State Party as soon as possible.

The Law dealt with and defined the issues of (1) internal waters, (2) territorial waters, (3) archipelagic waters, (4) the contiguous zone, (5) EEZs, and (6) the continental shelf. The issue of varying claims of territorial waters was raised in the UN in 1967 and again in 1973 when the *Third United Nations Conference on the Law of the Sea* was convened in New York. In an attempt to reduce the possibility of groups of nation-states dominating the negotiations, the conference used a consensus process rather than majority vote. The convention introduced a number of provisions. The most significant issues covered were setting limits, navigation, archipelagic status and transit regimes, EEZs, continental shelf jurisdiction, deep seabed mining, the exploitation regime, protection of the marine environment, scientific research, and settlement of disputes. The convention set the limit of various areas, measured from a carefully defined baseline—typically a sea baseline follows the low-water line, but when the coastline is deeply indented, has fringing islands or is highly unstable, straight baselines may be used.

Thus, *internal waters* cover all water and waterways on the landward side of the baseline. The coastal state is free to set laws, regulate use, and use any resource. Foreign vessels have no right of passage within internal waters. On the other hand, *territorial* waters stretch out to 12 nautical miles (14 miles, 22 km) from the baseline and the coastal state is free to set laws, regulate use, and use any resource. Vessels were given the right of *innocent passage* through any territorial waters, with strategic straits allowing the passage of military craft as *transit passage*, in that naval vessels are allowed to maintain (non-threatening) postures that would be illegal in territorial waters. *Innocent passage* was defined by the convention as passing through waters in an expeditious and continuous manner, which is not *prejudicial to the peace, good order, or the security* of the coastal state. Fishing, polluting, weapons practice, and spying were not classed as *innocent*, and submarines as well as any other underwater vehicles were required to navigate on the surface and to show their flag. Nations had the right to temporarily suspend innocent passage in specific areas of their territorial seas, if doing so was essential for the protection of national security.

Archipelagic waters were set by drawing a baseline between the outermost points of the outermost islands, subject to these points being sufficiently close to one another and all waters within the this baseline were designated as *archipelagic waters*. The state has full sovereignty over these waters (the same sovereignty as over *internal waters*), but foreign vessels do have right of innocent passage through archipelagic waters (the same innocent passage as for territorial waters).

The *contiguous zone* lies beyond the 12-nautical-mile (14 miles, 22 km) limit and lies within a further 12 nautical miles from the territorial sea baseline limit in which a state (coastal nation) can continue to enforce laws in four specific areas: (1) customs, (2) taxation, (3) immigration, and (4) pollution, if the infringement started within the state's territory or territorial waters, or if this infringement is about to occur within the state's territory or territorial waters.

Exclusive economic zones extend from the edge of the territorial sea out to 200 nautical miles (230 miles) from the baseline and, within this area, the coastal nation has sole exploitation rights over all natural resources. The *EEZs* were introduced to halt the increasingly heated clashes over fishing rights, although at that time the importance of offshore oil and gas was being realized. The success of an offshore oil platform in the Gulf of Mexico in 1947 (Chapter 2) was soon repeated elsewhere in the world, and by 1970 oil platforms operating in waters 13,000 ft had become technically feasible. As a result, the Law of the Sea gave foreign nations the freedom of navigation and overflight, subject to the regulation of the coastal nations. Foreign nations were also given the tights to lay submarine pipes and cables, subject to agreement with the host coastal nation.

The Law defined the *continental shelf* as the natural prolongation of the land territory to the outer edge of the continental margin or 200 nautical miles (230 miles) from the baseline of the coastal state, whichever is greater. The continental shelf of a coastal nation may exceed 200 nautical miles (230 miles) until the natural prolongation ends. However, it may never exceed 350 nautical miles (400 miles, 650 km) from the baseline; or it may never exceed 100 nautical miles (120 miles, 190 km) beyond the 8,250 ft (2500 m) isobath (the line connecting the depth of 8,250 ft). Coastal nations have the right to harvest mineral and non-living material in the subsoil of its continental shelf, to the exclusion of others. The coastal nations also have exclusive control over living resources attached to the continental shelf, but not to creatures living in the water column beyond the EEZ.

The Third Conference on the Law of the Sea opened in 1973 with a brief organizational session, followed in 1974 by a second session held in Caracas, Venezuela. A first draft was submitted to delegates in 1975. Over the next 7 years, the text underwent several major revisions. But on 30 April 1982, an agreement had been reached and the final text of the new convention was put to a vote. The vote, which took place at United Nations Headquarters in New York, marked the end of over a decade of intense and often strenuous negotiations, involving the participation of more than 160 countries from all regions of the world and all legal and political systems. The Convention was adopted with 130 States voting in favor, four against and 17 abstaining. Later that same year, on 10 December, the Convention was opened for signature at Montego Bay, Jamaica, and received a record number of signatures—119—on the first day.

The United Nations Convention on the Law of the Sea entered into force on 16 November 1994, one year after it had reached the 60 ratifications necessary. Today the Convention is fast approaching universal participation, with 138 States, including the European Union, having become parties. New advances in technology have allowed exploration and production activities to go further offshore and deeper into the oceans. Life on the seabed, which was once thought of as existing only in the shallow waters of the continental shelf, has now been found at depths of more than 4,000 ft.

In the years to come, the oil and gas industry will be affected by the continuing work of the Commission on the Limits of the Continental Shelf, as States seek to establish the outer limits of their continental shelves beyond 200 nautical miles (230 miles). As deep seabed mineral activities move from prospecting to exploration and eventually to exploitation, the International Seabed Authority will have to give ever-greater attention to environmental considerations, in addition to benefit-sharing, as it continues to administer the resources of the Area for humankind.

It is argued that virtually all aspects related to maritime oil transport are covered by International law, but that there are considerable loopholes in the international legal framework for offshore oil development. Many countries already engaging in offshore oil extraction have developed their own national or regional laws and standards.

10.4 COASTAL ZONE MANAGEMENT

The continental margins, the geographic region contiguous to and lying seaward of a coastline (Chapter 2), have become increasingly important to the crude oil and natural gas industry over the past century. The continental margins consist of three portions: (1) the continental shelf which has shallow water depths rarely deeper than 650 ft) and extends seaward from the shoreline to distances ranging from 12.3 miles to 249 miles, (2) the continental slope where the bottom drops off to depths of up to 3.1 miles, and (3) the continental rise which dips very shallowly seaward from the base of the continental slope and is in part composed of down-washed sediments deposited at the base of the slope. The continental margins are of great importance for many reasons, not least of which is that they are presently the source of increasing amounts of crude oil and natural gas supplies.

In the United States, as interest in the commercial development of natural gas and oil increased in the 1940s, control over these resources became a major issue, especially in the offshore regions. While these legislative efforts have are used here as example, the countries of the European Union and other countries have followed suit with similar legislation.

The Congress of the United States eventually resolved this issue by passage of the Submerged Lands Act (SLA) in 1953, which established the jurisdiction of the Federal Government to (and ownership of) submerged lands located on a majority of the continental margin. States were given jurisdiction over any natural resources within 3 nautical miles (3.45 miles) of the coastline excepting Texas and the west coast of Florida where the SLA extends the States' Gulf of Mexico jurisdiction to 9 nautical miles (10.35 miles).

Passage of the SLA prepared the way for passage of the OCSLA, also in 1953, which defined the Outer Continental Shelf (OCS), separate from geologic definitions, as any submerged land outside state jurisdiction and reaffirmed federal jurisdiction over these waters and all resources they contain. Moreover, the OCSLA outlined the responsibilities of the federal government for managing and maintaining offshore lands subject to environmental constraints and safety concerns. The Act (OCSLA) authorizes the United States DOI to lease the defined areas for development and to formulate regulations pertaining thereto as necessary. Between 1978 and

1998 the OCSLA was amended six times to account for changing issues and remains the cornerstone of offshore legislation.

On a worldwide basis, the 1994 International Law of the Sea granted the same 200 nautical miles (230 miles) to all countries. Prior to this, countries had claimed jurisdiction to offshore areas in bilateral agreement with neighboring countries. For example, since 1978, the United States and Mexico signed two treaties in order to fully define jurisdictional boundaries in the Gulf of Mexico. In some instances the International Law of the Sea provides that jurisdiction over natural resources extends beyond the 200-nautical mile (230-mile) boundary to the edge of the geological continental margin based on geological factors such as sediment thickness and water depth. For this reason the boundaries associated with Alaska, parts of the East Coast and the Gulf of Mexico extend beyond 200 nautical miles (230 miles), but the Pacific coast has the standard boundary limits (Figure 10.1).

The issue of the availability of outer continental shelf (OCS) lease areas has been a source of debate, even controversy, over the years. On the one hand, there are arguments that one side are those who argue that countries (for example, the United States) need to open such areas to crude oil and natural gas production in order to meet future energy needs or other policy objectives such as reduced dependence on foreign (imported) oil. On the other hand, there are other arguments related to protection of the ocean and coastal environments from further pollution as well as avoiding the potential negative effects on fishing or tourism with the caveat that the negative impacts of exploration and development needed to extract the natural gas and oil likely outweigh the potential benefits.

10.5 THE FUTURE

As the 21st century progresses, a larger proportion of crude oil and natural gas will be produced from reservoirs that lie under the oceans. From the perspective of the United States, the production of crude oil and natural gas from beneath the seabed is a major activity in the Gulf of Mexico, offshore southern California, and in some regions of Alaska. The same reasoning applies to offshore resources currently being produced by the European Union and other coastal countries.

Offshore construction design standards developed with industry participation have contributed to the growth of the industry by reducing project risks and creating a common language used by project participants. These efforts will continue and the rig or platform of the mid-to-late 21st century (yes, crude oil and natural gas will be a major energy source for at least the next fifty years) will be a vast improvement over those used at this time. The profile of the either unit on the horizon might appear to be similar but the advancement of technology will contribute to vast improvement and energy-efficient operations within the next 50 years. This will be accompanied by continuous examination of (and change to) the various pieces of legislation that relate to offshore crude oil and natural gas exploration and production.

In fact, the European Union—because of the Deepwater Horizon incident in the Gulf of Mexico in 2010 (Hopkins, 2012)—is instituting new rules aimed at bringing offshore oil and gas drilling under a common policy. The likelihood of a major offshore accident in European waters remains unacceptably high and the legislation will help prevent such future crises from happening in all marine waters which fall under jurisdiction of the member countries of the European Union. The regulation establishes rules to cover exploration and production activities from design to the final removal of a crude oil or natural gas installation (Watkins, 2011).

The new approach taken by the European Union will lead will lead to a risk assessment that upgrades continuously by taking into account new technology and any accompanying risks. The key requirements for preventing and responding to a major accident are : (1) licensing, which will ensure that only operators with sufficient technical and financial capacities necessary to control the safety of offshore activities and environmental protection are allowed to explore and produce in European Union waters, (2) independent verifiers, who will ensure that the plans presented by the operator that are critical for safety on the installation will be verified before and periodically after the installation, (3) obligatory emergency planning, which will require companies to prepare a major hazard report for their installation, containing a risk assessment, and an emergency response plan before exploration or production begins, (4) inspections by independent national authorities responsible for the safety of installations, which will verify safety provisions, environmental protection, and emergency preparedness of rigs and platforms and the operations conducted on them, (5) emergency response, which will require companies to prepare emergency response plans based on the rig or platform risk assessments and keep response equipment available, (6) liability, which requires that oil and gas companies will be fully liable for environmental damages caused to the protected marine species and natural habitat within the geographic zone of all European Union marine waters including the EEZ and the continental shelf where the coastal

European Union members exercises jurisdiction, and (7) transparency by which comparable information will be made available to citizens via web sites about industry performance standards and the activities of the national authorities.

The offshore oil and natural gas industry within the United States and many other countries has significant economic benefits as a contributor to employment, the national economy, government revenues, and domestic energy production (Burden et al., 2011; Hillegeist et al., 2013). Current offshore oil and gas production in the United States is essentially limited to the Central, Western and a small amount of the eastern Gulf of Mexico with limited additional production off Alaska and California. Various developments in the area of ocean affairs and the law of the sea can be foreseen in the coming years—new advances in technology have allowed exploration and production to move further offshore and deeper into the oceans with attention to the relative technological advances (Carré et al., 2002).

However, deeper and colder waters create problems related to with hydrates production, wax deposition, and asphaltene deposition and treatment of the reservoir fluids is already needed to maintain fluid flow from the reservoir to the production platform. This will be derived from a more intimate knowledge of crude oil properties and behavior in terms of phase separation diagrams and the influence of pressurized natural gas on the properties of the crude oil (Speight, 2009, 2014).

Furthermore, advances in technology and safety management systems have also contributed to an improved OCS environmental protection and safety. In company with the technological advances, significant developments have occurred in the offshore service vessel fleet, where new deep-draft, very large, high-horsepower anchor handling/tug/supply vessels have evolved to move these large new drilling rigs, handle their anchors, chain and mooring lines, and meet all kinds of service demands of the new generation of deepwater rigs and production platforms. These types of advances will continue as long as offshore crude oil and natural gas production is viable.

However, the health of the oceans will continue to remain a vital issue related to offshore exploration and production. To ignore such an issue would contribute to the continued degradation of the marine environment, to shortcomings in environmental responses and to constraints in the availability of natural resources that could in turn affect food security and lead to conflict situations. Many of the issues currently facing the international community today, and those likely to linger in the future, move beyond national borders and can only be countered effectively by nations acting in concert at the national, international, and regional levels.

10.5.1 Safety Management

The major issue currently facing offshore crude oil and natural gas development is to reduce the risk to the environment—the central has to be designed in a manner such as to limit risk by reducing the frequency of malfunctions and minimizing the consequences. More specifically for the future this means: (1) minimizing the likelihood of a loss of control of production and particularly of release of environmentally hazardous materials, (2) reducing the probability of ignition/explosion where there is a release, (3) containing the consequences of any fire, explosion, or escape of toxic substances, and (4) to ensure that the means of evacuation for all contingencies are available (Cox and Cheyne, 2000; API, 2010; Vinnem, 2010; Skogdalen et al., 2011).

In order to bring this about, safety imperatives must be integrated in the preliminary design stage into the overall facility layout—in particular, ensuring separation of the crude oil and natural gas treatment plant from ignition sources. The organization of each project therefore draws on traditional risk management methods such as quality control, risk assessment, safety reviews and audits (Sutton, 2012). In short, safety will continue to be monitored and controlled throughout the entire lifetime of an offshore installation.

Human error is one of the major causes of accidents, and obviously will continue to be addressed. In most cases, the risk may have been underestimated suggesting the need for appreciation of this danger in the organization of the work and equipment. The organization of work and training, and the dissemination of information will continue to be critical to accident prevention and to minimizing the consequences of error. In fact, it must be realized by all concerned that safety management is not only a matter of solving technical problems and also of enacting and enforcing the necessary regulations.

10.5.2 Environmental Risk Management

Planned emissions and unplanned incidents in the marine environment have the potential to result in adverse environmental effects if a discharge of hazardous material or release of energy comes into contact with sensitive receptors. Environmental risk management is the process of systematically identifying credible environmental hazards, analyzing the likelihood of occurrence and severity of the potential consequences, and managing the resulting level of risk (Wills, 2000; Stoklosa, 1998, 1999).

Companies involved in offshore crude oil and natural gas production will continue to endeavor to prevent or control environmental problems resulting from their activities and set clear environmental targets. The environmental targets will relate predominantly to reducing (1) gas flaring of gas, (2) emissions of hydrocarbons, (3) oil-content of effluent, and (4) minimizing the environmental impact of the operations such as the of oil spills.

In order to combat the occurrence of spills of crude oil, and because of the enormous costs that could be involved in such pills, the Unites States government has enacted legislation related to unlimited liability for gross (or wilful) negligence. All tankers trading in US waters are required to demonstrate to local authorities (with Certificates of Financial Responsibility) that they carry adequate insurance to cover maximum financial risk. In contrast to the IMO, the same liability rules for vessels do also apply for offshore oil installations – unlimited financial liability of the ship owner or the company managing an offshore platform is considered in the United States as an incentive to display for responsible conduct. With adequate liability at risk, oil companies and ship owners will be motivated to design, construct, and operate their projects as safely as possible. Insurance companies will be less likely to take the risk to insure sub-standard vessels or offshore platforms with unlimited liability at stake. However, the issue of unlimited liability is according to the crude oil and natural gas industry and many legal experts and impractical solution and a limited liability which is realistic and which would provide for sufficient compensation after an accident is to be preferred.

In addition, the initiative to reduce flaring has the aim to significantly reduce the emissions of carbon dioxide. Reducing flaring is not a simple task as it means limiting the emissions of associated gas in the process where the use or reinjection is not easy. New projects they will study and limit, as far as possible, flaring of gas which will mean a significant reduction of greenhouse gas emissions, especially in countries where there is also significant flaring of natural gas. The main pollutants generated by offshore exploration and production activities are sulfur-containing gases. Measure will be taken to ensure that the gases to be flared are purified so that compliance with the appropriate standards is ensured.

The problems posed by liquid effluents will also be addressed with more vigor. Water is a by-product of oil production, and the water naturally contains hydrocarbon emulsions and it is vital that the effluent is cleaned up before being discharged. Effluent containing variable amounts of oil may be tolerated in some areas but standards for the allowable oil content will be lowered to more stringent levels and more strictly enforced. This will affect production from depleted reservoirs present difficulties because large quantities of water are used in the production process.

Site rehabilitation at the end of platform life or field life and in particular, the decommissioning of offshore installations, is currently the focus of considerable attention. After development wells have been drilled and transportation or pipeline infrastructure is in place, drilling platforms are removed, often used in nearby areas for additional drilling, scuttled, or dismantled for scrap (Neff et al., 1987). Little or no reclamation of the seabed around offshore well heads is attempted, so these locations remain contaminated by drilling muds and materials discharged from drilling platforms for the indefinite future. Little research has been completed concerning the distribution or dissolution of such materials after removal of drilling platforms. International regulations that have been for the most part indicative of the end result will become stricture in order to increase the protection that will be offered to the environment.

Overall the evaluation of ecological hazards must fit into decision making when comparisons of risk are necessary for a wide range of crude oil and natural gas exploration and production activities and naturally occurring events. The process is not, of course, entirely objective and scientific in nature since social expectations and the pressures of special interest groups can have a major influence on the perceptions of risk that cannot be ignored (Rundmo, 1997).

An important aspect of the environmental risk management process will continue to be—perhaps even more so than in current risk management processes—the involvement of industry, government, special interest groups, and community stakeholders in the assessment of activities that may impact the marine environment. In doing so, clear objectives and assessment criteria will need to adopted from the beginning so that there is constructive input that can be used to identify all potential (perhaps non-emotional) hazards and a constructive approach is taken to manage all of the risks that are to be assessed.

In order to obtain a more confident estimation of risk, the need to seek a balance to perform risk assessments in the most efficient manner will be recognized. Furthermore, the objectives of risk management are also likely to change in the future with a concomitant re-assessment of risk and management strategies, which will be result of political imperatives, newly discovered or introduced threats, or the perceived value of particular environmental resources. It will be essential that there is an understanding of the key concept of assessment and measurement endpoints and the criteria used to judge the significance of potential risks to the environment will be even more imperative, such as in the analysis of produced formation water discharge from offshore platforms (Terrens and Tait, 1994).

There will also be a more definitive move the recognizing when a hazard will have obvious unacceptable impacts. In addition, there will also be more definitive moves to recognize when the potential effects of an environmental hazard are near the threshold of what might be considered acceptable risk by clearly defining (1) the expected effect, (2) the threshold of acceptability, and (3) the likelihood that the threshold would be exceeded. This will (or should) remove all of the various aspects of emotional decision making. Thus, the interpretation of the acceptability of adverse impacts that might occur from human actions will be viewed in the perspective of the naturally occurring variability of ecological systems. While the importance of this perspective is already recognized (Bartell et al., 1992; Suter, 1993; Beer and Ziolkowski, 1995) government and the public will be more acceptable to the potential consequences of marine environmental hazards.

Finally, all countries involved in offshore oil and gas production will be required to formally acknowledge and abide by the Law of the Sea Treaty—this is an important aspect of crude oil and natural gas production as exploration and production activities move into deeper and deeper waters. The component of the Treaty that protects the right of both commercial and military ships and aircraft to move freely through and over straits used for international navigation, to engage in *innocent passage* through territorial seas, and to allow passage through *EEZs*, also is important to the energy security of many countries (especially the United States) as the sources of crude oil and natural gas become more global (Greene, 2010; Greene and Hopson, 2010; Speight, 2011).

In the future, seeking a balance between the use of natural resources and the preservation of environmental values will be a more focused aspect of offshore crude oil and natural gas exploration and production is not an easy task. The issue will be recognized as involving the range of environmental values may not be well under stood or agreed among government agencies, the public, and special interest groups. Such an integrated and inclusive approach toward managing risk will be imperative.

However, determining the impact of new programs or of any operational can only be achieved by having proper performance indicators that represent the attendant risks. This will require a careful assessment of risks due to: (1) risk due equipment failure, (2) risk due to major hazards, (3) risk due to incidents that may represent challenges for emergency preparedness and not forgetting, (4) occupational injury risk, (5) occupational illness risk, and (6) risk perception and cultural factors. While debate tends to be on-going as to the type of indicators should be used to determine risk and identifying those indicators that are meaningful (Skogdalen et al., 2011), future efforts will focus on the current indicators for major hazards, particularly for preventing such incidents as blowouts. Nevertheless, it must be membered that any specific indicator might provide early warning for major-hazard risk at a global level, but not always for individual facilities since there is lack of uniformity in indicators used worldwide—*site specificity* is the operative phrase.

10.5.3 Recovery and Processing Options

It might be said that *the most difficult predictions to make are predictions of what might transpire in the future*! Of all the future development, the issues of crude oil recovery and processing options represents the most difficult to predict of all. The processing options for offshore crude oil straight from the well are limited because of the unavailability of space on production platforms.

For example, there are continuing efforts to evaluate the potential for enhanced oil recovery in in offshore fields, most of them are at the early stages of evaluation or may not be economically attractive with the current technology. As a result, it is expected that commercial applications of enhanced oil recovery methods will likely not be applied with any high degree of frequency for several years. Constraints related to surface facilities constraints and environmental regulations (e.g., chemical additives for enhanced oil recovery also represent major hurdles for large enhanced oil recovery applications in offshore fields. It is more than likely that waterflooding and gas injection and their combined processes (e.g., water–assisted gas injection) will continue to support offshore production in the near term.

In addition, cleaning crude oil to meet pipeline specifications is an obvious area that will have to continue and may advance in terms of efficiency of the cleaning processes rather than the installation of new processes on to the platform.

The concept of installing a simple visbreaker on a platform as a controlled treatment option would *clean up* such crude oils by removing undesirable constituents (which, if not removed, would cause problems further down the refinery sequence) as coke or sediment (Chapter 6)

(Radovanović and Speight, 2011; Speight, 2012). However there always remains the issue of available space and the issue of dealing with any waste products from the visbreaker.

In terms of natural gas cleaning, membrane systems represent the most promising option for installation on a platform. In fact, in the short term, opportunities exist throughout the petroleum refining and gas processing industries for the use of membranes in the separation of low molecular weight hydrocarbon mixtures, as well as the removal of water and nitrogen from these mixtures. In the long term, however, these applications will likely require membranes with higher selectivity than the current membranes can achieve and such membranes could replace the current gas processing options used on a platform. The development of viable membranes could expand the platform-related membrane market. Over the next two decades, the use of membranes to purify natural gas for pipeline transport could easily grow to become a large membrane gas separation application. The need for efficient gas pretreatment options prior to pipeline shipment is an essential part of offshore natural gas production operations. Thus, applications of membrane technology is likely to grow significantly, particularly in applications such as offshore platforms where membrane technology offers advantages over the more conventional (and space-demanding) gas processing technology (Baker, 2002).

10.5.4 Materials of Construction

The materials of construction used in offshore units typically are fabricated rom steel and other metals or metal alloys. However, the use of composite materials has shown several advantages and, in recognition of these potential advantages, the use of composites will show considerable growth in the development activities of several offshore structural systems, particularly those that are water depth sensitive such as mooring and risers. Indeed, there will be a move to expand the use of composite materials which can offer (in some applications) several advantages for marine construction because of the low density, corrosion resistance, and excellent fatigue performance.

In addition, the use of composites allows for greater design flexibility by tailoring the properties to meet specific design requirements, thus promoting better system-oriented solutions. In some cases, units fabricated from composite components are often more expensive than the steel-fabricated counterpart but on a performance-related basis, the economic incentive to use composite components can often be demonstrated based on the capability to reduce system costs and improve the life cycle efficiency.

REFERENCES

API, 2010. Process Safety Performance Indicators for the Refining & Petrochemical Industries. Report No. RP-754. American Petroleum Institute, Petroleum Institute, Washington, DC.

Bartell, S.M., Gardner, R.H., O'Neill, R.V., 1992. Ecological Risk Estimation. Lewis Publishers, Stockport, United Kingdom.

Beer, T., Ziolkowski, F., 1995. Environmental Risk Assessment: An Australian Perspective. Supervising Scientist Report Number 102, Commonwealth Environment Protection Agency, Canberra, Australia.

Burden, P., Goldsmith, S., Cuyno, L., McCoy, T., 2011. Potential National-Level Benefits of Alaska OCS Development. Prepared for Shell Exploration and Production by Northern Economics, Anchorage, Alaska.

Carré, G., Pradié, E., Christie, A., Delabroy, L., Greeson, B., Watson, G., Fett, D., Piedras, J., Jenkins, R., Schmidt, D., Kolstad, E., Stimatz, G., Taylor, G., 2002. High Expectations from Deepwater Wells. Schlumberger Oilfield Review, Winter 2002/2003, pp. 36–51.

Cox, S.J., Cheyne, A.J.T., 2000. Assessing the safety culture in offshore environments. Safety Science 34, 111–129.

Dudley, N., Stolton, S., 2002. To Dig or Not to Dig: Criteria for Determining the Suitability or Acceptability of Mineral Exploration, Extraction, and Transport from Ecological and Social Perspectives. World Wildlife Fund, Washington, DC.

Ebinger, C., Banks, J.P., Schackmann, A., 2014. Offshore Oil and Gas Governance in the Arctic: A Leadership Role for the U.S. Policy Brief No. 14-01. Brookings Institution, Washington, DC, March.

Greene, D.L., 2010. Measuring energy security: Can the United States achieve oil independence? Energy Policy 38, 1614–1621.

Greene, D.L., Hopson, J., 2010. The Costs of Oil Dependence. Oak Ridge National Laboratory, Oak Ridge, Tennessee.

Hedberg, H.D., 1976. Ocean boundaries and petroleum resources. Science 191, 1009–1018.

Hillegeist, P., Shafer, S., Gross, M., 2013. The Economic Benefits off Increasing U.S. Access to Offshore Oil and Natural Gas Resources in the Atlantic. Quest Offshore Inc., for the American Petroleum Institute (API) and the National Ocean Industries Association (NOIA), Washington, DC, December.

Hopkins, A., 2012. Safety Indicators for Offshore Drilling- A Working Paper for the CSB Inquiry of the Macondo Blowout, US Chemical Safety and Hazard Investigation Board, Washington, DC.

IMO, 2007. Implications of the United Nations Convention on the Law of the Sea for the International Maritime Organization. Report No. LEG/MISC.5. International Maritime Organization, London, United Kingdom.

Mannan, M.S., Mentzer, R.A., Rocha-Valadez, A., Mims, A., 2014a. Offshore drilling risks – 1: Study: Risk indicators have varying impact on mitigation. Oil & Gas Journal, May 5.

Mannan, M.S., Mentzer, R.A., Rocha-Valadez, A., Mims, A., 2014b. Evaluating offshore drilling risks – 2 (Conclusion): Global, regional statistics show continuing improvement. Oil & Gas Journal 2, June.

Mearns, K., Flin, R., Gordon, R., Fleming, M., 1998. Measuring safety climate on offshore installations. Work and Stress 12, 238–254.

OCSLA, 1953. Outer Continental Shelf Lands Act – the Act Of August 7, 1953, Chapter 345, As Amended [As Amended Through Public Law 106-580, Dec. 29, 2000]. An Act to Provide for the Jurisdiction of the United States over the Submerged Lands of the Outer Continental Shelf and to Authorize the Secretary of the Interior to Lease Such Lands For Certain Purposes. Enacted by the Senate and House of Representatives of the United States of America in Congress, Washington, DC.

Radovanović, L., Speight, J.G., 2011. Visbreaking: A Technology of the Future. Proceedings. First International Conference – Process Technology and Environmental Protection (PTEP 2011). University of Novi Sad, Technical Faculty "Mihajlo Pupin," Zrenjanin, Republic of Serbia, December 7, pp. 335–338.

Rundmo, T., 1994. Associations between safety and contingency measures and occupational accidents on offshore petroleum platforms. Scandinavian Journal of Work Environment and Health 20, 128–131.

Rundmo, T., 1997. Associations between risk perception and safety. Safety Science 24, 197–209.

Skogdalen, J.E., Utne, I.B., Vinnem, J.E., 2011. Developing safety indicators for preventing offshore oil and gas deepwater drilling blowouts. Safety Science 49, 1187–1199.

Speight, J.G., 2009. Enhanced Recovery Methods for Heavy Oil and Tar Sands. Gulf Publishing Company, Houston, Texas.

Speight, J.G., 2011. An Introduction to Petroleum Technology, Economics, and Politics. Scrivener Publishing, Salem, Massachusetts.

Speight, J.G., 2012. Visbreaking: A technology of the past and the future. Scientia Iranica 19 (3), 569–573, 2012C.

Speight, J.G., 2013. Shale Gas Production Processes. Gulf Professional Publishing, Elsevier, Oxford, United Kingdom.

Speight, J.G., 2014. The Chemistry and Technology of Petroleum, 5th Edition. CRC Press, Taylor & Francis Group, Boca Raton, Florida.

Stoklosa, R.T., 1998. The relevance of risk assessment in environmental approvals and decision making. The APPEA Journal 38 (1), 715–723.

Stoklosa, R.T., 1999. Practical application of environmental risk management—Gorgon LNG Project Case Study. The APPEA Journal 39 (1), 606–621.

Suter, II, G.W., 1993. Ecological Risk Assessment. Lewis Publishers, Stockport, United Kingdom.

Sutton, I., 2012. Offshore Safety Management. Elsevier, Amsterdam, Netherlands.

Terrens, G.W., Tait, R.D., 1994. Effects on the Marine Environment of Produced Formation Water Discharges from Offshore Development in Bass Strait, Australia. Paper No. SPE 27149. Proceedings. 2nd International Conference on Health, Safety and Environment in Oil and Gas Exploration and Production, Jakarta, Indonesia.

UN, 1994. United Nations Convention on the Law of the Sea III (entered into force November 16, 1994).

UN, 2012. Suspension Of Innocent Passage. Oceans and Law of the Sea. United Nations. Division for Ocean Affairs and Law of the Sea. United Nations, Geneva, Switzerland. http://www.un.org/depts/los/convention_agreements/innocent_passages_suspension. htm, accessed July 4, 2014.

Vann, A., 2010. Offshore Oil and Gas Development: Legal Framework. CRS Report for Congress, Congressional Research Service, United States Congress, Washington, DC.

Vinnem, J.E., 2010. Risk indicators for major hazards on offshore installations. Safety Science 48, 778–787.

Von Sydow, 2014. Energy Law & Policy: Yearbook 2013. Claeys and Casteels, Deventer, Netherlands.

Watkins, E., 2011. European Union Seeks unified policy for offshore drilling. Oil & Gas Journal, November 14.

Wills, J.W.G., 2000. Muddied Waters - A Survey of Offshore Oilfield Drilling Wastes and Disposal Techniques to Reduce the Ecological Impact of Sea Dumping. Ekologicheskaya Vahkta Sakhalina, Yuzhno-Sakhalinsk, Russia.

World Offshore Oil and Gas Projects

(http://www.offshoreoilandgas.gov.bc.ca/world-offshore-oil-and-gas/)

Table A.1 Africa and the Middle East

- Abana, Gulf of Guinea, Nigeria
- Agbami Discovery Well, Niger Delta, Nigeria
- Al Shaheen Oil Field, Qatar
- BBLT, Block 14 Compliant Piled Tower (CPT), Angola
- Bonga Deepwater Project, Niger Delta, Nigeria
- Ceiba, Rio Muni Basin, Equatorial Guinea
- Chinguetti Oil Field, Mauritania
- Dalia Field Development of Block 17, Angola
- Ekpe Phase II, Gulf of Guinea, Nigeria
- Espoir Field, Ivory Coast
- Gimboa Field, Angola
- Girassol, Luanda, Angola
- Greater Plutonio, Block 18, Deepwater Drillship Pride, Angola
- Kizomba Deepwater Project, Angola
- Mossel Bay, Bredasdorp Basin, South Africa
- Okume Complex, Equatorial Guinea
- Rosa Field, Angola
- Sanha / Bomboco Development, Angola
- Scarab and Saffron Gas Fields, Eastern Mediterranean, Egypt
- South Pars, Qatar North Field, Iran
- Xikomba Oil Field Deepwater Development, Angola
- Yoho Oil Field, Nigeria
- Zafiro, Gulf of Guinea, Equatorial Guinea

Handbook of Offshore Oil and Gas Operations. http://dx.doi.org/10.1016/B978-1-85617-558-6.00011-9

Table A.2 Asia-Pacific

- Bayu-Undan, Timor Sea, Australia
- Buffalo, Timor Sea, Australia
- Gorgon, Northern Carnarvon Basin, Australia
- Kikeh Floating Production, Storage and Offloading Development, Malaysia
- Laminaria, Timor Sea, Australia
- Langsa Oil Pool, Straits of Malacca, Indonesia
- Liuhua 11-1, South China Sea, China
- Lufeng 22-1, South China Sea, China
- Malampaya, South China Sea, Philippines
- Mutineer-Exeter MODEC Venture II FPSO, Carnarvon Basin, Australia
- Natuna Gas Field, Indonesia
- Otway Basin (Minerva, Geographe, Thylacine and Casino) Fields, Australia
- PM-3 Commercial Arrangement Area (CAA)
- Stag, North West Shelf, Australia
- Stybarrow Oil Field, Australia
- Tui, Amokura and Pateke Reserves Area Oil Project, Taranaki Basin, New Zealand
- West Seno, Makassar Strait Deepwater Development, Indonesia
- Wonnich, Carnarvon Basin, Australia

Table A.3 Central Asia

- Azeri-Chirag-Gunashli (ACG) Oil Field, Caspian Sea, Azerbaijan
- Blue Stream Natural Gas Pipeline, Russia/Turkey
- Kashagan, Caspian Sea, Kazakhstan
- Sakhalin II, Sea of Okhotsk, Russia
- Shah Deniz South Caspian Sea, Azerbaijan

Table A.4 North America

- Baldpate, Gulf of Mexico, USA
- Bombax Pipeline Development, Trinidad and Tobago
- Brutus, Gulf of Mexico, USA
- Cameron Highway Oil Transport System, Gulf of Mexico, USA
- Cantarell Oil Field, Gulf of Mexico, Mexico
- Canyon Express Gas Field, Mississippi Canyon, USA
- Constitution/Ticonderoga Field, Spar Technology, Gulf of Mexico, Afghanistan
- Deep Panuke Gas Field, Canada
- Devils Tower Gas Field, Gulf of Mexico, USA
- Genesis, Gulf of Mexico, USA
- Glider, Gulf of Mexico, USA
- Greater Angostura, Eastern Venezuela Basin, Trinidad and Tobago
- Gyrfalcon, Gulf of Mexico, USA
- Hibernia, Jeanne d'Arc Basin, Canada
- Hickory, Gulf of Mexico, USA
- Holstein Oil and Gas Development, Gulf of Mexico, USA
- Hoover Diana, Gulf of Mexico, USA
- Horn Mountain Field, Gulf of Mexico, USA
- Independence Hub, Gulf of Mexico, USA
- King Field, Gulf of Mexico, USA
- Llano, Gulf of Mexico, USA
- Mad Dog Field, Gulf of Mexico, USA
- Magnolia Field, Gulf of Mexico, USA
- Manatee Field, Gulf of Mexico, USA
- Marco Polo Field Gulf of Mexico, USA
- Mardi Gras Oil and Gas Transportation System, Gulf of Mexico, USA
- Mars, Gulf of Mexico, USA
- Matterhorn Field, Gulf of Mexico, USA
- Mensa, Gulf of Mexico, USA
- Morpeth, Gulf of Mexico, USA
- Na Kika Oil and Gas Fields, Gulf of Mexico, USA
- Nansen and Boomvang Gas Fields, Gulf of Mexico, USA
- Neptune, Gulf of Mexico, USA
- Petronius, Gulf of Mexico, USA
- Ram Powell, Gulf of Mexico, USA
- Red Hawk, Gulf of Mexico, USA
- Sable Offshore Energy Project, Sable Island, Canada
- Serrano and Oregano, Gulf of Mexico, USA
- Tahoe, Gulf of Mexico, USA
- Tanzanite, Gulf of Mexico, USA
- Terra Nova, Jeanne d'Arc Basin, Canada
- Thunder Horse Field, Gulf of Mexico, USA
- Troika, Gulf of Mexico, USA
- Typhoon, Gulf of Mexico, USA
- Ursa, Gulf of Mexico, USA
- White Rose Oil and Gas Field, Jeanne d'Arc Basin, Canada

Table A.5 North Atlantic

- Corrib Gas Field, Republic of Ireland
- Foinaven Oil Field, United Kingdom
- Liverpool Bay Oil and Gas Fields, United Kingdom
- Rivers Fields, East Irish Sea, United Kingdom
- Schiehallion Oil Field, United Kingdom

Table A.6 North Sea

- Alba Phase II, North Sea Northern, United Kingdom
- Alvheim North Sea Northern, Norway
- Åsgard, North Sea Northern, Norway
- Balder, North Sea Northern, Norway
- Banff, North Sea Central, United Kingdom
- Blake Flank, North Sea Northern, United Kingdom
- Boulton, North Sea Southern, United Kingdom
- Brigantine, North Sea Southern, United Kingdom
- Britannia, North Sea Central, United Kingdom
- Bruce Phase II, North Sea Northern, United Kingdom
- Buzzard Field North Sea Central, United Kingdom
- Caister Murdoch Phase 3, North Sea Southern, United Kingdom
- Captain, North Sea Central, United Kingdom
- Clair Field, Shetlands, United Kingdom
- Cook Field, North Sea Central, United Kingdom
- Curlew, North Sea Central, United Kingdom
- Dunbar Phase II, North Sea Central, United Kingdom
- Easington Catchment Area (ECA), North Sea Southern, United Kingdom
- Eastern Trough Area Project (ETAP), North Sea Central, United Kingdom
- Ekofisk II, North Sea Central, Norway
- Elgin-Franklin, North Sea Central, United Kingdom
- Erskine, North Sea Central, United Kingdom
- Ettrick Field, United Kingdom `NEW`
- Gannet, North Sea Central, United Kingdom
- Gjøa Field, North Sea Northern, Norway
- Glitne, North Sea Northern, Norway
- Goldeneye Gas Platform, North Sea Northern, United Kingdom
- Gullfaks, North Sea Northern, Norway
- Hanze F2A, Dutch North Sea, Netherlands
- Harding Area Gas Project, North Sea, United Kingdom
- Jade Oil and Gas Platform, North Sea Central, United Kingdom
- Janice, North Sea Central, United Kingdom
- Jotun, North Sea Northern, Norway
- Jura Field, North Sea Northern, United Kingdom
- K5F Gas Field, Netherlands

Table A.6 North Sea *(cont.)*

- Kristin Deepwater Project, Norwegian Sea, Norway
- Leadon, North Sea Northern, United Kingdom
- Lukoil's Kravtsovskoye (D-6) Oil Field, Russia
- MacCulloch, North Sea Central, United Kingdom
- Magnus EOR, Shetlands, United Kingdom
- Mikkel Deepwater Project, Norwegian Sea, Norway
- Mittelplate Redevelopment, Germany
- Njord, North Sea Northern, Norway
- Norne, North Sea Northern, Norway
- NUGGETS, North Sea Northern, United Kingdom
- Ormen Lange, North Sea Northern, Norway
- Oseberg Sør, North Sea Northern, Norway
- Pierce, North Sea Central, United Kingdom
- Prirazlomnoye Oilfield – Barents Sea, Russia
- R Block Development, North Sea Central, United Kingdom
- Ross, North Sea Central, United Kingdom
- Shearwater, North Sea Central, United Kingdom
- Shtokman Gas Condensate Deposit Barents Sea, Russia
- Sigyn Gas Field, Norwegian North Sea, Norway
- Siri, North Sea Northern, Denmark
- Skinfaks Development and Rimfaks Expansion Project, Norway
- Snøhvit Gas Field, Barents Sea, Norway
- Snorre, North Sea Central, Norway
- South Arne, Danish North Sea, Denmark
- Tordis IOR Project, Norway
- Triton, North Sea Central, United Kingdom
- Troll West, North Sea Northern, Norway
- Tyrihans, Norwegian Sea, Norway
- Valhall Flank Water Injection Platform, Norwegian North Sea, Norway
- Viking B, North Sea Southern, United Kingdom
- Visund, North Sea Northern, Norway
- Vixen, North Sea Southern, United Kingdom

Table A.7 South America

- Barracuda and Caratinga Fields, Campos Basin, Brazil
- Bijupira and Salema Fields, Campos Basin, Brazil
- Carina Aries Natural Gas Production, Offshore Block CMA-1, Argentina
- Espadarte, Campos Basin, Brazil
- Marlim Oil Field, Campos Basin, Brazil
- Marlim Sul, Campos Basin, Brazil
- PROCAP 2000, Campos Basin, Brazil
- Roncador, Campos Basin, Brazil

Deepwater Natural Gas and Oil Qualified Fields

The following table lists the nicknames of *qualified* deepwater natural gas and oil fields and leases discovered in the Gulf of Mexico. Leases are listed as having discoveries when a well on the lease is qualified as capable of producing under 30 CFR 550.115 or 30 CFR 550.16. These new producible leases are then designated as either a new field or assigned to an existing field. The designated field name is listed for those leases that have qualified as capable of producing by BOEM. This table provides a listing of field nicknames, location (lease, area, and block), the water depth and the current operator, field name code, field discovery date, field first production date, lease qualification date, lease first production date, and lease expiration date from field.

Deepwater Qualified Fields in the Gulf of Mexico as of 07-01-2014 10:36:58 a.m.

(Leases in field water depths over 1,000 ft.)

https://www.data.boem.gov/homepg/data_center/other/tables/deeptbl2.asp

Handbook of Offshore Oil and Gas Operations. http://dx.doi.org/10.1016/B978-1-85617-558-6.00012-X

Field nickname	Lease	Area	Block	Field water depth (ft)	Operator	Field name code	Field discovery date	Field first production date	Lease qualification date	Lease first production date	Lease expiration date from field
Aconcagua	G19935	MC	305	7,050	ATP Oil & Gas Corporation	MC305	02/21/1999	10/2002	09/20/2002	09/2002	12/2013
Allegheny	G07049	GC	254	3,250	Eni US Operating Co. Inc.	GC254	01/01/1985	10/1999	02/23/1994	09/1999	N/A
Allegheny	G07049	GC	254	3,250	Newfield Exploration Gulf Coast LLC	GC254	01/01/1985	10/1999	02/23/1994	09/1999	N/A
Allegheny	G08010	GC	298	3,250	Eni US Operating Co. Inc.	GC254	01/01/1985	10/1999	04/10/2000	04/2000	N/A
Allegheny	G08010	GC	298	3,250	Newfield Exploration Gulf Coast LLC	GC254	01/01/1985	10/1999	04/10/2000	04/2000	N/A
Amberjack	G05825	MC	109	1,046	Stone Energy Corporation	MC109	11/13/1983	10/1991	08/17/1984	10/1991	N/A
Amberjack	G05826	MC	110	1,046	BP Exploration Inc.	MC109	11/13/1983	10/1991	N/A	N/A	12/1988
Amberjack	G09777	MC	108	1,046	BP Exploration & Production Inc.	MC109	11/13/1983	10/1991	07/01/1992	11/1993	N/A

Amberjack	G09777	MC	108	1,046	Stone Energy Corporation	MC109	11/13/1983	10/1991	07/01/1992	11/1993	N/A
Amberjack	G18192	MC	110	1,046	Apache Corporation	MC109	11/13/1983	10/1991	08/18/1999	05/2000	N/A
Angus	G15545	GC	112	1,839	Marubeni Oil & Gas (USA) Inc.	GC112	06/08/1997	09/1999	02/25/1998	10/1999	N/A
Angus	G15546	GC	113	1,839	Marubeni Oil & Gas (USA) Inc.	GC112	06/08/1997	09/1999	09/11/1999	09/1999	N/A
Angus	G15546	GC	113	1,839	Shell Offshore Inc.	GC112	06/08/1997	09/1999	09/11/1999	09/1999	N/A
Appaloosa	G18244	MC	459	2,833	Eni US Operating Co. Inc.	MC503	12/11/2007	12/2010	09/15/2008	N/A	03/2013
Appaloosa	G18245	MC	460	2,833	Eni US Operating Co. Inc.	MC503	12/11/2007	12/2010	09/15/2008	06/2011	N/A
Appaloosa	G27277	MC	503	2,833	Eni US Operating Co. Inc.	MC503	12/11/2007	12/2010	02/19/2008	12/2011	N/A
Appaloosa	G27277	MC	503	2,833	LLOG Exploration Offshore, L.L.C.	MC503	12/11/2007	12/2010	02/19/2008	12/2011	N/A
Appomattox	G26253	MC	392	7,257	Shell Gulf of Mexico Inc.	MC392	12/21/2009	N/A	01/10/2012	N/A	N/A

(Continued)

Field nickname	Lease	Area	Block	Field water depth (ft)	Operator	Field name code	Field discovery date	Field first production date	Lease qualification date	Lease first production date	Lease expiration date from field
Ariel	G07944	MC	429	6,132	BP Exploration & Production Inc.	MC429	11/20/1995	04/2004	05/13/1996	04/2004	N/A
Arnold	G13084	EW	963	1,682	Marathon Oil Company	EW963	06/12/1996	05/1998	07/15/1996	05/1998	N/A
Aspen	G20051	GC	243	3,039	Nexen Petroleum U.S.A. Inc.	GC243	01/27/2001	12/2002	10/31/2001	12/2002	N/A
Aspen	G20051	GC	243	3,039	Walter Oil & Gas Corporation	GC243	01/27/2001	12/2002	10/31/2001	12/2002	N/A
Atlantis	G15604	GC	699	6,285	BP Exploration & Production Inc.	GC743	05/12/1998	10/2007	10/10/2004	04/2013	N/A
Atlantis	G15606	GC	742	6,285	BP Exploration & Production Inc.	GC743	05/12/1998	10/2007	N/A	04/2009	N/A
Atlantis	G15607	GC	743	6,285	BP Exploration & Production Inc.	GC743	05/12/1998	10/2007	10/06/2004	10/2007	N/A

Atlantis	G15608	GC	744	6,285	BP Exploration & Production Inc.	GC743	05/12/1998	10/2007	N/A	N/A	N/A
Atlas	G23458	LL	50	8,944	Anadarko Petroleum Corporation	LL050	05/29/2003	07/2007	04/17/2006	07/2007	02/2012
Atlas Nw	G23450	LL	5	8,807	Anadarko Petroleum Corporation	LL005	01/13/2004	07/2007	04/17/2006	07/2007	02/2012
Auger	G07493	GB	427	2,847	Shell Offshore Inc.	GB426	05/01/1987	04/1994	08/21/1987	05/2007	N/A
Auger	G07498	GB	471	2,847	Shell Offshore Inc.	GB426	05/01/1987	04/1994	08/21/1987	N/A	N/A
Auger	G08241	GB	426	2,847	Shell Offshore Inc.	GB426	05/01/1987	04/1994	05/16/1990	05/1994	N/A
Auger	G08248	GB	470	2,847	Shell Offshore Inc.	GB426	05/01/1987	04/1994	05/27/1992	04/1994	N/A
Baccarat	G25123	GC	178	1,404	W & T Offshore, Inc.	GC178	05/11/2004	08/2005	06/10/2004	08/2005	06/2009
Baha	G08580	AC	600	7,620	Shell Offshore Inc.	AC600	05/23/1996	N/A	09/26/1996	N/A	12/2001
Balboa	G22288	EB	597	3,352	Apache Deepwater LLC	EB597	07/02/2001	12/2010	08/29/2001	12/2010	N/A

(Continued)

Field nickname	Lease	Area	Block	Field water depth (ft)	Operator	Field name code	Field discovery date	Field first production date	Lease qualification date	Lease first production date	Lease expiration date from field
Baldpate	G07461	GB	259	1,605	Hess Corporation	GB260	11/01/1991	09/1998	05/18/1995	01/1999	N/A
Baldpate	G07462	GB	260	1,605	Hess Corporation	GB260	11/01/1991	09/1998	09/14/1992	09/1998	N/A
Baldpate	G09216	GB	215	1,605	Hess Corporation	GB260	11/01/1991	09/1998	05/16/1995	06/2003	N/A
Baldpate	G14224	GB	216	1,605	Hess Corporation	GB260	11/01/1991	09/1998	04/09/1997	05/1999	N/A
Bass Lite	G18603	AT	426	6,623	Apache Deepwater LLC	AT426	01/25/2001	02/2008	01/19/2006	02/2008	N/A
Big Foot	G16942	WR	29	5,444	Chevron U.S.A. Inc.	WR029	12/02/2005	N/A	05/09/2006	N/A	N/A
Bison	G07028	GC	166	2,348	Exxon Mobil Corporation	GC166	03/01/1986	N/A	06/02/1986	N/A	12/1994
Blind Faith	G16641	MC	696	6,952	Chevron U.S.A. Inc.	MC696	05/30/2001	11/2007	05/17/2005	11/2007	N/A
Blind Faith	G21182	MC	695	6,952	Chevron U.S.A. Inc.	MC696	05/30/2001	11/2007	11/25/2008	11/2008	N/A
Bonsai	G16910	AT	398	3,619	BP Exploration & Production Inc.	AT398	08/14/2005	N/A	09/09/2006	N/A	02/2007
Boomvang East	G09191	EB	688	3,756	Anadarko Petroleum Corporation	EB688	05/01/1988	06/2002	05/13/1992	06/2002	04/2010

Boomvang North	G09183	EB	642	3,397	Anadarko Petroleum Corporation	EB643	12/13/1997	08/2002	12/07/1999	06/2002	N/A
Boomvang North	G09184	EB	643	3,397	Anadarko Petroleum Corporation	EB643	12/13/1997	08/2002	10/01/1999	08/2002	N/A
Boomvang North	G19027	EB	598	3,397	Anadarko Petroleum Corporation	EB643	12/13/1997	08/2002	11/02/2004	10/2004	N/A
Boomvang North	G19028	EB	599	3,397	Anadarko Petroleum Corporation	EB643	12/13/1997	08/2002	12/04/2002	10/2004	N/A
Boomvang West	G09183	EB	642	3,749	Anadarko Petroleum Corporation	EB642	10/31/1999	06/2002	12/07/1999	06/2002	N/A
Boomvang West	G20729	EB	686	3,749	Anadarko US Offshore Corporation	EB642	10/31/1999	06/2002	N/A	02/2004	09/2008
Boomvang West	G21374	EB	641	3,749	Anadarko Petroleum Corporation	EB642	10/31/1999	06/2002	03/26/2003	03/2003	N/A
Boris	G16727	GC	282	2,367	BHP Billiton Petroleum (GOM) Inc.	GC282	09/29/2001	02/2003	01/02/2002	02/2003	N/A

(Continued)

Field nickname	Lease	Area	Block	Field water depth (ft)	Operator	Field name code	Field discovery date	Field first production date	Lease qualification date	Lease first production date	Lease expiration date from field
Boris	G16727	GC	282	2,367	Energy Resource Technology GOM, Inc.	GC282	09/29/2001	02/2003	01/02/2002	02/2003	N/A
Boris	G26302	GC	238	2,367	BHP Billiton Petroleum (GOM) Inc.	GC282	09/29/2001	02/2003	N/A	07/2011	N/A
Boris	G26302	GC	238	2,367	Energy Resource Technology GOM, Inc.	GC282	09/29/2001	02/2003	N/A	07/2011	N/A
Brutus	G07995	GC	158	2,939	Shell Offshore Inc.	GC158	03/01/1989	08/2001	05/13/1992	09/2001	N/A
Brutus	G07998	GC	202	2,939	Shell Offshore Inc.	GC158	03/01/1989	08/2001	N/A	08/2001	N/A
Brutus	G12210	GC	201	2,939	LLOG Exploration Offshore, L.L.C.	GC158	03/01/1989	08/2001	11/20/2000	11/2000	N/A
Brutus	G12210	GC	201	2,939	Shell Offshore Inc.	GC158	03/01/1989	08/2001	11/20/2000	11/2000	N/A
Brutus	G24154	GC	157	2,939	LLOG Exploration Offshore, L.L.C.	GC158	03/01/1989	08/2001	09/09/2003	05/2005	N/A

Field	ID	Area	Block	Depth	Operator	Lease					
Bullwinkle	G05889	GC	65	1,336	SandRidge Energy Offshore, LLC	GC065	10/01/1983	07/1989	02/23/1984	07/1989	N/A
Bullwinkle	G05890	GC	66	1,336	Marathon Oil Company	GC065	10/01/1983	07/1989	07/29/1988	N/A	12/1990
Bullwinkle	G05900	GC	109	1,336	SandRidge Energy Offshore, LLC	GC065	10/01/1983	07/1989	12/13/1984	08/1989	N/A
Bullwinkle	G05901	GC	110	1,336	Marathon Oil Company	GC065	10/01/1983	07/1989	11/21/1984	N/A	12/1990
Bullwinkle	G07005	GC	64	1,336	Shell Offshore Inc.	GC065	10/01/1983	07/1989	04/15/1985	02/1992	05/1998
Bullwinkle	G07014	GC	108	1,336	Eni Petroleum Co. Inc.	GC065	10/01/1983	07/1989	04/02/1987	N/A	06/1991
Bullwinkle	G14023	GC	110	1,336	Shell Offshore Inc.	GC065	10/01/1983	07/1989	01/23/1996	01/1996	03/2005
Bullwinkle	G14668	GC	108	1,336	SandRidge Energy Offshore, LLC	GC065	10/01/1983	07/1989	04/02/2006	04/2006	N/A
Callisto	G21190	MC	875	7,789	Anadarko Petroleum Corporation	MC876	03/08/2001	01/2011	N/A	01/2011	N/A

(Continued)

Field nickname	Lease	Area	Block	Field water depth (ft)	Operator	Field name code	Field discovery date	Field first production date	Lease qualification date	Lease first production date	Lease expiration date from field
Callisto	G21191	MC	876	7,789	Anadarko Petroleum Corporation	MC876	03/08/2001	01/2011	06/24/2008	N/A	N/A
Camden Hills	G19939	MC	348	7,223	ATP Oil & Gas Corporation	MC348	08/04/1999	10/2002	09/14/2000	10/2002	N/A
Camden Hills	G19939	MC	348	7,223	Shell Gulf of Mexico Inc.	MC348	08/04/1999	10/2002	09/14/2000	10/2002	N/A
Cascade	G16965	WR	206	8,148	Petrobras America Inc.	WR206	04/14/2002	02/2012	04/21/2006	N/A	N/A
Cheyenne	G23480	LL	399	8,986	Anadarko Petroleum Corporation	LL399	07/20/2004	10/2007	04/17/2006	10/2007	06/2012
Cheyenne	G23481	LL	400	8,986	Anadarko Petroleum Corporation	LL399	07/20/2004	10/2007	01/23/2012	02/2012	N/A
Chinook	G16987	WR	425	8,842	Petrobras America Inc.	WR469	06/19/2003	09/2012	N/A	N/A	N/A
Chinook	G16997	WR	469	8,842	Petrobras America Inc.	WR469	06/19/2003	09/2012	03/24/2006	09/2012	N/A
Claymore	G16865	AT	140	3,739	Anadarko US Offshore Corporation	AT140	05/05/2006	N/A	06/16/2006	N/A	12/2007
Clipper	G15571	GC	299	3,429	ATP Oil & Gas Corporation	GC299	10/05/2005	03/2013	02/10/2006	N/A	09/2009

Clipper	G15571	GC	299	3,429	Shell Gulf of Mexico Inc.	GC299	10/05/2005	03/2013	02/10/2006	N/A	09/2009
Clipper	G22939	GC	300	3,429	Bennu Oil & Gas, LLC	GC299	10/05/2005	03/2013	01/12/2007	03/2013	N/A
Clipper	G22939	GC	300	3,429	Murphy Exploration & Production Company – USA	GC299	10/05/2005	03/2013	01/12/2007	03/2013	N/A
Cognac	G02638	MC	194	1,022	Shell Offshore Inc.	MC194	07/01/1975	09/1979	03/31/1977	09/1980	N/A
Cognac	G02639	MC	195	1,022	Shell Offshore Inc.	MC194	07/01/1975	09/1979	03/01/1982	03/1982	N/A
Cognac	G02642	MC	150	1,022	Shell Offshore Inc.	MC194	07/01/1975	09/1979	11/08/1977	11/1979	N/A
Cognac	G02643	MC	151	1,022	Shell Offshore Inc.	MC194	07/01/1975	09/1979	03/31/1977	09/1979	N/A
Condor	G28077	GC	448	3,266	Deep Gulf Energy LP	GC448	01/23/2008	06/2011	02/26/2008	06/2011	N/A
Constitution	G21811	GC	679	4,998	Anadarko Petroleum Corporation	GC680	10/31/2001	03/2006	05/23/2005	06/2006	N/A
Constitution	G21811	GC	679	4,998	Eni US Operating Co. Inc.	GC680	10/31/2001	03/2006	05/23/2005	06/2006	N/A

(Continued)

Field nickname	Lease	Area	Block	Field water depth (ft)	Operator	Field name code	Field discovery date	Field first production date	Lease qualification date	Lease first production date	Lease expiration date from field
Constitution	G21811	GC	679	4,998	Noble Energy, Inc.	GC680	10/31/2001	03/2006	05/23/2005	06/2006	N/A
Constitution	G22987	GC	680	4,998	Anadarko Petroleum Corporation	GC680	10/31/2001	03/2006	05/23/2005	03/2006	N/A
Cooper	G07485	GB	387	2,210	Newfield Exploration Gulf Coast LLC	GB388	03/16/1989	09/1995	07/10/1990	05/1996	09/2005
Cooper	G07486	GB	388	2,210	Newfield Exploration Gulf Coast LLC	GB388	03/16/1989	09/1995	06/21/1991	09/1995	09/2005
Cooper	G08232	GB	344	2,210	Energy Resource Technology GOM, Inc.	GB388	03/16/1989	09/1995	07/31/1997	07/1997	07/2008
Cooper	G08232	GB	344	2,210	Hess Corporation	GB388	03/16/1989	09/1995	07/31/1997	07/1997	07/2008
Cooper	G08232	GB	344	2,210	Newfield Exploration Gulf Coast LLC	GB388	03/16/1989	09/1995	07/31/1997	07/1997	07/2008
Cottonwood	G15860	GB	244	2,089	Petrobras America Inc.	GB244	08/15/2001	02/2007	12/10/2001	02/2007	N/A

Field	Lease	Area	Block	Water Depth	Operator					
Coulomb	G08496	MC	657	7,558	Shell Offshore Inc.	11/01/1987	06/2004	03/31/1988	06/2004	N/A
Coulomb	G19974	MC	613	7,558	Shell Offshore Inc.	11/01/1987	06/2004	N/A	07/2004	N/A
Crosby	G09895	MC	898	4,148	Shell Offshore Inc.	01/04/1998	12/2001	03/09/2005	03/2007	N/A
Crosby	G09896	MC	899	4,148	Shell Offshore Inc.	01/04/1998	12/2001	03/19/1998	12/2001	N/A
Crosby	G09904	MC	942	4,148	Shell Deepwater Development Inc.	01/04/1998	12/2001	06/08/1999	N/A	04/2001
Crosby	G16661	MC	941	4,148	Bennu Oil & Gas, LLC	01/04/1998	12/2001	04/26/1999	10/2010	N/A
Crosby	G16661	MC	941	4,148	Shell Offshore Inc.	01/04/1998	12/2001	04/26/1999	10/2010	N/A
Crosby	G16661	MC	941	4,148	Statoil USA E&P Inc.	01/04/1998	12/2001	04/26/1999	10/2010	N/A
Crosby	G24130	MC	942	4,148	Bennu Oil & Gas, LLC	01/04/1998	12/2001	N/A	02/2012	N/A
Crosby	G24130	MC	942	4,148	Statoil USA E&P Inc.	01/04/1998	12/2001	N/A	02/2012	N/A
Cyclops	G09995	AT	8	3,135	Shell Deepwater Development Inc.	04/26/1997	N/A	08/14/1997	N/A	04/1998

(Continued)

Field nickname	Lease	Area	Block	Field water depth (ft)	Operator	Field name code	Field discovery date	Field first production date	Lease qualification date	Lease first production date	Lease expiration date from field
Dalmatian	G10440	DC	48	5,876	Murphy Exploration & Production Company - USA	DC048	09/29/2008	N/A	12/01/2008	04/2014	N/A
Dalmatian North	G10437	DC	4	5,822	Murphy Exploration & Production Company - USA	DC004	04/30/2010	N/A	06/17/2010	06/2014	N/A
Dalmatian South	G23488	DC	134	6,394	Murphy Exploration & Production Company - USA	DC134	09/27/2012	N/A	03/18/2013	N/A	N/A
Daniel Boone	G20083	GC	646	4,230	W & T Offshore, Inc.	GC646	01/10/2004	08/2008	04/21/2004	09/2009	N/A
Deep Blue	G21813	GC	723	5,040	Noble Energy, Inc.	GC723	05/03/2010	N/A	01/01/2011	N/A	05/2012
Devils Tower	G16644	MC	728	5,343	Eni US Operating Co. Inc.	MC773	12/13/1999	05/2004	09/16/2002	11/2005	N/A
Devils Tower	G16647	MC	772	5,343	Chevron U.S.A. Inc.	MC773	12/13/1999	05/2004	09/12/2002	06/2005	N/A
Devils Tower	G16647	MC	772	5,343	Eni US Operating Co. Inc.	MC773	12/13/1999	05/2004	09/12/2002	06/2005	N/A

Devils Tower	G19996	MC	773	5,343	Eni US Operating Co. Inc.	MC773	12/13/1999	05/2004	09/14/2000	05/2004	N/A
Devils Tower	G24107	MC	771	5,343	Deep Gulf Energy II, LLC	MC773	12/13/1999	05/2004	08/17/2004	12/2005	N/A
Devils Tower	G24107	MC	771	5,343	Eni US Operating Co. Inc.	MC773	12/13/1999	05/2004	08/17/2004	12/2005	N/A
Diamond	G06951	MC	445	2,095	Anadarko US Offshore Corporation	MC445	12/05/1992	10/1993	10/16/1993	10/1993	09/1999
Diana	G08211	EB	945	4,646	Exxon Mobil Corporation	EB945	08/01/1990	05/2000	01/24/1994	06/2000	N/A
Diana	G08212	EB	946	4,646	Exxon Mobil Corporation	EB945	08/01/1990	05/2000	04/23/1996	05/2000	N/A
Diana	G08214	EB	989	4,646	Exxon Mobil Corporation	EB945	08/01/1990	05/2000	02/23/2000	06/2000	N/A
Diana South	G09249	AC	65	4,852	Exxon Mobil Corporation	AC065	03/24/1997	03/2004	09/24/1997	03/2004	N/A
Dionysus	G04498	VK	864	1,482	Conoco Inc.	VK864	10/01/1981	N/A	04/26/1982	N/A	10/1985
Double Corona	G06992	GC	27	1,593	Eni Petroleum Co. Inc.	GC027	07/01/1989	N/A	09/29/1989	N/A	12/1990
Dulcimer	G17351	GB	367	1,123	Mariner Energy, Inc.	GB367	02/09/1998	04/1999	N/A	04/1999	03/2002

(Continued)

Field nickname	Lease	Area	Block	Field water depth (ft)	Operator	Field name code	Field discovery date	Field first production date	Lease qualification date	Lease first production date	Lease expiration date from field
East Anstey	G09837	MC	607	6,536	BP Exploration & Production Inc.	MC607	11/12/1997	11/2003	03/16/1998	N/A	04/2011
East Anstey	G09838	MC	608	6,536	BP Exploration & Production Inc.	MC607	11/12/1997	11/2003	N/A	11/2003	04/2011
East Anstey	G19966	MC	562	6,536	BP Exploration & Production Inc.	MC607	11/12/1997	11/2003	07/17/2007	06/2012	N/A
East Anstey	G27278	MC	519	6,536	Noble Energy, Inc.	MC607	11/12/1997	11/2003	06/24/2009	06/2012	N/A
Einset	G08471	VK	873	3,584	Shell Offshore Inc.	VK873	03/01/1988	12/2001	06/07/1988	N/A	05/1991
Einset	G19908	VK	873	3,584	Shell Offshore Inc.	VK873	03/01/1988	12/2001	12/18/2000	12/2001	07/2008
El Toro	G05893	GC	69	1,437	Conoco Inc.	GC069	09/13/1984	N/A	N/A	N/A	12/1988
El Toro	G13159	GC	69	1,437	Devon SFS Operating, Inc.	GC069	09/13/1984	N/A	04/23/1996	N/A	12/2000
El Toro	G13159	GC	69	1,437	Shell Deepwater Development Inc.	GC069	09/13/1984	N/A	04/23/1996	N/A	12/2000

Europa	G07969	MC	890	3,871	Shell Offshore Inc.	MC935	04/22/1994	02/2000	N/A	01/2000	N/A
Europa	G07975	MC	934	3,871	Shell Offshore Inc.	MC935	04/22/1994	02/2000	06/19/1995	02/2000	N/A
Europa	G07976	MC	935	3,871	Shell Offshore Inc.	MC935	04/22/1994	02/2000	11/01/1994	07/2006	N/A
Falcon	G19025	EB	579	3,454	Marubeni Oil & Gas (USA) Inc.	EB579	03/27/2001	03/2003	03/18/2003	03/2003	03/2011
Falcon	G19030	EB	623	3,454	Marubeni Oil & Gas (USA) Inc.	EB579	03/27/2001	03/2003	03/15/2003	03/2003	N/A
Flying Dutchman	G22971	GC	511	3,831	Marathon Oil Company	GC511	04/17/2010	N/A	04/18/2011	N/A	04/2011
Fourier	G08823	MC	522	6,884	BP Exploration & Production Inc.	MC522	07/01/1989	11/2003	03/20/1992	11/2003	N/A
Fourier	G08831	MC	566	6,884	BP Exploration & Production Inc.	MC522	07/01/1989	11/2003	05/11/2008	05/2008	N/A
Fourier	G09821	MC	520	6,884	BP Exploration & Production Inc.	MC522	07/01/1989	11/2003	05/06/1997	12/2003	N/A

(Continued)

Field nickname	Lease	Area	Block	Field water depth (ft)	Operator	Field name code	Field discovery date	Field first production date	Lease qualification date	Lease first production date	Lease expiration date from field
Freedom	G24133	MC	992	6,081	Marathon Oil Company	MC948	08/15/2008	N/A	08/19/2013	N/A	N/A
Freedom	G24133	MC	992	6,081	Noble Energy, Inc.	MC948	08/15/2008	N/A	08/19/2013	N/A	N/A
Freedom	G28030	MC	948	6,081	Noble Energy, Inc.	MC948	08/15/2008	N/A	03/16/2009	N/A	N/A
Friesian	G16762	GC	599	3,838	Plains Exploration & Production Company	GC599	10/08/2006	N/A	01/26/2007	N/A	05/2011
Friesian	G16772	GC	643	3,838	Plains Exploration & Production Company	GC599	10/08/2006	N/A	03/17/2009	N/A	05/2011
Front Runner	G21790	GC	338	3,327	Murphy Exploration & Production Company – USA	GC339	01/23/2001	12/2004	11/06/2001	12/2004	N/A
Front Runner	G21790	GC	338	3,327	Walter Oil & Gas Corporation	GC339	01/23/2001	12/2004	11/06/2001	12/2004	N/A
Front Runner	G21791	GC	339	3,327	Murphy Exploration & Production Company – USA	GC339	01/23/2001	12/2004	09/14/2001	03/2005	N/A

Front Runner	G22950	GC	382	3,327	Murphy Exploration & Production Company – USA	GC339	01/23/2001	12/2004	10/25/2002	04/2005	N/A
Fuji	G08879	GC	505	4,256	Texaco Exploration and Production Inc.	GC506	01/30/1995	N/A	01/09/1998	N/A	12/1999
Fuji	G08880	GC	506	4,256	Texaco Exploration and Production Inc.	GC506	01/30/1995	N/A	02/12/1996	N/A	12/1999
Fuji	G11064	GC	461	4,256	BP Exploration & Oil Inc.	GC506	01/30/1995	N/A	12/15/1998	N/A	04/1999
Gb302	G09223	GB	302	2,346	N/A	GB302	02/01/1991	01/2008	05/20/1991	N/A	09/1995
Gb302	G24479	GB	302	2,346	Anadarko Petroleum Corporation	GB302	02/01/1991	01/2008	02/24/2006	01/2008	N/A
Gb302	G24479	GB	302	2,346	Shell Offshore Inc.	GB302	02/01/1991	01/2008	02/24/2006	01/2008	N/A

(Continued)

Field nickname	Lease	Area	Block	Field water depth (ft)	Operator	Field name code	Field discovery date	Field first production date	Lease qualification date	Lease first production date	Lease expiration date from field
Gb302	G24479	GB	302	2,346	Walter Oil & Gas Corporation	GB302	02/01/1991	01/2008	02/24/2006	01/2008	N/A
Gb302	G27632	GB	258	2,346	Anadarko Petroleum Corporation	GB302	02/01/1991	01/2008	03/16/2007	06/2008	N/A
Gb302	G27632	GB	258	2,346	W & T Offshore, Inc.	GB302	02/01/1991	01/2008	03/16/2007	06/2008	N/A
Geauxpher	G19127	GB	462	2,789	Apache Deepwater LLC	GB462	01/25/2007	05/2009	12/30/2008	05/2009	N/A
Geauxpher	G26664	GB	506	2,789	Energy Resource Technology GOM, Inc.	GB462	01/25/2007	05/2009	09/08/2008	09/2008	N/A
Gemini	G08805	MC	291	3,529	Texaco Exploration and Production Inc.	MC292	09/07/1995	05/1999	02/26/1998	N/A	05/1998
Gemini	G08806	MC	292	3,529	Chevron U.S.A. Inc.	MC292	09/07/1995	05/1999	05/28/1999	05/1999	N/A
Gemini	G08806	MC	292	3,529	Noble Energy, Inc.	MC292	09/07/1995	05/1999	05/28/1999	05/1999	N/A
Gemini	G13114	MC	248	3,529	Chevron U.S.A. Inc.	MC292	09/07/1995	05/1999	06/04/2001	N/A	06/2001

Gemini	G27268	MC	248	3,529	Noble Energy, Inc.	MC292	09/07/1995	05/1999	07/27/2007	11/2008	N/A
Genesis	G05909	GC	161	2,786	Chevron U.S.A. Inc.	GC205	09/01/1988	01/1999	03/04/1991	02/2000	N/A
Genesis	G05911	GC	205	2,786	Chevron U.S.A. Inc.	GC205	09/01/1988	01/1999	01/05/1989	01/1999	N/A
Genesis	G07996	GC	160	2,786	Chevron U.S.A. Inc.	GC205	09/01/1988	01/1999	07/14/1992	09/1999	N/A
Genesis	G15565	GC	248	2,786	Shell Offshore Inc.	GC205	09/01/1988	01/1999	N/A	07/2004	N/A
Gladden	G18292	MC	800	3,116	W & T Offshore, Inc.	MC800	04/29/2008	02/2011	07/16/2008	02/2011	N/A
Gomez	G07956	MC	755	2,941	Exxon Mobil Corporation	MC755	03/19/1986	03/2006	06/19/1986	N/A	04/1992
Gomez	G14016	MC	711	2,941	ATP Oil & Gas Corporation	MC755	03/19/1986	03/2006	12/30/1997	03/2006	06/2013
Gomez	G14017	MC	755	2,941	Anadarko E&P Company LP	MC755	03/19/1986	03/2006	06/11/1998	N/A	04/2001
Gomez	G24104	MC	754	2,941	ATP Oil & Gas Corporation	MC755	03/19/1986	03/2006	06/08/2009	02/2011	12/2012
Gomez	G24105	MC	755	2,941	ATP Oil & Gas Corporation	MC755	03/19/1986	03/2006	10/19/2007	10/2007	10/2013

(Continued)

Field nickname	Lease	Area	Block	Field water depth (ft)	Operator	Field name code	Field discovery date	Field first production date	Lease qualification date	Lease first production date	Lease expiration date from field
Gomez	G24105	MC	755	2,941	Nexen Petroleum U.S.A. Inc.	MC755	03/19/1986	03/2006	10/19/2007	10/2007	10/2013
Goose	G22899	MC	751	1,589	Statoil USA E&P Inc.	MC751	12/15/2002	11/2012	04/24/2003	N/A	04/2008
Goose	G33175	MC	751	1,589	LLOG Exploration Offshore, L.L.C.	MC751	12/15/2002	11/2012	10/04/2011	11/2012	N/A
Great White	G17561	AC	813	7,918	Shell Offshore Inc.	AC857	05/18/2002	03/2010	10/10/2006	N/A	N/A
Great White	G17565	AC	857	7,918	Shell Offshore Inc.	AC857	05/18/2002	03/2010	10/10/2006	03/2010	N/A
Great White	G17571	AC	901	7,918	Shell Offshore Inc.	AC857	05/18/2002	03/2010	10/21/2010	10/2010	N/A
Great White	G24593	AC	812	7,918	Shell Offshore Inc.	AC857	05/18/2002	03/2010	N/A	05/2011	N/A
Gretchen	G20034	GC	114	2,506	Devon Energy Production Company, L.P.	GC114	12/18/1999	N/A	03/10/2000	N/A	07/2008
Gretchen	G20034	GC	114	2,506	Shell Offshore Inc.	GC114	12/18/1999	N/A	03/10/2000	N/A	07/2008

Gunnison	G15927	GB	625	3,064	Anadarko Petroleum Corporation	GB668	05/09/2000	12/2003	12/15/2004	07/2006	N/A
Gunnison	G17406	GB	667	3,064	Anadarko Petroleum Corporation	GB668	05/09/2000	12/2003	09/14/2001	12/2003	N/A
Gunnison	G17407	GB	668	3,064	Anadarko Petroleum Corporation	GB668	05/09/2000	12/2003	09/14/2000	12/2003	N/A
Gunnison	G17408	GB	669	3,064	Anadarko Petroleum Corporation	GB668	05/09/2000	12/2003	01/11/2001	12/2003	N/A
Hadrian South	G21450	KC	963	7,508	Exxon Mobil Corporation	KC964	09/21/2008	N/A	01/08/2010	N/A	07/2011
Hadrian South	G21451	KC	964	7,508	Exxon Mobil Corporation	KC964	09/21/2008	N/A	03/18/2009	N/A	N/A
Hal	G20403	WR	848	7,657	Statoil Gulf of Mexico LLC	WR848	01/15/2008	N/A	04/04/2008	N/A	08/2008
Harrier	G20744	EB	758	4,114	Marubeni Oil & Gas (USA) Inc.	EB759	01/28/2003	01/2004	N/A	N/A	09/2008
Harrier	G20745	EB	759	4,114	Marubeni Oil & Gas (USA) Inc.	EB759	01/28/2003	01/2004	04/24/2003	01/2004	09/2008

(Continued)

Field nickname	Lease	Area	Block	Field water depth (ft)	Operator	Field name code	Field discovery date	Field first production date	Lease qualification date	Lease first production date	Lease expiration date from field
Hawkes	G21764	MC	508	4,082	Exxon Mobil Corporation	MC509	11/20/2001	N/A	05/06/2004	N/A	04/2009
Hawkes	G21765	MC	509	4,082	Exxon Mobil Corporation	MC509	11/20/2001	N/A	10/01/2002	N/A	04/2009
Heidelberg	G24194	GC	859	5,357	Anadarko Petroleum Corporation	GC859	01/23/2009	N/A	05/21/2012	N/A	N/A
Holstein	G11080	GC	644	4,341	Freeport-McMoRan Oil & Gas LLC	GC644	01/09/1999	12/2004	03/29/1999	12/2004	N/A
Holstein	G11081	GC	645	4,341	Freeport-McMoRan Oil & Gas LLC	GC644	01/09/1999	12/2004	04/06/2005	04/2005	N/A
Holstein	G18423	GC	688	4,341	Freeport-McMoRan Oil & Gas LLC	GC644	01/09/1999	12/2004	12/25/2008	12/2008	N/A
Hoover	G10380	AC	25	4,807	Exxon Mobil Corporation	AC025	01/30/1997	09/2000	09/11/1997	09/2000	N/A
Hoover	G10381	AC	26	4,807	Exxon Mobil Corporation	AC025	01/30/1997	09/2000	12/15/2001	N/A	N/A
Hornet	G22947	GC	379	3,878	Hess Corporation	GC379	12/14/2001	N/A	05/17/2002	N/A	06/2012

Ida/Fastball	G21160	VK	1003	4,914	W & T Offshore, Inc.	VK1003	04/15/2007	10/2009	01/27/2009	10/2009	N/A
Jack	G17015	WR	758	6,963	Chevron U.S.A. Inc.	WR759	07/09/2004	N/A	01/24/2006	N/A	N/A
Jack	G17016	WR	759	6,963	Chevron U.S.A. Inc.	WR759	07/09/2004	N/A	04/15/2005	N/A	N/A
Jolliet	G04518	GC	184	1,718	MC Offshore Petroleum, LLC	GC184	07/01/1981	11/1989	04/16/1982	11/1989	N/A
Jolliet	G05910	GC	185	1,718	Texaco Exploration and Production Inc.	GC184	07/01/1981	11/1989	04/01/1985	N/A	09/1996
Jubilee	G18577	AT	349	8,778	Anadarko Petroleum Corporation	AT349	03/30/2003	10/2007	04/17/2006	10/2007	N/A
Jubilee	G23473	LL	309	8,778	Anadarko Petroleum Corporation	AT349	03/30/2003	10/2007	04/17/2006	11/2007	03/2012
Julia	G20351	WR	584	7,104	Exxon Mobil Corporation	WR627	04/07/2007	N/A	10/14/2008	N/A	N/A
Julia	G20361	WR	627	7,104	Exxon Mobil Corporation	WR627	04/07/2007	N/A	08/23/2007	N/A	N/A

(Continued)

Field nickname	Lease	Area	Block	Field water depth (ft)	Operator	Field name code	Field discovery date	Field first production date	Lease qualification date	Lease first production date	Lease expiration date from field
K2	G11075	GC	562	4,034	Anadarko Petroleum Corporation	GC562	08/14/1999	05/2005	11/02/1999	05/2005	N/A
K2	G11076	GC	563	4,034	Anadarko Petroleum Corporation	GC562	08/14/1999	05/2005	03/09/2006	N/A	05/2010
K2	G11076	GC	563	4,034	Eni US Operating Co. Inc.	GC562	08/14/1999	05/2005	03/09/2006	N/A	05/2010
K2	G16753	GC	561	4,034	Anadarko Petroleum Corporation	GC562	08/14/1999	05/2005	N/A	N/A	N/A
K2	G21801	GC	518	4,034	Anadarko Petroleum Corporation	GC562	08/14/1999	05/2005	03/18/2006	01/2006	N/A
K2	G21801	GC	518	4,034	Eni US Operating Co. Inc.	GC562	08/14/1999	05/2005	03/18/2006	01/2006	N/A
K2	G21807	GC	606	4,034	Anadarko Petroleum Corporation	GC562	08/14/1999	05/2005	11/03/2009	11/2009	N/A
Kaskida	G19530	KC	244	5,775	BP Exploration & Production Inc.	KC292	05/22/2006	N/A	11/08/2007	N/A	04/2011

Kaskida	G25792	KC	292	5,775	BP Exploration & Production Inc.	KC292	05/22/2006	N/A	07/27/2006	N/A	N/A
Kepler	G07937	MC	383	5,741	BP Exploration & Production Inc.	MC383	08/01/1987	04/2004	11/06/1987	04/2004	N/A
King Kong	G05097	GC	472	3,813	Eni US Operating Co. Inc.	GC472	02/01/1989	02/2002	04/03/1989	02/2002	12/2012
King Kong	G05922	GC	473	3,813	Eni US Operating Co. Inc.	GC472	02/01/1989	02/2002	06/24/1992	04/2003	12/2012
King Kong	G20074	GC	516	3,813	Eni US Operating Co. Inc.	GC472	02/01/1989	02/2002	04/05/2002	04/2002	12/2012
King'S Peak	G09789	MC	173	6,509	ATP Oil & Gas Corporation	DC133	03/01/1993	09/2002	11/11/2002	N/A	03/2013
King'S Peak	G09790	MC	217	6,509	ATP Oil & Gas Corporation	DC133	03/01/1993	09/2002	10/22/1997	10/2002	03/2013
King'S Peak	G10444	DC	133	6,509	ATP Oil & Gas Corporation	DC133	03/01/1993	09/2002	11/18/1993	09/2002	08/2009

(Continued)

Field nickname	Lease	Area	Block	Field water depth (ft)	Operator	Field name code	Field discovery date	Field first production date	Lease qualification date	Lease first production date	Lease expiration date from field
King'S Peak	G10445	DC	177	6,509	ATP Oil & Gas Corporation	DC133	03/01/1993	09/2002	10/30/1997	N/A	06/2009
King/Horn Mt.	G08484	MC	84	5,300	Freeport-McMoRan Oil & Gas LLC	MC084	01/01/1993	04/2002	11/18/1993	06/2003	N/A
King/Horn Mt.	G08797	MC	85	5,300	Freeport-McMoRan Oil & Gas LLC	MC084	01/01/1993	04/2002	11/04/1997	04/2002	N/A
King/Horn Mt.	G10977	MC	129	5,300	Freeport-McMoRan Oil & Gas LLC	MC084	01/01/1993	04/2002	11/05/1997	04/2009	N/A
King/Horn Mt.	G18194	MC	126	5,300	Freeport-McMoRan Oil & Gas LLC	MC084	01/01/1993	04/2002	05/16/2000	01/2003	N/A
King/Horn Mt.	G19925	MC	127	5,300	Freeport-McMoRan Oil & Gas LLC	MC084	01/01/1993	04/2002	05/03/2000	11/2002	N/A
La Femme	G16608	MC	427	5,782	Newfield Exploration Gulf Coast LLC	MC427	12/02/2004	N/A	06/17/2005	N/A	08/2006

Ladybug	G15891	GB	409	1,358	Union Oil Company of California	GB409	05/13/1997	09/2001	11/25/1997	09/2001	N/A
Lena	G03205	MC	281	1,005	Exxon Mobil Corporation	MC281	05/01/1976	01/1984	05/29/1980	05/1984	N/A
Lena	G03605	MC	324	1,005	Exxon Mobil Corporation	MC281	05/01/1976	01/1984	08/18/1980	01/1984	N/A
Lena	G03818	MC	280	1,005	Exxon Mobil Corporation	MC281	05/01/1976	01/1984	08/01/1980	04/1985	N/A
Lena	G03820	MC	325	1,005	Exxon Mobil Corporation	MC281	05/01/1976	01/1984	09/15/1985	09/1985	N/A
Leo	G06960	MC	502	2,537	BP America Production Company	MC546	02/01/1986	10/2009	N/A	N/A	08/1990
Leo	G06965	MC	546	2,537	BP America Production Company	MC546	02/01/1986	10/2009	05/02/1986	N/A	08/1990
Leo	G14642	MC	546	2,537	Eni Petroleum Co. Inc.	MC546	02/01/1986	10/2009	03/11/1999	N/A	04/2002
Leo	G24084	MC	502	2,537	Eni US Operating Co. Inc.	MC546	02/01/1986	10/2009	10/27/2009	10/2009	N/A
Leo	G25098	MC	546	2,537	Eni US Operating Co. Inc.	MC546	02/01/1986	10/2009	10/26/2009	10/2009	N/A

(Continued)

Field nickname	Lease	Area	Block	Field water depth (ft)	Operator	Field name code	Field discovery date	Field first production date	Lease qualification date	Lease first production date	Lease expiration date from field
Leo	G25098	MC	546	2,537	LLOG Exploration Offshore, L.L.C.	MC546	02/01/1986	10/2009	10/26/2009	10/2009	N/A
Leo	G27277	MC	503	2,537	Eni US Operating Co. Inc.	MC546	02/01/1986	10/2009	02/19/2008	12/2011	N/A
Leo	G27277	MC	503	2,537	LLOG Exploration Offshore, L.L.C.	MC546	02/01/1986	10/2009	02/19/2008	12/2011	N/A
Leo	G32334	MC	547	2,537	LLOG Exploration Offshore, L.L.C.	MC546	02/01/1986	10/2009	06/14/2010	12/2011	N/A
Llano	G07485	GB	387	2,322	Newfield Exploration Gulf Coast LLC	GB387	10/03/1994	05/1996	07/10/1990	05/1996	09/2005
Llano	G10350	GB	386	2,322	Shell Gulf of Mexico Inc.	GB387	10/03/1994	05/1996	10/01/1998	05/2004	N/A
Llano	G15879	GB	341	2,322	Shell Offshore Inc.	GB387	10/03/1994	05/1996	05/06/2002	11/2003	N/A
Llano	G17358	GB	385	2,322	Shell Gulf of Mexico Inc.	GB387	10/03/1994	05/1996	01/17/2002	04/2004	N/A

Logan	G26419	WR	969	7,532	Statoil Gulf of Mexico LLC	WR969	09/20/2011	N/A	07/16/2013	N/A	N/A
Lost Ark	G17255	EB	421	2,754	Noble Energy, Inc.	EB421	01/31/2001	06/2002	08/28/2001	06/2002	01/2008
Lost Ark	G17259	EB	464	2,754	Noble Energy, Inc.	EB421	01/31/2001	06/2002	04/15/2004	12/2007	10/2009
Lucius	G21444	KC	875	7,106	Anadarko Petroleum Corporation	KC875	01/23/2010	N/A	03/13/2010	N/A	N/A
Lucius	G26771	KC	874	7,106	Anadarko Petroleum Corporation	KC875	01/23/2010	N/A	03/28/2013	N/A	N/A
Lucius	G26771	KC	874	7,106	Union Oil Company of California	KC875	01/23/2010	N/A	03/28/2013	N/A	N/A
Macaroni	G11553	GB	602	3,688	Shell Offshore Inc.	GB602	01/21/1996	08/1999	N/A	08/1999	N/A
Mad Dog	G09981	GC	825	4,780	BP Exploration & Production Inc.	GC826	11/24/1998	01/2005	10/17/2008	N/A	N/A
Mad Dog	G09982	GC	826	4,780	BP Exploration & Production Inc.	GC826	11/24/1998	01/2005	03/16/1999	08/2008	N/A

(Continued)

Field nickname	Lease	Area	Block	Field water depth (ft)	Operator	Field name code	Field discovery date	Field first production date	Lease qualification date	Lease first production date	Lease expiration date from field
Mad Dog	G15609	GC	781	4,780	BP Exploration & Production Inc.	GC826	11/24/1998	01/2005	10/21/2008	N/A	N/A
Mad Dog	G15610	GC	782	4,780	BP Exploration & Production Inc.	GC826	11/24/1998	01/2005	N/A	01/2005	N/A
Mad Dog	G16786	GC	738	4,780	BP Exploration & Production Inc.	GC826	11/24/1998	01/2005	02/14/2012	N/A	N/A
Madison	G10379	AC	24	4,856	Exxon Mobil Corporation	AC024	06/25/1998	02/2002	10/02/1998	02/2002	N/A
Magnolia	G11573	GB	783	4,659	ConocoPhillips Company	GB783	05/03/1999	12/2004	05/24/1999	12/2004	N/A
Magnolia	G11574	GB	784	4,659	ConocoPhillips Company	GB783	05/03/1999	12/2004	N/A	N/A	N/A
Magnolia	G20797	GB	782	4,659	BP Exploration & Production Inc.	GB783	05/03/1999	12/2004	08/23/2007	N/A	06/2009
Magnolia	G20797	GB	782	4,659	Callon Entrada Company	GB783	05/03/1999	12/2004	08/23/2007	N/A	06/2009

Field	Lease	Area	Block	Depth	Operator	Block					
Manta Ray	G05820	EW	1006	1,851	MOBIL OIL EXPLORATION & PRODUCING SOUTHEAST INC.	EW1006	01/26/1988	03/1999	N/A	N/A	03/1988
Manta Ray	G10968	EW	1006	1,851	Marathon Oil Company	EW1006	01/26/1988	03/1999	10/23/1996	03/1999	N/A
Manta Ray	G10968	EW	1006	1,851	Walter Oil & Gas Corporation	EW1006	01/26/1988	03/1999	10/23/1996	03/1999	N/A
Marathon	G05907	GC	153	1,687	Marathon Oil Company	GC153	04/01/1984	N/A	08/15/1984	N/A	12/1990
Marathon	G07022	GC	152	1,687	Marathon Oil Company	GC153	04/01/1984	N/A	02/18/1988	N/A	05/1989
Marco Polo	G18402	GC	608	4,290	Anadarko Petroleum Corporation	GC608	04/21/2000	07/2004	08/06/2002	07/2004	N/A
Marco Polo	G18402	GC	608	4,290	BHP Billiton Petroleum (GOM) Inc.	GC608	04/21/2000	07/2004	08/06/2002	07/2004	N/A
Marlin	G06894	VK	915	3,406	Freeport-McMoRan Oil & Gas LLC	VK915	06/01/1993	11/1999	08/17/1993	11/1999	N/A

(Continued)

Field nickname	Lease	Area	Block	Field water depth (ft)	Operator	Field name code	Field discovery date	Field first production date	Lease qualification date	Lease first production date	Lease expiration date from field
Marlin	G08469	VK	871	3,406	Freeport-McMoRan Oil & Gas LLC	VK915	06/01/1993	11/1999	02/25/2001	02/2001	N/A
Marmalard	G22868	MC	300	5,939	LLOG Exploration Offshore, L.L.C.	MC300	05/03/2012	N/A	08/21/2012	N/A	N/A
Marmalard	G24064	MC	255	5,939	LLOG Exploration Offshore, L.L.C.	MC300	05/03/2012	N/A	07/25/2013	N/A	N/A
Mars–Ursa	G05868	MC	809	3,341	Shell Offshore Inc.	MC807	04/01/1989	07/1996	11/15/1996	03/1999	N/A
Mars–Ursa	G05871	MC	853	3,341	Shell Offshore Inc.	MC807	04/01/1989	07/1996	N/A	01/2000	N/A
Mars–Ursa	G06981	MC	808	3,341	Shell Offshore Inc.	MC807	04/01/1989	07/1996	04/14/2009	04/2009	N/A
Mars–Ursa	G07957	MC	762	3,341	Shell Offshore Inc.	MC807	04/01/1989	07/1996	N/A	07/2007	N/A
Mars–Ursa	G07958	MC	763	3,341	Shell Offshore Inc.	MC807	04/01/1989	07/1996	04/24/1992	07/1997	N/A
Mars–Ursa	G07962	MC	806	3,341	Shell Offshore Inc.	MC807	04/01/1989	07/1996	05/26/1992	12/1996	N/A
Mars–Ursa	G07963	MC	807	3,341	Shell Offshore Inc.	MC807	04/01/1989	07/1996	04/24/1992	07/1996	N/A

Mars–Ursa	G08852	MC	764	3,341	Shell Offshore Inc.	MC807	04/01/1989	07/1996	09/30/1998	04/2000	N/A
Mars–Ursa	G09873	MC	810	3,341	Shell Offshore Inc.	MC807	04/01/1989	07/1996	09/19/1996	N/A	N/A
Mars–Ursa	G09881	MC	850	3,341	Shell Offshore Inc.	MC807	04/01/1989	07/1996	10/03/2009	10/2009	N/A
Mars–Ursa	G09882	MC	851	3,341	Shell Offshore Inc.	MC807	04/01/1989	07/1996	10/29/2008	10/2008	N/A
Mars–Ursa	G09883	MC	854	3,341	Shell Offshore Inc.	MC807	04/01/1989	07/1996	03/23/1992	03/2001	N/A
Mars–Ursa	G12166	MC	765	3,341	Shell Offshore Inc.	MC807	04/01/1989	07/1996	09/06/2000	11/2002	N/A
Mars–Ursa	G14653	MC	766	3,341	Shell Offshore Inc.	MC807	04/01/1989	07/1996	09/10/2003	12/2003	N/A
Marshall	G10322	EB	948	4,376	Exxon Mobil Corporation	EB949	07/30/1998	10/2001	10/21/2001	10/2001	N/A
Marshall	G10323	EB	949	4,376	Exxon Mobil Corporation	EB949	07/30/1998	10/2001	10/02/1998	10/2001	N/A
Matterhorn	G07927	MC	243	2,780	Conoco Inc.	MC243	09/01/1990	11/2003	02/03/1993	N/A	06/1995
Matterhorn	G19931	MC	243	2,780	W & T Energy VI, LLC	MC243	09/01/1990	11/2003	N/A	11/2003	N/A
Matterhorn	G32301	MC	199	2,780	LLOG Exploration Offshore, L.L.C.	MC243	09/01/1990	11/2003	03/03/2010	06/2012	N/A

(Continued)

Field nickname	Lease	Area	Block	Field water depth (ft)	Operator	Field name code	Field discovery date	Field first production date	Lease qualification date	Lease first production date	Lease expiration date from field
Mckinley	G09932	GC	416	4,019	Texaco Exploration and Production Inc.	GC416	07/14/1998	N/A	10/16/1998	N/A	04/2000
Medusa	G14005	MC	496	2,136	Marubeni Oil & Gas (USA) Inc.	MC582	08/23/1998	06/2003	05/14/2001	06/2003	N/A
Medusa	G16614	MC	538	2,136	Murphy Exploration & Production Company – USA	MC582	08/23/1998	06/2003	04/12/2000	11/2003	N/A
Medusa	G16623	MC	582	2,136	Murphy Exploration & Production Company – USA	MC582	08/23/1998	06/2003	04/12/2000	03/2004	N/A
Medusa	G16624	MC	583	2,136	Walter Oil & Gas Corporation	MC582	08/23/1998	06/2003	09/17/2004	06/2005	N/A
Mensa	G05862	MC	686	5,286	Shell Offshore Inc.	MC731	12/01/1986	07/1997	07/12/1997	07/1997	N/A
Mensa	G05863	MC	687	5,286	Shell Offshore Inc.	MC731	12/01/1986	07/1997	11/20/1998	11/1998	N/A

Mensa	G07954	MC	730	5,286	Shell Offshore Inc.	MC731	12/01/1986	07/1997	04/27/1992	11/1997	N/A
Mensa	G07955	MC	731	5,286	Shell Offshore Inc.	MC731	12/01/1986	07/1997	06/01/1987	N/A	07/2007
Merganser	G21825	AT	36	7,938	Anadarko Petroleum Corporation	AT037	11/28/2001	08/2007	05/13/2010	07/2010	N/A
Merganser	G21826	AT	37	7,938	Anadarko Petroleum Corporation	AT037	11/28/2001	08/2007	01/29/2002	08/2007	05/2011
Mica	G08801	MC	167	4,321	Exxon Mobil Corporation	MC211	05/01/1990	06/2001	06/06/2001	06/2001	04/2009
Mica	G08803	MC	211	4,321	Exxon Mobil Corporation	MC211	05/01/1990	06/2001	01/25/1994	06/2001	N/A
Mission Deep	G20114	GC	955	7,120	Anadarko E&P Company LP	GC955	12/13/1999	N/A	12/12/2007	N/A	04/2009
Moccasin	G13816	GB	254	1,920	Chevron U.S.A. Inc.	GB254	07/23/1993	N/A	08/16/1993	N/A	11/2000
Moccasin	G22367	KC	736	6,580	Chevron U.S.A. Inc.	KC736	08/04/2011	N/A	06/01/2012	N/A	N/A
Mondo Nw	G10486	LL	1	8,351	Anadarko Petroleum Corporation	LL001	01/08/2005	08/2007	04/17/2006	07/2007	05/2010

(Continued)

Field nickname	Lease	Area	Block	Field water depth (ft)	Operator	Field name code	Field discovery date	Field first production date	Lease qualification date	Lease first production date	Lease expiration date from field
Mondo Nw	G10487	LL	2	8,351	Anadarko Petroleum Corporation	LL001	01/08/2005	08/2007	08/10/2009	03/2010	N/A
Monet	G19965	MC	561	6,295	Noble Energy, Inc.	MC561	06/03/2008	N/A	12/09/2008	N/A	01/2009
Morpeth	G06925	EW	966	1,712	ORYX ENERGY COMPANY	EW921	07/30/1989	10/1998	02/03/1989	N/A	03/1990
Morpeth	G06935	EW	1010	1,712	Texaco Exploration and Production Inc.	EW921	07/30/1989	10/1998	02/14/1989	N/A	06/1990
Morpeth	G12142	EW	921	1,712	Eni US Operating Co. Inc.	EW921	07/30/1989	10/1998	08/19/1993	03/1999	N/A
Morpeth	G12145	EW	965	1,712	Eni US Operating Co. Inc.	EW921	07/30/1989	10/1998	12/05/1996	10/1998	N/A
Morpeth	G18184	EW	966	1,712	Apache Deepwater LLC	EW921	07/30/1989	10/1998	06/16/1999	10/2000	N/A
Morpeth	G18184	EW	966	1,712	Chevron U.S.A. Inc.	EW921	07/30/1989	10/1998	06/16/1999	10/2000	N/A

Mosquito Hawk	G12644	GB	269	1,049	Texaco Exploration and Production Inc.	GB269	03/06/1996	N/A	04/23/1996	N/A	09/1998
N.Thunder Horse	G09866	MC	776	5,668	BP Exploration & Production Inc.	MC776	07/23/2000	06/2009	09/07/2004	02/2009	N/A
N.Thunder Horse	G09867	MC	777	5,668	BP Exploration & Production Inc.	MC776	07/23/2000	06/2009	02/15/2009	02/2009	N/A
N.Thunder Horse	G09868	MC	778	5,668	BP Exploration & Production Inc.	MC776	07/23/2000	06/2009	09/29/1999	07/2010	N/A
N.Thunder Horse	G19997	MC	775	5,668	BP Exploration & Production Inc.	MC776	07/23/2000	06/2009	01/09/2010	01/2010	N/A
N.Thunder Horse	G21778	MC	734	5,668	Murphy Exploration & Production Company - USA	MC776	07/23/2000	06/2009	11/30/2006	07/2009	N/A

(Continued)

Field nickname	Lease	Area	Block	Field water depth (ft)	Operator	Field name code	Field discovery date	Field first production date	Lease qualification date	Lease first production date	Lease expiration date from field
N.Thunder Horse	G27312	MC	819	5,668	Murphy Exploration & Production Company – USA	MC776	07/23/2000	06/2009	04/03/2007	N/A	N/A
Nansen	G14205	EB	602	3,692	Anadarko Petroleum Corporation	EB602	09/25/1999	01/2002	11/16/1999	01/2002	N/A
Nansen	G17266	EB	558	3,692	Anadarko Petroleum Corporation	EB602	09/25/1999	01/2002	09/09/2006	02/2008	11/2010
Nansen	G20725	EB	646	3,692	Anadarko Petroleum Corporation	EB602	09/25/1999	01/2002	10/19/2000	01/2002	N/A
Nansen	G22295	EB	689	3,692	Anadarko Petroleum Corporation	EB602	09/25/1999	01/2002	08/22/2002	05/2003	N/A
Nansen	G22296	EB	690	3,692	Anadarko Petroleum Corporation	EB602	09/25/1999	01/2002	10/25/2001	05/2002	N/A
Neptune	G05778	VK	825	1,864	Anadarko Petroleum Corporation	VK825	11/01/1987	03/1997	01/14/1988	11/1998	11/2013
Neptune	G05784	VK	869	1,864	Exxon Mobil Corporation	VK825	11/01/1987	03/1997	07/22/1988	N/A	08/1990

Neptune	G06888	VK	826	1,864	Noble Energy, Inc.	VK825	11/01/1987	03/1997	08/21/1989	03/1997	N/A
Neptune	G13065	VK	869	1,864	Anadarko Petroleum Corporation	VK825	11/01/1987	03/1997	N/A	03/2004	11/2013
Neptune (At)	G08035	AT	574	6,205	BHP Billiton Petroleum (GOM) Inc.	AT575	09/26/1995	07/2008	02/10/2005	07/2008	N/A
Neptune (At)	G08036	AT	575	6,205	BHP Billiton Petroleum (GOM) Inc.	AT575	09/26/1995	07/2008	11/29/1995	08/2008	N/A
Neptune (At)	G08037	AT	617	6,205	BHP Billiton Petroleum (GOM) Inc.	AT575	09/26/1995	07/2008	02/10/2005	N/A	07/2013
Neptune (At)	G08038	AT	618	6,205	BHP Billiton Petroleum (GOM) Inc.	AT575	09/26/1995	07/2008	02/10/2005	12/2009	N/A
Ness	G22970	GC	507	4,001	Hess Corporation	GC507	12/27/2001	N/A	02/01/2002	N/A	06/2013
Nile	G08785	VK	914	3,535	BP Exploration & Production Inc.	VK914	04/30/1997	04/2001	08/20/1997	04/2001	07/2009
Nirvana	G07926	MC	162	3,454	BP Exploration & Oil Inc.	MC162	11/30/1994	N/A	04/28/1995	N/A	12/1997

(Continued)

Field nickname	Lease	Area	Block	Field water depth (ft)	Operator	Field name code	Field discovery date	Field first production date	Lease qualification date	Lease first production date	Lease expiration date from field
Nirvana	G21746	MC	162	3,454	Deep Gulf Energy LP	MC162	11/30/1994	N/A	01/13/2010	N/A	06/2010
Nirvana	G21746	MC	162	3,454	Shell Offshore Inc.	MC162	11/30/1994	N/A	01/13/2010	N/A	06/2010
Northwestern	G15852	GB	200	1,391	Hess Corporation	GB200	05/14/1998	11/2000	09/09/1998	11/2000	N/A
Northwestern	G17300	GB	158	1,391	Hess Corporation	GB200	05/14/1998	11/2000	09/19/2000	01/2002	N/A
Northwestern	G17307	GB	201	1,391	Hess Corporation	GB200	05/14/1998	11/2000	11/13/2002	11/2002	N/A
Ochre	G24040	MC	66	1,144	Mariner Energy, Inc.	MC066	10/25/2002	01/2004	05/30/2003	01/2004	06/2010
Oregano	G11546	GB	559	3,398	Shell Offshore Inc.	GB559	03/27/1999	10/2001	08/27/1999	10/2001	N/A
Pegasus	G25142	GC	385	3,514	Eni US Operating Co. Inc.	GC385	04/28/2005	12/2008	03/30/2007	12/2008	N/A
Petronius	G10944	VK	830	1,795	Chevron U.S.A. Inc.	VK786	07/14/1995	07/2000	10/04/2001	09/2001	N/A
Petronius	G12119	VK	786	1,795	Chevron U.S.A. Inc.	VK786	07/14/1995	07/2000	08/14/1995	07/2000	N/A
Petronius	G13988	VK	742	1,192	Chevron U.S.A. Inc.	VK742	08/08/1997	02/2002	03/12/1998	02/2002	N/A
Phoenix	G05913	GC	235	2,016	Placid Oil Company	GC236	10/01/1984	07/2001	01/08/1985	N/A	06/1988

Phoenix	G15562	GC	236	BHP Billiton Petroleum (GOM) Inc.	GC236	10/01/1984	07/2001	10/20/1999	07/2001	06/2011
Phoenix	G15562	GC	236	Energy Resource Technology GOM, Inc.	GC236	10/01/1984	07/2001	10/20/1999	07/2001	06/2011
Phoenix	G15563	GC	237	BHP Billiton Petroleum (GOM) Inc.	GC236	10/01/1984	07/2001	10/20/1999	08/2001	N/A
Phoenix	G15563	GC	237	Energy Resource Technology GOM, Inc.	GC236	10/01/1984	07/2001	10/20/1999	08/2001	N/A
Pilsner	G02648	EB	161	Union Oil Company of California	EB205	05/02/2001	12/2001	07/07/1976	02/1987	N/A
Pilsner	G17237	EB	205	Union Oil Company of California	EB205	05/02/2001	12/2001	10/11/2001	11/2001	06/2008
Pluto	G07952	MC	718	Apache Deepwater LLC	MC718	10/20/1995	12/1999	01/30/1996	09/2006	05/2011
Pluto	G07952	MC	718	BP Exploration & Production Inc.	MC718	10/20/1995	12/1999	01/30/1996	09/2006	05/2011

(Continued)

Field nickname	Lease	Area	Block	Field water depth (ft)	Operator	Field name code	Field discovery date	Field first production date	Lease qualification date	Lease first production date	Lease expiration date from field
Pluto	G13687	MC	674	2,802	Apache Deepwater LLC	MC718	10/20/1995	12/1999	02/02/1996	12/1999	05/2011
Pluto	G13687	MC	674	2,802	BP Exploration & Production Inc.	MC718	10/20/1995	12/1999	02/02/1996	12/1999	05/2011
Pompano	G04256	MC	28	1,436	Atlantic Richfield Company	VK990	05/01/1981	10/1994	04/13/1981	N/A	09/1984
Pompano	G06898	VK	989	1,436	Stone Energy Corporation	VK990	05/01/1981	10/1994	12/30/1985	11/1994	N/A
Pompano	G06899	VK	990	1,436	Stone Energy Corporation	VK990	05/01/1981	10/1994	07/23/1985	08/1995	N/A
Pompano	G07923	MC	27	1,436	Stone Energy Corporation	VK990	05/01/1981	10/1994	02/23/1989	10/1994	N/A
Pompano	G08483	MC	72	1,436	Energy XXI GOM, LLC	VK990	05/01/1981	10/1994	07/13/1992	05/1996	N/A
Pompano	G08483	MC	72	1,436	Stone Energy Corporation	VK990	05/01/1981	10/1994	07/13/1992	05/1996	N/A
Pompano	G09771	MC	28	1,436	Stone Energy Corporation	VK990	05/01/1981	10/1994	02/13/1989	10/1994	N/A
Pompano I	G13997	MC	29	2,040	Newfield Exploration Company	MC029	03/04/1998	01/2002	01/14/2002	01/2002	N/A

Pompano I	G13997	MC	29	2,040	Stone Energy Corporation	MC029	03/04/1998	01/2002	01/14/2002	01/2002	N/A
Popeye	G05896	GC	72	2,019	Shell Offshore Inc.	GC072	07/10/1985	01/1996	05/21/1986	01/1996	N/A
Popeye	G05897	GC	73	2,019	Shell Offshore Inc.	GC072	07/10/1985	01/1996	09/13/1985	N/A	06/2005
Popeye	G05904	GC	116	2,120	Shell Offshore Inc.	GC116	02/01/1985	01/1996	04/29/1985	01/1996	N/A
Popeye	G05905	GC	117	2,019	Shell Offshore Inc.	GC072	07/10/1985	01/1996	04/16/2002	04/2002	N/A
Poseidon	G08016	GC	691	4,489	BP Exploration & Oil Inc.	GC691	02/27/1996	N/A	06/07/1996	N/A	09/1997
Prince	G06921	EW	958	1,526	EnVen Energy Ventures, LLC	EW958	07/20/1994	09/2001	03/28/1995	12/2001	N/A
Prince	G13091	EW	1003	1,526	EnVen Energy Ventures, LLC	EW958	07/20/1994	09/2001	10/13/1994	09/2001	N/A
Ptolemy	G06371	GB	412	1,390	Union Exploration Partners, Ltd.	GB412	07/01/1984	N/A	08/10/1984	N/A	12/1988

(Continued)

Field nickname	Lease	Area	Block	Field water depth (ft)	Operator	Field name code	Field discovery date	Field first production date	Lease qualification date	Lease first production date	Lease expiration date from field
Puma	G16808	GC	823	4,155	BP Exploration & Production Inc.	GC823	11/05/2003	N/A	05/11/2005	N/A	07/2009
Pyrenees	G32409	GB	293	2,079	W & T Offshore, Inc.	GB293	04/14/2009	02/2012	N/A	02/2012	N/A
Q	G26281	MC	961	7,926	Statoil USA E&P Inc.	MC961	06/17/2005	10/2007	04/28/2005	10/2007	N/A
Ram–Powell	G05784	VK	869	3,238	Exxon Mobil Corporation	VK956	05/01/1985	09/1997	07/22/1988	N/A	08/1990
Ram–Powell	G06892	VK	911	3,238	Shell Offshore Inc.	VK956	05/01/1985	09/1997	01/29/1999	01/1999	N/A
Ram–Powell	G06893	VK	912	3,238	Shell Offshore Inc.	VK956	05/01/1985	09/1997	07/24/1985	09/1997	N/A
Ram–Powell	G06896	VK	956	3,238	Shell Offshore Inc.	VK956	05/01/1985	09/1997	03/19/1986	12/1997	N/A
Ram–Powell	G08474	VK	955	3,238	Shell Offshore Inc.	VK956	05/01/1985	09/1997	03/22/2002	03/2002	N/A
Ram–Powell	G08475	VK	957	3,238	Shell Offshore Inc.	VK956	05/01/1985	09/1997	02/07/1998	02/1998	N/A
Ram–Powell	G08784	VK	913	3,238	Shell Offshore Inc.	VK956	05/01/1985	09/1997	N/A	07/2004	N/A
Ram–Powell	G13065	VK	869	3,238	Anadarko Petroleum Corporation	VK956	05/01/1985	09/1997	N/A	03/2004	11/2013

Raptor	G23243	EB	668	3,710	Marubeni Oil & Gas (USA) Inc.	EB668	09/13/2003	06/2004	10/29/2003	06/2004	09/2011
Red Hawk	G21408	GB	877	5,329	Anadarko Petroleum Corporation	GB877	08/15/2001	07/2004	10/05/2002	07/2004	10/2009
Rigel	G18207	MC	252	5,091	Dominion Exploration & Production, Inc.	MC252	11/29/1999	03/2006	07/03/2003	N/A	06/2007
Rigel	G21164	MC	296	5,091	Eni US Operating Co. Inc.	MC252	11/29/1999	03/2006	02/17/2004	03/2006	12/2011
Rigel	G24055	MC	209	5,091	LLOG Exploration Offshore, L.L.C.	MC252	11/29/1999	03/2006	09/09/2013	N/A	N/A
Rigel	G24062	MC	253	5,091	ConocoPhillips Company	MC252	11/29/1999	03/2006	03/03/2013	N/A	N/A
Rigel	G24062	MC	253	5,091	LLOG Exploration Offshore, L.L.C.	MC252	11/29/1999	03/2006	03/03/2013	N/A	N/A
Rockefeller	G10325	EB	992	4,865	Exxon Mobil Corporation	EB992	11/28/1995	08/2009	07/14/1997	N/A	11/1998

(Continued)

Field nickname	Lease	Area	Block	Field water depth (ft)	Operator	Field name code	Field discovery date	Field first production date	Lease qualification date	Lease first production date	Lease expiration date from field
Rockefeller	G23257	EB	992	4,865	Exxon Mobil Corporation	EB992	11/28/1995	08/2009	08/18/2009	08/2009	01/2013
Rocky	G05901	GC	110	1,960	Marathon Oil Company	GC110	08/07/1987	01/1996	11/21/1984	N/A	12/1990
Rocky	G05907	GC	153	1,960	Marathon Oil Company	GC110	08/07/1987	01/1996	08/15/1984	N/A	12/1990
Rocky	G05908	GC	154	1,960	Marathon Oil Company	GC110	08/07/1987	01/1996	04/19/1988	N/A	12/1990
Rocky	G14023	GC	110	1,960	Shell Offshore Inc.	GC110	08/07/1987	01/1996	01/23/1996	01/1996	03/2005
Rocky	G16698	GC	155	1,960	Marubeni Oil & Gas (USA) Inc.	GC110	08/07/1987	01/1996	04/24/2000	07/2002	N/A
Rocky	G24160	GC	199	1,960	Noble Energy, Inc.	GC110	08/07/1987	01/1996	11/22/2004	04/2006	N/A
Sable	G05912	GC	228	1,694	Texaco Exploration and Production Inc.	GC228	07/01/1985	N/A	09/20/1985	N/A	09/1996
Saint Malo	G18753	WR	677	6,951	Union Oil Company of California	WR678	10/13/2003	N/A	05/28/2008	N/A	N/A
Saint Malo	G21245	WR	678	6,951	Union Oil Company of California	WR678	10/13/2003	N/A	12/12/2006	N/A	N/A

Salsa	G06353	GB	171	1,195	Marathon Oil Company	GB171	04/18/1984	08/1998	07/22/1988	N/A	02/1990
Salsa	G09216	GB	215	1,195	Hess Corporation	GB171	04/18/1984	08/1998	05/16/1995	06/2003	N/A
Salsa	G13369	GB	171	1,195	Anadarko US Offshore Corporation	GB171	04/18/1984	08/1998	05/08/1996	N/A	06/1999
Salsa	G14221	GB	172	1,195	Shell Offshore Inc.	GB171	04/18/1984	08/1998	03/11/1996	08/1998	N/A
San Jacinto	G23526	DC	618	7,805	Eni US Operating Co. Inc.	DC618	04/24/2004	10/2007	02/14/2005	09/2007	01/2012
San Patricio	G13211	AT	153	4,785	Exxon Mobil Corporation	AT153	08/09/2001	N/A	11/29/2001	N/A	02/2002
Sangria	G16702	GC	177	1,487	Mariner Gulf of Mexico LLC	GC177	08/22/1999	06/2002	12/02/1999	06/2002	03/2005
Serrano	G08252	GB	516	3,340	Shell Offshore Inc.	GB516	07/23/1996	12/2001	11/21/1996	06/2002	06/2005
Serrano	G11528	GB	472	3,340	Shell Offshore Inc.	GB516	07/23/1996	12/2001	01/01/2000	02/2002	05/2010
Serrano	G20792	GB	515	3,340	Marathon Oil Company	GB516	07/23/1996	12/2001	10/30/2008	12/2011	07/2013
Seventeen Hands	G21752	MC	299	5,881	Eni US Operating Co. Inc.	MC299	05/04/2001	03/2006	05/08/2002	03/2006	09/2011

(Continued)

Field nickname	Lease	Area	Block	Field water depth (ft)	Operator	Field name code	Field discovery date	Field first production date	Lease qualification date	Lease first production date	Lease expiration date from field
Shaft	G21785	GC	141	1,016	LLOG Exploration Offshore, L.L.C.	GC141	06/29/2008	02/2010	08/12/2008	07/2010	N/A
Shenzi	G16764	GC	609	4,304	BHP Billiton Petroleum (GOM) Inc.	GC654	12/02/2002	10/2007	N/A	06/2012	N/A
Shenzi	G16765	GC	610	4,304	BHP Billiton Petroleum (GOM) Inc.	GC654	12/02/2002	10/2007	01/30/2007	N/A	N/A
Shenzi	G20084	GC	653	4,304	BHP Billiton Petroleum (GOM) Inc.	GC654	12/02/2002	10/2007	11/09/2006	09/2009	N/A
Shenzi	G20085	GC	654	4,304	BHP Billiton Petroleum (GOM) Inc.	GC654	12/02/2002	10/2007	11/09/2006	03/2009	N/A
Shenzi	G21810	GC	652	4,304	BHP Billiton Petroleum (GOM) Inc.	GC654	12/02/2002	10/2007	04/05/2006	10/2007	N/A
Shiner Deep	G19154	GB	700	4,542	Anadarko E&P Company LP	GB700	09/13/2003	N/A	05/05/2004	N/A	11/2007
Slammer	G21188	MC	849	3,600	Noble Energy, Inc.	MC849	03/19/2002	N/A	05/17/2002	N/A	06/2009

Son Of Bluto 2	G22877	MC	431	6,426	LLOG Exploration Offshore, L.L.C.	MC431	02/06/2012	N/A	04/26/2012	N/A	N/A
Spiderman/Amazo	G23528	DC	620	8,082	Anadarko Petroleum Corporation	DC621	11/29/2003	09/2007	04/18/2006	11/2007	03/2013
Spiderman/Amazo	G23529	DC	621	8,082	Anadarko Petroleum Corporation	DC621	11/29/2003	09/2007	04/18/2006	09/2007	N/A
Stampede	G26313	GC	468	3,494	Hess Corporation	GC468	06/24/2006	N/A	10/01/2008	N/A	N/A
Stampede	G26315	GC	512	3,494	Hess Corporation	GC468	06/24/2006	N/A	05/01/2009	N/A	N/A
Stones	G17001	WR	508	8,860	Shell Offshore Inc.	WR508	03/10/2005	N/A	03/23/2009	N/A	N/A
Supertramp	G10971	MC	26	1,272	BP Exploration & Oil Inc.	MC026	05/27/1994	N/A	06/28/1994	N/A	06/1997
Sw Horseshoe	G19001	EB	430	2,285	Hess Corporation	EB430	05/03/2000	05/2006	12/04/2001	05/2006	N/A
Sw Horseshoe	G19001	EB	430	2,285	Walter Oil & Gas Corporation	EB430	05/03/2000	05/2006	12/04/2001	05/2006	N/A

(Continued)

Field nickname	Lease	Area	Block	Field water depth (ft)	Operator	Field name code	Field discovery date	Field first production date	Lease qualification date	Lease first production date	Lease expiration date from field
Swordfish	G15441	VK	917	4,374	BP Exploration & Production Inc.	VK917	12/08/2001	10/2006	01/16/2002	10/2006	N/A
Swordfish	G15441	VK	917	4,374	Noble Energy, Inc.	VK917	12/08/2001	10/2006	01/16/2002	10/2006	N/A
Swordfish	G15444	VK	961	4,676	BP Exploration & Production Inc.	VK962	11/15/2001	10/2005	10/01/2004	10/2005	01/2012
Swordfish	G15444	VK	961	4,676	Noble Energy, Inc.	VK962	11/15/2001	10/2005	10/01/2004	10/2005	01/2012
Swordfish	G15445	VK	962	4,676	BP Exploration & Production Inc.	VK962	11/15/2001	10/2005	01/16/2002	10/2005	N/A
Swordfish	G15445	VK	962	4,676	Noble Energy, Inc.	VK962	11/15/2001	10/2005	01/16/2002	10/2005	N/A
Tahiti/Cae/Tong	G16759	GC	596	4,320	Chevron U.S.A. Inc.	GC640	03/29/2002	04/2007	04/27/2005	07/2007	N/A
Tahiti/Cae/Tong	G16783	GC	727	4,320	Anadarko Petroleum Corporation	GC640	03/29/2002	04/2007	04/10/2006	01/2014	N/A
Tahiti/Cae/Tong	G16794	GC	771	4,320	Anadarko Petroleum Corporation	GC640	03/29/2002	04/2007	04/23/2007	N/A	06/2007

Tahiti/Cae/Tong	G18421	GC	683	4,320	Anadarko Petroleum Corporation	GC640	03/29/2002	04/2007	03/15/2007	01/2007	N/A
Tahiti/Cae/Tong	G20082	GC	640	4,320	Chevron U.S.A. Inc.	GC640	03/29/2002	04/2007	05/02/2005	04/2007	N/A
Tahiti/Cae/Tong	G24179	GC	726	4,320	Anadarko Petroleum Corporation	GC640	03/29/2002	04/2007	N/A	03/2012	N/A
Tahiti/Cae/Tong	G24184	GC	770	4,320	Anadarko Petroleum Corporation	GC640	03/29/2002	04/2007	N/A	N/A	N/A
Tahoe/Se Tahoe	G06886	VK	783	1,328	W & T Energy VI, LLC	VK783	12/01/1984	01/1994	03/29/1985	01/1994	N/A
Tahoe/Se Tahoe	G07910	VK	827	2,104	Shell Deepwater Production Inc.	VK827	11/08/1998	N/A	01/07/1999	N/A	04/2001
Tahoe/Se Tahoe	G13060	VK	784	1,328	W & T Energy VI, LLC	VK783	12/01/1984	01/1994	04/25/1996	10/1996	N/A
Telemark	G13198	AT	63	4,412	Bennu Oil & Gas, LLC	AT063	02/11/2000	03/2010	07/25/2000	03/2010	N/A
Thunder Horse	G09868	MC	778	6,077	BP Exploration & Production Inc.	MC778	04/01/1999	06/2008	09/29/1999	07/2010	N/A

(Continued)

Field nickname	Lease	Area	Block	Field water depth (ft)	Operator	Field name code	Field discovery date	Field first production date	Lease qualification date	Lease first production date	Lease expiration date from field
Thunder Horse	G14657	MC	821	6,077	BP Exploration & Production Inc.	MC778	04/01/1999	06/2008	03/14/2012	N/A	N/A
Thunder Horse	G14658	MC	822	6,077	BP Exploration & Production Inc.	MC778	04/01/1999	06/2008	N/A	06/2008	N/A
Thunder Ridge	G16645	MC	737	6,108	Murphy Exploration & Production Company – USA	MC737	10/27/2006	N/A	11/28/2006	N/A	05/2007
Ticonderoga	G21817	GC	768	5,257	Anadarko Petroleum Corporation	GC768	09/10/2004	02/2006	10/18/2004	02/2006	N/A
Ticonderoga	G21817	GC	768	5,257	Noble Energy, Inc.	GC768	09/10/2004	02/2006	10/18/2004	02/2006	N/A
Tiger	G20863	AC	818	9,004	Chevron U.S.A. Inc.	AC818	02/28/2004	N/A	05/09/2007	N/A	10/2008
Tiger	G24159	GC	195	1,844	Deep Gulf Energy LP	GC195	05/25/2006	01/2007	12/30/2006	12/2006	06/2009
Timber Wolf	G21770	MC	555	4,749	Exxon Mobil Corporation	MC555	10/30/2001	N/A	10/01/2002	N/A	05/2010
Tobago	G19409	AC	815	9,436	Shell Offshore Inc.	AC859	04/17/2004	12/2010	12/09/2010	12/2010	N/A

Tobago	G20871	AC	859	9,436	Shell Offshore Inc.	AC859	04/17/2004	12/2010	10/10/2006	10/2011	N/A
Troika	G11043	GC	244	2,795	Marathon Oil Company	GC244	05/30/1994	11/1997	08/04/1994	03/1998	N/A
Troika	G11043	GC	244	2,795	Shell Offshore Inc.	GC244	05/30/1994	11/1997	08/04/1994	03/1998	N/A
Troika	G12209	GC	200	2,795	Shell Offshore Inc.	GC244	05/30/1994	11/1997	10/31/1997	10/1997	N/A
Troika	G12210	GC	201	2,795	LLOG Exploration Offshore, L.L.C.	GC244	05/30/1994	11/1997	11/20/2000	11/2000	N/A
Troika	G12210	GC	201	2,795	Shell Offshore Inc.	GC244	05/30/1994	11/1997	11/20/2000	11/2000	N/A
Tubular Bells	G16636	MC	683	4,522	Hess Corporation	MC682	10/17/2003	N/A	04/01/2008	N/A	N/A
Tubular Bells	G21776	MC	682	4,522	Hess Corporation	MC682	10/17/2003	N/A	03/15/2007	N/A	N/A
Tubular Bells	G22898	MC	725	4,522	Hess Corporation	MC682	10/17/2003	N/A	04/23/2008	N/A	N/A
Tubular Bells	G24101	MC	726	4,522	Hess Corporation	MC682	10/17/2003	N/A	04/09/2008	N/A	N/A
Valley Forge	G25103	MC	707	1,538	LLOG Exploration Offshore, L.L.C.	MC707	06/20/2007	09/2007	10/09/2007	07/2008	N/A

(*Continued*)

Field nickname	Lease	Area	Block	Field water depth (ft)	Operator	Field name code	Field discovery date	Field first production date	Lease qualification date	Lease first production date	Lease expiration date from field
Virgo	G10942	VK	823	1,142	W & T Energy VI, LLC	VK823	01/01/1993	12/1999	05/29/1997	12/1999	N/A
Virgo	G16549	VK	822	1,142	W & T Energy VI, LLC	VK823	01/01/1993	12/1999	N/A	06/2000	N/A
Vito	G22919	MC	984	4,025	Shell Offshore Inc.	MC940	03/29/2010	N/A	01/13/2011	N/A	N/A
Vito	G31534	MC	940	4,025	Shell Offshore Inc.	MC940	03/29/2010	N/A	11/30/2010	N/A	N/A
Vortex	G16890	AT	261	8,344	Anadarko Petroleum Corporation	AT261	12/02/2002	10/2007	03/31/2006	10/2007	09/2012
Wide Berth	G31732	GC	490	3,714	Apache Deepwater LLC	GC490	10/25/2009	04/2012	N/A	04/2012	N/A
Wrigley	G26261	MC	506	3,682	W & T Offshore, Inc.	MC506	02/06/2005	07/2007	10/12/2005	07/2007	N/A
Zinc	G02961	MC	398	1,493	Exxon Mobil Corporation	MC354	08/01/1977	08/1993	11/20/1979	N/A	05/2010
Zinc	G02962	MC	399	1,493	Exxon Mobil Corporation	MC354	08/01/1977	08/1993	04/10/1979	N/A	05/2010
Zinc	G02963	MC	354	1,493	Exxon Mobil Corporation	MC354	08/01/1977	08/1993	04/10/1979	08/1993	05/2010
Zinc	G02964	MC	355	1,493	Exxon Mobil Corporation	MC354	08/01/1977	08/1993	01/22/1980	08/1993	05/2010

G02972	MC	113	1,857	Placid Oil Company	MC113	01/01/1976	N/A	03/03/1976	N/A	11/1977	N/A
G02974	MC	68	1,214	ATOFINA Petrochemicals, Inc.	MC068	12/09/1975	01/2001	N/A	N/A	11/1977	N/A
G03208	VK	862	1,048	Exxon Mobil Corporation	VK862	10/01/1976	12/1995	09/10/1980	N/A	02/1982	N/A
G05783	VK	862	1,048	BP Exploration Inc.	VK862	10/01/1976	12/1995	09/18/1985	N/A	03/1989	N/A
G05832	MC	285	2,902	Texaco Inc.	MC285	09/01/1987	04/2012	11/01/1987	N/A	06/1989	N/A
G05879	GC	21	1,224	Odeco Oil & Gas Company	GC021	10/01/1984	N/A	01/14/1985	N/A	02/1991	N/A
G05882	GC	29	1,565	Placid Oil Company	GC029	01/01/1984	11/1988	04/12/1984	06/1989	12/1990	N/A
G05883	GC	39	1,976	Placid Oil Company	GC039	04/01/1984	N/A	07/25/1984	N/A	06/1988	N/A
G05894	GC	70	1,618	Conoco Inc.	GC070	06/01/1984	N/A	09/06/1984	N/A	12/1988	N/A
G06931	EW	999	1,565	Placid Oil Company	GC029	01/01/1984	11/1988	11/14/1985	10/1988	07/1990	N/A
G06953	MC	455	1,400	Union Exploration Partners, Ltd.	MC455	02/01/1986	N/A	04/23/1986	N/A	05/1989	N/A

(Continued)

Field nickname	Lease	Area	Block	Field water depth (ft)	Operator	Field name code	Field discovery date	Field first production date	Lease qualification date	Lease first production date	Lease expiration date from field
N/A	G06973	MC	665	2,525	Atlantic Richfield Company	MC709	02/01/1987	N/A	09/01/1987	N/A	10/1989
N/A	G06975	MC	709	2,525	Atlantic Richfield Company	MC709	02/01/1987	N/A	05/14/1987	N/A	10/1989
N/A	G06984	MC	929	2,250	Conoco Inc.	MC929	11/01/1987	N/A	04/08/1988	N/A	05/1989
N/A	G06994	GC	31	2,172	EP Operating Limited Partnership	GC075	05/01/1985	11/1988	01/08/1986	11/1988	04/1990
N/A	G07007	GC	75	2,172	ORYX ENERGY COMPANY	GC075	05/01/1985	11/1988	09/06/1985	N/A	08/1989
N/A	G07020	GC	147	1,275	Texaco Exploration and Production Inc.	GC147	05/01/1988	N/A	07/21/1988	N/A	06/1990
N/A	G07025	GC	162	2,583	ORYX ENERGY COMPANY	GC162	07/01/1989	N/A	06/20/1989	N/A	10/1990
N/A	G07075	AT	1	2,381	Texaco Exploration and Production Inc.	AT001	05/14/1986	N/A	08/08/1986	N/A	06/1989

G07424	EB	377	Shell Offshore Inc.	EB377	10/01/1985	N/A	01/14/1986	N/A	08/1989	N/A
G07483	GB	379	Conoco Inc.	GB379	07/01/1985	12/2004	10/01/1985	N/A	08/1991	N/A
G08222	GB	208	Eni Petroleum Co. Inc.	GB208	09/01/1991	11/2004	04/21/1992	N/A	04/1995	N/A
G08268	PI	525	BP Exploration & Oil Inc.	PI525	04/30/1996	N/A	08/14/1996	N/A	11/1996	N/A
G11066	GC	463	BP Exploration & Oil Inc.	GC463	12/01/1998	N/A	01/21/1999	N/A	04/1999	N/A
G13064	VK	862	Walter Oil & Gas Corporation	VK862	10/01/1976	12/1995	12/19/1995	12/1995	N/A	N/A
G15464	MC	68	Walter Oil & Gas Corporation	MC068	12/09/1975	01/2001	07/20/2000	01/2001	10/2005	N/A
G16650	MC	837	Walter Oil & Gas Corporation	EW878	07/03/2000	12/2001	07/02/2001	04/2004	05/2007	N/A
G16679	GC	39	Eni Petroleum Co. Inc.	GC039	04/01/1984	N/A	04/30/1999	N/A	08/2003	N/A

(Continued)

Field nickname	Lease	Area	Block	Field water depth (ft)	Operator	Field name code	Field discovery date	Field first production date	Lease qualification date	Lease first production date	Lease expiration date from field
N/A	G17456	GB	873	4,705	BP Exploration & Production Inc.	GB873	12/09/2006	N/A	03/03/2007	N/A	06/2007
N/A	G18169	EW	878	1,543	Walter Oil & Gas Corporation	EW878	07/03/2000	12/2001	08/02/2000	12/2001	N/A
N/A	G21163	MC	161	2,924	Walter Oil & Gas Corporation	MC161	07/16/2005	11/2007	10/20/2005	11/2007	N/A
N/A	G21750	MC	241	2,902	Shell Offshore Inc.	MC285	09/01/1987	04/2012	05/06/2009	04/2012	N/A
N/A	G21750	MC	241	2,902	Walter Oil & Gas Corporation	MC285	09/01/1987	04/2012	05/06/2009	04/2012	N/A
N/A	G23278	GB	205	1,330	Nexen Petroleum U.S.A. Inc.	GB205	07/25/2002	12/2002	09/18/2002	12/2002	01/2012
N/A	G24059	MC	214	5,802	Deep Gulf Energy II, LLC	MC214	09/17/2013	N/A	02/21/2014	N/A	N/A
N/A	G24182	GC	767	5,116	Anadarko Petroleum Corporation	GC767	08/13/2004	N/A	10/15/2004	N/A	06/2013
N/A	G24182	GC	767	5,116	Noble Energy, Inc.	GC767	08/13/2004	N/A	10/15/2004	N/A	06/2013

G24462	GB	208	1,267	McMoRan Oil & Gas LLC	GB208	09/01/1991	11/2004	10/16/2003	11/2004	11/2008	N/A
G24488	GB	378	2,047	LLOG Exploration & Production Company, L.L.C.	GB379	07/01/1985	12/2004	06/18/2004	12/2004	11/2011	N/A
G25118	GC	137	1,173	Walter Oil & Gas Corporation	GC137	03/02/2004	01/2005	01/13/2005	01/2005	N/A	N/A
G25610	EB	197	1,249	LLOG Exploration Offshore, L.L.C.	EB197	12/09/2004	N/A	03/24/2005	N/A	11/2012	N/A
G25673	GB	339	2,180	Deep Gulf Energy LP	GB339	11/22/2008	03/2010	06/09/2009	03/2010	N/A	N/A
G27243	VK	821	1,030	Walter Oil & Gas Corporation	VK821	04/24/2008	06/2008	10/31/2009	10/2009	N/A	N/A
G27779	SE	39	8,553	Anadarko Petroleum Corporation	SE039	04/09/2013	N/A	11/18/2013	N/A	N/A	N/A
G27982	EW	834	1,543	Walter Oil & Gas Corporation	EW878	07/03/2000	12/2001	04/26/2012	N/A	N/A	N/A

(*Continued*)

Field nickname	Lease	Area	Block	Field water depth (ft)	Operator	Field name code	Field discovery date	Field first production date	Lease qualification date	Lease first production date	Lease expiration date from field
N/A	G27988	EW	998	1,313	Walter Oil & Gas Corporation	EW998	02/27/2009	03/2011	N/A	03/2011	N/A
N/A	G28022	MC	698	7,221	Noble Energy, Inc.	MC698	11/15/2012	N/A	07/16/2013	N/A	N/A
N/A	G33178	MC	816	5,534	LLOG Exploration Offshore, L.L.C.	MC816	08/30/2013	N/A	03/18/2014	N/A	N/A
N/A	G33757	MC	782	6,573	Noble Energy, Inc.	MC782	N/A	N/A	01/14/2014	N/A	N/A

CONVERSION FACTORS

1 CONCENTRATION CONVERSIONS

1 part per million (1 ppm) = 1 microgram per liter (1 μg/L)

1 microgram per liter (1 μg/L) = 1 milligram per kilogram (1 mg/kg)

1 microgram per liter (μg/L) $\times$ 6.243 $\times$ 10^8 = 1 lb per cubic foot (1 lb/ft³)

1 microgram per liter (1 μg/L) $\times$ 10^{-3} = 1 milligram per liter (1 mg/L)

1 milligram per liter (1 mg/L) $\times$ 6.243 $\times$ 10^5 = 1 pound per cubic foot (1 lb/ft³)

I gram mole per cubic meter (1 g mol/m³) $\times$ 6.243 $\times$ 10^5 = 1 pound per cubic foot (1 lb/ft³)

10,000 ppm = 1% (w/w)

2 WEIGHT CONVERSION

1 ounce (1 oz) = 28.3495 grams (18.2495 g)

1 pound (1 lb) = 0.454 kilogram

1 pound (1 lb) = 454 grams (454 g)

1 kilogram (1 kg) = 2.20462 pounds (2.20462 lb)

1 stone (English, 1 st) = 14 pounds (14 lb)

1 ton (US; 1 short ton) = 2,000 lbs

1 ton (English; 1 long ton) = 2,240 lbs

1 metric ton = 2204.62262 pounds

1 tonne = 2204.62262 pounds

1 tonne = 7.49 barrels

1 US gallon of water = 8.34 lbs

1 imperial gallon of water − 10 lbs

1 barrel (water) = 350 lb at 60 °F

Density of water at 60 °F = 0.999 gram/cu cm = 62.367 Ib/cu ft = 8.337 Ib/gal

3 TEMPERATURE CONVERSIONS

°F = (°C $\times$ 1.8) + 32

°C = (°F − 32)/1.8

(°F − 32) $\times$ 0.555 = °C

Absolute zero = −273.15 °C

Absolute zero = −459.67 °F

Handbook of Offshore Oil and Gas Operations. http://dx.doi.org/10.1016/B978-1-85617-558-6.00013-1

4 AREA

1 square centimeter (1 cm²) = 0.1550 square inches
1 square meter 1 (m²) = 1.1960 square yards
1 hectare = 2.4711 acres
1 square kilometer (1 km²) = 0.3861 square miles
1 square inch (1 in.²) = 6.4516 square centimeters
1 square foot (1 ft²) = 0.0929 square meters
1 square yard (1 yd²) = 0.8361 square meters
1 acre = 4046.9 square meters
1 square mile (1 mi²) = 2.59 square kilometers
1 square mile (1 mi²) = 640 acres
1 yd³ = 27 ft³
1 yd³ = 0.765 m³

5 VOLUME AND PRESSURE CONVERSIONS

1 barrel (oil) = 42 gal = 5.6146 cu ft
1 m³ = 6.29 barrels
1 barrel = 0.159 m³
1 barrel per day = 1.84 cu cm per second
1 m³ = 35.3 ft³ (natural gas)
1 ft³ = 0.028 m³ (natural gas)
1 cubic foot = 28,317 cu cm = 7.4805 gal
1 gallon = 231 cu in. = 3,785.4 cu cm = 0.13368 cu ft
1 atmosphere = 760 mm Hg = 14.696 psia = 29.91 in. Hg
1 atmosphere = 1.0133 bars = 33.899 ft. H_2O
1 kiloPascal (kPa) × 9.8692 × 10^{-3} = 14.696 pounds per square inch (14.696 psi)

6 OTHER CONVERSIONS

1 Btu = 778.26 ft-lb
1 centipoise × 2.42 = Ib mass/(ft) (h), viscosity
1 centipoise × 0.000672 = Ib mass/(ft) (s), viscosity
1 inch = 2.54 cm
1 meter = 100 cm = 1,000 mm = 10 microns = 10 angstroms (A)
Million = 1 × 10^6
Billion = 1 × 10^9
Trillion = 1 × 10^{12}

GLOSSARY

Abiogenic produced from nonorganism materials.

Abiotic produced from nonorganism materials.

Abyssal of or relating to the bottom waters of the ocean.

Accumulate to amass or collect. When oil and gas migrate into porous formations, the quantity collected is called an accumulation.

Aerobic bacteria bacteria which can grow in the presence of oxygen.

Air gun a chamber filled with compressed air, often used offshore in seismic exploration. As the gun is trailed behind a boat, air is released, making a low-frequency popping noise, which penetrates the subsurface rock layers and is reflected by the layers. Sensitive hydrophones receive the reflections and transmit them to recording equipment on the boat.

Alluvial fan a large, sloping sedimentary deposit at the mouth of a canyon, laid down by intermittently flowing water, especially in arid climates, and composed of gravel and sand. The deposit tends to be coarse and unworked, with angular, poorly sorted grains in thin, overlapping sheets. A line of fans may eventually coalesce into an apron that grows broader and higher as the slopes above are eroded away.

Amphibole any group of common rock-forming silicate minerals.

Anaerobic bacteria bacteria which can grow in the absence of oxygen.

Angle of deflection in directional drilling, the angle at which a well diverts from vertical; usually expressed in degrees, with vertical being 0°.

Angle of dip the angle at which a formation dips downward from the horizontal.

Annular blowout preventer a large, specialized valve used to seal, control and monitor oil and gas wells; usually installed above the ram preventers that forms a seal in the annular space between the pipe and well bore or, if no pipe is present, over the well bore itself.

Anticlinal trap a hydrocarbon trap in which petroleum accumulates in the top of an anticline. See *anticline.*

Anticline rock layers folded in the shape of an arch. Anticlines sometimes trap oil and gas. Compare *syncline.*

Associated gas natural gas that over-lies and contacts crude oil in a reservoir. Where reservoir conditions are such that the production of associated gas does not substantially affect the recovery of crude oil in the reservoir, such gas may also be reclassified as nonassociated gas by a regulatory agency. Also called associated free gas. See *gas cap.*

Barrel (bbl) a measure of volume for petroleum products. One barrel is the equivalent of 35 imperial gallons or 42 U.S. gallons or 0.15899 cubic metres (9,702 cubic inches). One cubic meter equals 6.2897 barrels.

Basement rock igneous or metamorphic rock, which seldom contains petroleum. Ordinarily, it lies below sedimentary rock. When it is encountered in drilling, the well is usually abandoned.

Basin a local depression in the earth's crust in which sediments can accumulate to form thick sequences of sedimentary rock.

Bbl barrel.

Bed a specific layer of earth or rock that presents a contrast to other layers of different material lying above, below, or adjacent to it.

Handbook of Offshore Oil and Gas Operations. http://dx.doi.org/10.1016/B978-1-85617-558-6.00014-3

Bedrock solid rock just beneath the soil.

Benthic relating to the seabed.

Biogenic produced by living organisms.

Biotic produced by living organisms.

Bit the cutting or boring element used in drilling oil and gas wells. The bit consists of a cutting element and a circulating element. The cutting element is steel teeth, tungsten carbide buttons, industrial diamonds, or polycrystalline diamonds (PDCs). These teeth, buttons, or diamonds penetrate and gouge or scrape the formation to remove it. The circulating element permits the passage of drilling fluid and utilizes the hydraulic force of the fluid stream to improve drilling rates. In rotary drilling, several drill collars are joined to the bottom end of the drill pipe column, and the bit is attached to the end of the drill collars. Drill collars provide weight on the bit to keep it in firm contact with the bottom of the hole. Most bits used in rotary drilling are roller cone bits, but diamond bits are also used extensively.

Bitumen a substance of dark to black color consisting almost entirely of carbon and hydrogen with very little oxygen, nitrogen, or sulfur. Bitumens occur naturally and can also be obtained by chemical decomposition.

Blowout an uncontrolled flow of gas, oil, or other well fluids into the atmosphere. A blowout, or gusher, occurs when formation pressure exceeds the pressure applied to it by the column of drilling fluid. A kick warns of an impending blowout. See *kick*.

Blowout preventer (BOP) one of several valves installed at the wellhead to prevent the escape of pressure either in the annular space between the casing and the drill pipe or in open hole (i.e., hole with no drill pipe) during drilling or completion operations. Blowout preventers on land rigs are located beneath the rig at the land's surface; on jackup or platform rigs, at the water's surface; and on floating offshore rigs, on the seafloor.

Bottom-hole assembly (bottom-hole assembly) the portion of the drilling assembly below the drill pipe. It can be as simple as bit and drill collars or it can be very complex and made up of multiple components.

Bottom-hole pressure (bottom-hole pressure) the pressure at the bottom-hole. For nonflowing condition, it is caused by the hydrostatic pressure of the wellbore fluid and by any backpressure held at the surface, if any; when wellbore fluid is being circulated, bottom-hole pressure is the hydrostatic pressure plus the friction pressure drop in annulus.

Bottom-supported offshore drilling rig a type of mobile offshore drilling unit that has a part of its structure in contact with the seafloor when it is on site.

Bright spot a seismic phenomenon that shows up on a seismic, or record, section as a sound reflection that is much stronger than usual. A bright spot sometimes directly indicates natural gas in a trap.

Cap gas natural gas trapped in the upper part of a reservoir and remaining separate from any crude oil, salt water, or other liquids in the well.

Cap rock a disk-like plate of anhydrite, gypsum, limestone, or sulfur overlying most salt domes in the Gulf Coast region: impermeable rock overlying an oil or gas reservoir that tends to prevent migration of oil or gas out of the reservoir.

Carbonate rock a sedimentary rock composed primarily of calcium carbonate (limestone) or calcium magnesium carbonate (dolomite); sometimes makes up petroleum reservoirs.

Casing steel pipe placed in an oil or gas well to prevent the wall of the hole from caving in, to prevent movement of fluids from one formation to another and to aid in well control.

Cenozoic era the time period from 65 million years ago until the present. It is marked by rapid evolution of mammals and birds, flowering plants, grasses, and shrubs, and little change in invertebrates.

Chadacryst a crystal enclosed in another crystal.

Christmas tree the control valves, pressure gauges, and chokes assembled at the top of a well to control flow of oil and/or gas after the well has been drilled and completed.

Clastic rock a sedimentary rock composed of fragments of preexisting rocks. The principal distinction among clastic rocks is grain size; conglomerates, sandstones, and shale are clastic rocks.

Clinopyroxene A subgroup name for monoclinic pyroxene group of minerals.

Coal a carbonaceous, rocklike material that forms from the remains of plants that were subjected to biochemical processes, intense pressure, and high temperatures. It is used as fuel.

Coal bed methane natural gas, primarily methane that occurs naturally in the fractures and matrix of coal beds.

Concrete gravity rigid platform rig a rigid offshore drilling platform built of steel-reinforced concrete and used to drill development wells. The platform is floated to the drilling site in a vertical position. At the site, one or more tall caissons that serve as the foundation of the platform are flooded so that the platform comes to rest on bottom. Because of the enormous weight of the platform, the force of gravity alone keeps it in place.

Condensate a light hydrocarbon liquid obtained by condensation of hydrocarbon vapors. It consists of varying proportions of butane, propane, pentane, and heavier fractions, with little or no methane or ethane.

Condensate reservoir a reservoir in which both condensate and gas exist in one homogeneous phase. When fluid is drawn from such a reservoir and the pressure decreases below the critical level, a liquid phase (condensate) appears.

Contact in geology, any sharp or well-defined boundary between two different bodies of rock; a bedding plane or unconformity that separates formations. 3. In a petroleum reservoir, a horizontal boundary where different types of fluids meet and mix slightly; for example, a gas–oil or oil–water contact; also called an interface.

Continuous accumulations petroleum that occurs in extensive reservoirs and is not necessarily related to conventional structural or stratigraphic traps. These accumulations of oil and/or gas lack well-defined down-dip petroleum/water contacts and thus are not localized by the buoyancy of oil or natural gas in water.

Conventional accumulations petroleum that occurs in structural or stratigraphic traps, commonly bounded b y a down-dip water contact, and therefore affected by the buoyancy of petroleum in water.

Conventional mud a drilling fluid containing essentially clay and water; no special or expensive chemicals or conditioners are added.

Core a cylindrical sample taken from a formation for geological analysis. Usually a conventional core barrel is substituted for the bit and procures a sample as it penetrates the formation. *V:* to obtain a solid, cylindrical formation sample for analysis.

Core analysis laboratory analysis of a core sample to determine porosity, permeability, lithology, fluid content, angle of dip, geological age, and probable productivity of the formation.

Coring the process of cutting a vertical, cylindrical sample of the formations encountered as an oil well is drilled. The purpose of coring is to obtain rock samples, or cores, in such a manner that the rock retains the same properties that it had before it was removed from the formation.

Cretaceous of or relating to the geologic period from about 135 million to 65 million years ago at the end of the Mesozoic era, or to the rocks formed during this period, including the extensive chalk deposits for which it was named.

Cross section a geological or geophysical profile of a vertical section of the earth.

Crude oil unrefined liquid petroleum which ranges in gravity from 9° API to 55° API and in color from yellow to black, and may have a paraffin, asphalt, or mixed base. If a crude oil, or crude, contains a sizable amount of sulfur or sulfur compounds, it is called a sour crude; if it has little or no sulfur, it is called a sweet crude. In addition, crude oils may be referred to as heavy or light according to API gravity, the lighter oils having the higher gravities.

Crust the outer layer of the earth, varying in thickness from 5 to 30 miles (10–50 km). It is composed chiefly of oxygen, silicon, and aluminum.

Cubic foot (ft³) the volume of a cube, all edges of which measure 1 ft. Natural gas in the United States is usually measured in cubic feet, with the most common standard cubic foot being measured at 60°F and 14.65 pounds per square inch absolute, although base conditions vary from state to state.

Cubic meter (m³) a unit of volume measurement in the SI metric system, replacing the previous standard unit known as the barrel, which was equivalent to 35 imperial gallons or 42 U.S. gallons. The cubic meter equals approximately 6,2898 barrels.

Cumulative production volumes of oil and natural gas liquids that have been produced.

Deep water in offshore operations, water depths greater than normal for the time and current technology.

Delineation well a well drilled in an existing field to determine, or delineate, the extent of the reservoir.

Deplete to exhaust a supply. An oil and gas reservoir is depleted when most or all economically recoverable hydrocarbons have been produced.

Depth the distance to which a well is drilled, stipulated in a drilling contract as contract depth. Total depth is the depth after drilling is finished.

Derrick a large load-bearing structure, usually of bolted construction. In drilling, the standard derrick has four legs standing at the corners of the substructure and reaching to the crown block; an assembly of heavy beams used to elevate the derrick and provide space to install equipment such as blowout preventers and casingheads; the standard derrick must be assembled piece by piece, it has largely been replaced by the mast, which can be lowered and raised without disassembly.

Development (of a gas or oil field) all operations associated with the construction of facilities to enable the production of oil and gas.

Development drilling drilling that occurs after the initial discovery of hydrocarbons in a reservoir. Usually, several wells are required to adequately develop a reservoir.

Development well a well drilled in proven territory in a field to complete a pattern of production; an exploitation well.

Deviation departure of the wellbore from the vertical, measured by the horizontal distance from the rotary table to the target. The amount of deviation is a function of the drift angle and hole depth. The term is sometimes used to indicate the angle from which a bit has deviated from the vertical during drilling.

Diamond bit a drill bit that has small industrial diamonds embedded in its cutting surface. Cutting is performed by the rotation of the very hard diamonds over the rock surface.

Directional drilling intentional deviation of a wellbore from the vertical. Although wellbores are normally drilled vertically, it is sometimes necessary or advantageous to drill at an angle from the vertical; controlled directional drilling makes it possible to reach subsurface areas laterally remote from the point where the bit enters the earth. It often involves the use of deflection tools.

Discovery well the first oil or gas well drilled in a new field that reveals the presence of a hydrocarbon-bearing reservoir. Subsequent wells are development wells.

Dissolved gas natural gas that is in solution with crude oil in the reservoir.

Dissolved water water in solution in oil at a defined temperature and pressure.

Drill to bore a hole in the earth, usually to find and remove subsurface formation fluids such as oil and gas.

Drilling fluid circulating fluid, one function of which is to lift cuttings out of the wellbore and to the surface. It also serves to cool the bit and to counteract downhole formation pressure. Although a mixture of barite, clay, water, and other chemical additives is the most common drilling fluid, wells can also be drilled by using air, gas, water, or oil-base mud as the drilling mud. Also called circulating fluid, drilling mud. See *mud.*

Drilling mud specially compounded liquid circulated through the wellbore during rotary drilling operations. See *drilling fluid, Mud.*

Drillship A self-propelled floating offshore drilling unit that is a ship constructed to permit a well to be drilled from it.

Drill string the string of drill pipe with attached tool joints that transmits fluid and rotational power from the kelly to the drill collars and the bit. The term is also loosely applied to both drill pipe and drill collars.

Dry a hole is dry when the reservoir it penetrates is not capable of producing hydrocarbons in commercial amounts.

Dry gas gas whose water content has been reduced by a dehydration process; gas containing few or no hydrocarbons commercially recoverable as liquid product; also called lean gas.

Dry hole any well that does not produce oil or gas in commercial quantities. A dry hole may flow water, gas, or even oil, but not in amounts large enough to justify production.

Ecology science of the relationships between organisms and their environment.

Eh/redox a measure of the degree of oxygenation of a sediment.

Endowment the sum of cumulative production, remaining reserves, mean undiscovered recoverable volumes, and mean additions to reserves by field growth.

Environment the sum of the physical, chemical, and biological factors that surround an organism; the water, air, and land and the interrelationship that exists among and between water, air, and land and all living things.

EPIC (Engineering, Procurement, Installation, Commissioning) an EPIC or "turnkey" contract integrates the responsibility going from the conception to the final acceptance of one or more elements of a production system. It can be awarded for all, or part, of a field development.

Erosion the process by which material (such as rock or soil) is worn away or removed (as by wind or water).

Essexite a dark gray or black holo-crystalline plutonic igneous rock.

Estuary a coastal indentation or bay into which a river empties and where fresh water mixes with seawater.

Ethanol produced through fermentation of agricultural raw materials (biomass), ethanol is used for various applications: drinks, pharmaceuticals, cosmetics, solvents, chemicals and more and more often in fuels, either in the form of an additive to gasoline (ETBE: Ethyl Tertiary-Butyl Ether) or blended directly with hydrocarbon-based gasoline.

Exploration the search for reservoirs of oil and gas, including aerial and geophysical surveys, geological studies, core testing and drilling of wildcats.

Exploration well a well drilled either in search of an as-yet-undiscovered pool of oil or gas (a wildcat well) or to extend greatly the limits of a known pool. It involves a relatively high degree of risk. Exploratory wells may be classified as (1) wildcat, drilled in an unproven area; (2) field extension or step-out, drilled in an unproven area to extend the proved limits of a field; or (3) deep test, drilled within a field area but to unproven deeper zones.

Fault a break in the earth's crust along which rocks on one side have been displaced (upward, downward, or laterally) relative to those on the other side.

Fault plane a surface along which faulting has occurred.

Fault trap a subsurface hydrocarbon trap created by faulting, in which an impermeable rock layer has moved opposite the reservoir bed or where impermeable gouge has sealed the fault and stopped fluid migration.

Feed Front-End Engineering Design.

Feldspathoid low silica igneous minerals that would have formed feldspars if only more silica (SiO_2) was present in the original magma.

Field a geographical area in which a number of oil or gas wells produce from a continuous reservoir. A field may refer to surface area only or to underground productive formations as well. A single field may have several separate reservoirs at varying depths; a contiguous area consisting of a single reservoir or multiple reservoirs of petroleum, all grouped on, or related to, a single geologic structural and/or stratigraphic feature.

Fixed platform a structure made of steel or concrete, firmly fixed to the bottom of the body of water in which it rests.

Flare an arrangement of piping and burners used to dispose (by burning) of surplus combustible vapors, usually situated near a gasoline plant, refinery, or producing well. *V.* To dispose of surplus combustible vapors by igniting them in the atmosphere. Flaring is rarely used, because of the high value of gas and the stringent air pollution controls.

Flare gas gas or vapor that is flared.

Flexible flow line flexible pipe laid on the seabed for the transportation of production or injection fluids. It is generally an infield line, linking a subsea structure to another structure or to a production facility. Its length ranges from a few hundred meters to several kilometers.

Flexible riser riser constructed with flexible pipe (see Riser).

Floaters floating production units including floating platforms, and floating, production, storage and offloading units (FPSOs).

Floating offshore drilling rig a type of mobile offshore drilling unit that floats and is not in contact with the seafloor (except with anchors) when it is in the drilling mode; floating units include barge rigs, drill ships, and semisubmersibles.

Floating production and system off-loader a floating offshore oil production vessel that has facilities for producing, treating, and storing oil from several producing wells and which puts (offloads) the treated oil into a tanker ship for transport to refineries on land; some floating, production, storage and offloading units are also capable of drilling, in case they are termed floating production, drilling, and system off-loaders (FPDSOs).

Flowing well a well that produces oil or gas by its own reservoir pressure rather than by use of artificial means (such as pumps).

Fold a flexure of rock strata (e.g., an arch or a trough) produced by horizontal compression of the earth's crust. See *anticline, syncline.*

Formation a bed or deposit composed throughout of substantially the same kind of rock; often a lithological unit. Each formation is given a name, frequently as a result of the study of the formation outcrop at the surface and sometimes based on fossils found in the formation.

Fossil the remains or impressions of a plant or animal of past geological ages that have been preserved in or as rock.

FPSO unit (Floating, Production, Storage and Offloading unit) a converted or custom-built ship-shaped floater, employed to process oil and gas and for temporary storage of the oil prior to trans-shipment.

FPU (Floating Production Unit) a ship-shaped floater or a semi-submersible used to process and export oil and gas.

Free water water produced with oil. It usually settles out within five minutes when the well fluids become stationary in a settling space within a vessel.

FSHR (Free Standing Hybrid Riser) a deepwater riser configuration (see Riser) consisting of a vertical rigid pipe section between the seabed and a submerged buoy and a catenary flexible pipe jumper between the submerged buoy and the floater.

Future petroleum the sum of the remaining reserves, mean reserve growth, and the mean of the undiscovered volume. cumulative production does not contribute to the future petroleum. the terms future oil, future liquid volume, or future endowment are sometimes used as variations of future petroleum to reflect those resources that are yet to be produced.

Gabbro a large group of dark, often coarse-grained, mafic intrusive igneous rocks chemically equivalent to plutonic basalt; formed when molten magma is trapped beneath the surface of the Earth and slowly cools into a holo-crystalline mass.

Gas a compressible fluid that completely fills any container in which it is confined. Technically, a gas will not condense when it is compressed and cooled, because a gas can exist only above the critical temperature for its particular composition. Below the critical temperature, this form of matter is known as a vapor, because liquid can exist and condensation can occur. Sometimes the terms "gas" and "vapor" are used interchangeable. The latter, however, should be used for those streams in which condensation can occur and that originate from, or are in equilibrium with, a liquid phase.

Gas cap a free-gas phase overlying an oil zone and occurring within the same producing formation as the oil. See *associated gas, reservoir.*

Gas–cap drive drive energy supplied naturally (as a reservoir is produced) by the expansion of the gas cap. In such a drive, the gas cap expands to force oil into the well and to the surface.

Gas detection analyzer a device used to detect and measure any gas in the drilling mud as it is circulated to the surface.

Gasoline a volatile, flammable liquid hydrocarbon refined from crude oils and used universally as a fuel for internal combustion, spark-ignition engines.

Gas pipeline a transmission system for natural gas or other gaseous material. The total system comprises pipes and compressors needed to maintain the flowing pressure of the system.

Gas processing the separation of constituents from natural gas for the purpose of making salable products and also for treating the residue gas to meet required specifications.

Gas reservoir a geological formation containing a single gaseous phase. When produced, the surface equipment may or may not contain condensed liquid, depending on the temperature, pressure, and composition of the single reservoir phase.

Gas sand a stratum of sand or porous sandstone from which natural gas is obtained.

Gas-to-liquids (GTL) transformation of natural gas into liquid fuel (Fischer Tropsch technology).

Gas well a well that primarily produces gas.

GC (Gas-chromatography_ an analytical technique which is used to separate and detect compounds.

GCMS (Gas-chromatography-mass-spectrometry) an analytical technique which can identify individual peaks

Geochemistry study of the relative and absolute abundances of the elements of the earth and the physical and chemical processes that have produced their observed distributions.

Geologist a scientist who gathers and interprets data pertaining to the rocks of the earth's crust.

Geology the science of the physical history of the earth and its life, especially as recorded in the rocks of the crust.

Geophone an instrument placed on the surface that detects vibrations passing through the earth's crust. It is used in conjunction with seismography. Geophones are often called jugs.

Geophysical exploration measurement of the physical properties of the earth to locate subsurface formations that may contain commercial accumulations of oil, gas, or other minerals; to obtain information for the design of surface structures, or to make other practical applications. The properties most often studied in the oil industry are seismic characteristics, magnetism, and gravity.

Geophysics the physics of the earth, including meteorology, hydrology, oceanography, seismology, volcanology, magnetism, and radioactivity.

Geothermal pertaining to heat within the earth.

Giant field recoverable reserves >500 million barrels of oil equivalents.

GK6 technology this technology improves the ethylene production of the furnaces by 35% compared to the original capacity. This increase in capacity is achieved by replacing the existing coil. Moreover, this technology allows the furnace to operate on a range of feedstocks from naphtha to heavy oils, with high selectivity and long on-stream time.

Graben a block of the earth's crust that has slid downward between two faults. Compare *horst*.

Gravel island a man-made construction of gravel used as a platform to support drilling rigs and oil and gas production equipment.

Gravity the attraction exerted by the earth's mass on objects at its surface; the weight of a body.

Gravity survey an exploration method in which an instrument that measures the intensity of the earth's gravity is passed over the surface or through the water. In places where the instrument detects stronger- or weaker-than-normal gravity forces, a geologic structure containing hydrocarbons may exist.

Groundwater water that seeps through soil and fills pores of underground rock formations; the source of water in springs and wells.

Gusher an oil well that has come in with such great pressure that the oil jets out of the well like a geyser. In reality, a gusher is a blowout and is extremely wasteful of reservoir fluids and drive energy. In the early days of the oil industry, gushers were common and many times were the only indication that a large reservoir of oil and gas had been struck. See *blowout*.

HCR (Hybrid Catenary Riser) riser configuration comprised of two flexible sections of flexible pipe (at the top and the bottom) and a rigid section in the middle.

Hole in drilling operations, the wellbore or borehole.

Holo-crystalline a rock that is completely crystalline.

Horizontal drilling deviation of the borehole at least 80° from vertical so that the borehole penetrates a productive formation in a manner parallel to the formation. A single horizontal hole can effectively drain a reservoir and eliminate the need for several vertical boreholes.

Horst a block of the earth's crust that has been raised (relatively) between two faults. Compare *graben*.

Hybrid riser a riser (see Riser) configuration combining both flexible and rigid pipe technologies.

Hydrocarbons organic compounds of hydrogen and carbon only whose densities, boiling points, and freezing points increase as their molecular weights increase; the smallest molecules of hydrocarbons are gaseous; the largest are solids. Petroleum is a mixture of many different hydrocarbons.

Ice scour the abrasion of material in contact with moving ice in a sea, ocean, or other body of water.

Igneous rock a rock mass formed by the solidification of magma within the earth's crust or on its surface. It is classified by chemical composition and grain size. Granite is an igneous rock.

Ilmenite a weakly magnetic titanium-iron oxide mineral which is iron-black or steel-gray.

Impermeable preventing the passage of fluid. A formation may be porous yet impermeable if there is an absence of connecting passages between the voids within it. See *permeability*.

Interstice a pore space in a reservoir rock.

IPB (Integrated Production Bundle) a patented flexible riser assembly combining multiple functions of production and gas lift, which incorporates both active heating and passive insulation. Used for severe flow assurance requirements.

IRM (Inspection, Repair and Maintenance) routine inspection and servicing of offshore installations and subsea infrastructures.

Jumper short pipe (flexible or rigid) sometimes used to connect a flow line to a subsea structure or two subsea structures located close to one another.

Jackup drilling rig a mobile bottom-supported offshore drilling structure with columnar or open-truss legs that support the deck and hull. When positioned over the drilling site, the bottoms of the legs penetrate the seafloor. A jackup rig is towed or propelled to a location with its legs up. Once the legs are firmly positioned on the bottom, the deck and hull height are adjusted and leveled. Also called self-elevating drilling unit.

J-Lay a vertical lay system for rigid pipes.

Kelly the heavy square or hexagonal steel member suspended from the swivel through the rotary table and connected to the drill pipe to turn the drill stem as the rotary table turns.

Landman a person in the petroleum industry who negotiates with landowners for oil and gas leases, options, minerals, and royalties and with producers for joint operations relative to production in a field; also called a leaseman.

Lava magma that reaches the surface of the earth.

Limestone a sedimentary rock rich in calcium carbonate that sometimes serves as a reservoir rock for petroleum.

Liquefied natural gas (LNG) a liquid composed chiefly of natural gas (i.e., mostly methane). Natural gas is liquefied to make it easy to transport if a pipeline is not feasible (as across a body of water). LNG must be put under low temperature and high pressure or under extremely low (cryogenic) temperature and close to atmospheric pressure to liquefy.

Liquid petroleum a term used to represent crude oil combined with natural gas liquids.

LNG (Liquefied Natural Gas) natural gas, liquefied through the reduction of its temperature to $-162°C$ $(-260°F)$, thus reducing its volume by 600 times, allowing its transport by LNG tanker.

Log a systematic recording of data, such as a driller's log, mud log, electrical well log, or radioactivity log. Many different logs are run in wells to discern various characteristics of downhole formation. V: to record data.

Macrofauna benthic invertebrates that live on or in the sediment and that are retained on a mesh with an aperture of 0.5 mm.

Magma the hot fluid matter within the earth's crust that is capable of intrusion or extrusion and that produces igneous rock when cooled.

Magnetic survey an exploration method in which an instrument that measures the intensity of the natural magnetic forces existing in the earth's subsurface is passed over the surface or through the water. The instrumentation detects deviations in magnetic forces, and such deviations may indicate the existence of underground formations that favor the entrapment of hydrocarbons.

Major field recoverable reserves >100 million barrels of oil equivalents.

Median a statistical measure of the midmost value, such that half the values in a set are greater and half are less than the median.

Meiofauna benthic invertebrates that live in the sediment and that are retained on a mesh with an aperture of 0.062 mm.

Methane a light, gaseous, flammable paraffinic hydrocarbon, CH4, that has a boiling point of $-25°F$ (and is the chief component of natural gas and an important basic hydrocarbon for petrochemical manufacture).

MFO (Mixed function oxygenases) an enzyme system, located within the cells, which can assist in the metabolism and excretion of contaminants.

Microgabbro gabbro with finer grain crystals (<1 mm).

Mineral rights the rights of ownership, conveyed by deed, of gas, oil, and other minerals beneath the surface of the earth.

MMSCFD million standard cubic feet per day.

Mobile offshore drilling unit a drilling rig that is used exclusively to drill offshore exploration and development wells and that floats upon the surface of the water when being used.

MODU (*plural*: MODUs) <u>m</u>obile <u>o</u>ffshore <u>d</u>rilling <u>u</u>nit often used in conjunction with semi-submersibles and floating, production, storage and offloading units, which do not have drilling rigs.

Moonpool opening in a vessel or a platform deck through which drilling, subsea pipe-laying, or construction is conducted. Several vessels are equipped with a moonpool allowing the use of the VLS for flexible pipe and Reel-Lay or J-Lay systems.

MSCC (Milli Second Catalytic Cracker) a catalytic cracking unit of FCC type (FCC stands for *fluid catalytic cracking*).

Mud the liquid circulated through the wellbore during rotary drilling and workover operations. In addition to its function of bringing cuttings to the surface, drilling mud cools and lubricates the bit and the drill stem, protects against blowouts by holding back subsurface pressures, and deposits a mud cake on the wall of the borehole to prevent loss of fluids to the formation; originally a suspension of earth solids (especially clay) in water, the mud used in modern drilling operations is a more complex, three-phase mixture of liquids, reactive solids, and inert solids—the liquid phase may be fresh water, diesel oil, or crude oil and may contain one or more conditioners. See *drilling fluid*.

Mud line (mudline) the sea bed, unless otherwise specified by the driller.

Natural gas a highly compressible, highly expansible mixture of hydrocarbons with a low specific gravity and occurring naturally in a gaseous form. Besides hydrocarbon gases, natural gas may contain appreciable quantities of nitrogen, helium, carbon dioxide, hydrogen sulfide, and water vapor. Although gaseous at normal temperatures and pressures, the gases making up the mixture that is natural gas are variable in form and may be found either as gases or as liquids under suitable conditions of temperature and pressure.

Natural gas liquids (NGL) those portions of reservoir gas that are liquefied at the surface in lease separators, field facilities, or gas processing plants; natural gas liquids are heavier homologs of methane (including ethane, propane, butane, pentane, natural gasoline, and condensate).

Nephelinemagnetite a sodium potassium aluminosilicate and magnetite—one of the two common naturally occurring iron oxides (chemical formula Fe_3O_4) and a member of the spinel group of minerals.

Nonporous containing no interstices; having no pores and therefore unable to hold fluids.

Norite a coarse-grained basic igneous rock dominated by essential calcic plagioclase and orthopyroxene. Norites also can contain up to 50% clinopyroxene and can be considered orthopyroxene dominated gabbro.

Offshore that geographic area that lies seaward of the coastline. In general, the term "coastline" means the line of ordinary low water along that portion of the coast that is in direct contact with the open sea or the line marking the seaward limit of inland waters.

Offshore drilling drilling for oil or gas in an ocean, gulf, or sea. A drilling unit for offshore operations may be a mobile floating vessel with a ship or barge hull, a semisubmersible or submersible base, a self-propelled or towed structure with jacking legs (jackup drilling rig), or a permanent structure used as a production platform when drilling is completed. In general, wildcat wells are drilled from mobile floating vessels or from jackups, while development wells are drilled from platforms or jackups.

Offshore oil and gas installation subsea or surface (platform) oil and gas drilling facilities.

Offshore production platform an immobile offshore structure from which wells are produced.

Offshore rig any of various types of drilling structures designed for use in drilling wells in oceans, seas, bays, gulfs, and so forth. Offshore rigs include platforms, jackup drilling rigs, semisubmersible drilling rigs, submersible drilling rigs, and drill ships.

Oikocrysts small, randomly orientated, crystals are enclosed within larger crystals of another mineral; the term is most commonly applied to igneous rock textures. The smaller enclosed crystals are known as chadacrysts, while the larger crystals are known as oikocrysts.

Oil a simple or complex liquid mixture of hydrocarbons that can be refined to yield gasoline, kerosene, diesel fuel, and various other products.

Oil-based mud a drilling or workover fluid in which oil is the continuous phase and which contains a few percent water.

Oil field (oilfield) a field producing oil and gas is termed an oil field when the petroleum contained within has a gas/oil ratio (gor) of less than 20,000 cubic feet per barrel. If the gor >20,000 cubic feet per barrel, the field is called a gas field.

Oil in place crude oil that is estimated to exist in a reservoir but that has not been produced.

Oil patch (slang) the oilfield.

Oil pool a loose term for an underground reservoir where oil occurs. Oil is actually found in the pores of rocks, not in a pool.

Oil seep a surface location where oil appears, the oil having permeated its subsurface boundaries and accumulated in small pools or rivulets. Also called oil spring.

Oil shale a shale containing hydrocarbons that cannot be recovered by an ordinary oil well but that can be extracted by mining and processing.

Oil slick a film of oil floating on water; considered a pollutant.

Oil spill *n* a quantity of oil that has leaked or fallen onto the ground or onto the surface of a body of water.

Oil well a well from which oil is obtained.

Oil zone a formation or horizon of a well from which oil may be produced. The oil zone is usually immediately under the gas zone and on top of the water zone if all three fluids are present and segregated.

Olefins family of molecules including in particular ethylene and propylene, which constitutes the raw material allowing for the manufacture of many plastics.

Olivine a name for a series between two end members, fayalite and forsterite; fayalite is the iron rich member $(Fe_2SiO_4.)$ whereas forsterite is the magnesium rich member (Mg_2SiO_4).

Onshore oil and gas installation onshore oil and gas exploration/production.

Organic compounds chemical compounds that contain carbon atoms, either in straight chains or in rings, and hydrogen atoms. They may also contain oxygen, nitrogen, or other atoms.

Organic rock rock materials produced by plant or animal life (coal, petroleum, limestone, and so on).

Orthopyroxene an essential constituent of various types of igneous rocks and metamorphic rocks

Outcrop part of a formation exposed at the earth's surface. *V*: to appear on the earth's surface (as a rock).

PAH polycyclic aromatic hydrocarbons, which contain more than one fused benzene ring; see PNA.

Peat an organic material that forms by the partial decomposition and disintegration of vegetation in tropical swamps and other wet, humid areas. It is believed to be the precursor of coal.

Pegmatite a holo-crystalline, intrusive igneous rock composed of interlocking phaneritic crystals usually larger than 2.5 cm in size; such rocks are referred to as *pegmatitic*.

Permeability a measure of the ease with which a fluid flows through the connecting pore spaces of rock or cement—the unit of measurement is the millidarcy; fluid conductivity of a porous medium; the ability of a fluid to flow within the interconnected pore network of a porous medium.

Permeable rock a porous rock formation in which the individual pore spaces are connected, allowing fluids to flow through the formation.

Petroleum a substance occurring naturally in the earth in solid, liquid, or gaseous state and composed mainly of mixtures of chemical compounds of carbon and hydrogen, with or without other nonmetallic elements such as sulfur, oxygen, and nitrogen.

Petroleum geology the study of oil and gas-bearing rock formations. It deals with the origin, occurrence, movement, and accumulation of hydrocarbon fuels.

Petroleum reservoir a rock formation that holds oil and gas.

Petroleum rock sandstone, limestone, dolomite, fractured shale, and other porous rock formations where accumulations of oil and gas may be found.

Petroleum window the conditions of temperature and pressure under which petroleum will form; also called oil window.

pH a measure of the acidity or alkalinity of a solution.

Pip (Pipe-in-pipe) steel pipes assembly consisting of a standard production pipe surrounded by a so-called carrier pipe. The gap between the carrier and production pipes is filled with an insulation material. As the insulation is protected from the external pressure by the carrier pipe, a high thermal performance material can be used.

Plagioclase a large group of dark, often phaneritic (coarse-grained), mafic intrusive igneous rocks, chemically equivalent to plutonic basalt; formed when molten magma is trapped beneath the surface of the Earth and slowly cools into a holo-crystalline mass.

Plate tectonics movement of great crustal plates of the earth on slow currents in the *plastic* mantle.

Platform the structure that supports production and drilling operations. The types of offshore platforms can be either floating or fixed, depending on the location, water depth, climate and the facility's size.

Play the extent of a petroleum-bearing formation; the activities associated with petroleum development in an area.

Plutonic rock an intrusive igneous rock that is crystallized from magma slowly cooling below the surface of the Earth.

PNA polynuclear aromatic hydrocarbons, which contain more than one fused benzene ring; see PAH.

Polypropylene due to its exceptional shock resistance properties, polypropylene is a plastic material used in a wide range of industries including automobile parts, household goods, fibers and films.

Pool a reservoir or group of reservoirs. The term is a misnomer in that hydrocarbons seldom exist in pools, but, rather, in the pores of rock. *V*: to combine small or irregular tracts into a unit large enough to meet state spacing regulations for drilling.

Porosity 1. The condition of being porous (such as a rock formation). 2. The ratio of the volume of empty space to the volume of solid rock in a formation, indicating how much fluid rock can hold.

Potential the maximum volume of oil or gas that a well is capable of producing, calculated from well test data.

Producer 1. A well that produces oil or gas in commercial quantities. 2. An operating company or individual in the business of producing oil; commonly called the operator.

Production 1. The phase of the petroleum industry that deals with bringing the well fluids to the surface and separating them and with storing, gauging, and otherwise preparing the product for the pipeline. 2. The amount of oil or gas produced in a given period.

Production casing a casing string that serves to isolate the reservoir from undesired fluids in the producing formation, and from other zones penetrated by the wellbore.

Proved reserves of crude oil according to API standard definitions, proved reserves of crude oil as of December 31 of any given year are the estimated quantities of all liquids statistically defined as crude oil that geological and engineering data demonstrate with reasonable certainty to be recoverable in future years from known reservoirs under existing economic and operating conditions.

Proved reserves of natural gas according to API standard definitions, proved reserves of natural gas as of December 31 of any given year are the estimated quantities of natural gas that geological and engineering data demonstrate with reasonable certainty to be recoverable in future years from known natural gas reservoirs under existing economic and operating conditions.

Pyroxene a group of rock-forming inosilicate minerals found in many igneous and metamorphic rocks; variable composition, among which calcium-, magnesium-, and iron-rich varieties predominate.

Quartzgabbro a coarse-grained igneous rock dominated by plagioclase (50%, v/v), orthopyroxene (15%), clinopyroxene (25%) and quartz (<5%) with minor biotite and accessory apatite and opaques.

Reeled pipe an installation method based on the onshore assembly of long sections of rigid steel pipeline, approximately 0.5 mile long, which are welded together as they are spooled onto a vessel-mounted reel for transit and subsequent cost-effective unreeling onto the seabed. Minimum welding is done at sea.

Remaining reserves recoverable oil and ngl volumes that were originally present and have not yet been produced.

Reserve the estimated quantities of petroleum expected to be commercially recovered from known accumulations relative to a specified date, under prevailing economic conditions, operating practices, and government regulations.

Reserve growth (field growth) the increases of estimated petroleum volume that commonly occur as oil and gas fields are developed and produced.

Reserves part of the identified (discovered) resources and include only recoverable materials; the unproduced but recoverable oil or gas in a formation that has been proved by production.

Reservoir a subsurface, porous, permeable or naturally fractured rock body in which oil or gas is stored. Most reservoir rocks are limestone, dolomites, sandstones, or a combination of these. The four basic types of hydrocarbon reservoirs are oil, volatile oil, dry gas, and gas condensate. An oil reservoir generally contains three fluids—gas, oil, and water-with oil the dominant product. In the typical oil reservoir, these fluids become vertically segregated because of their different densities. Gas, the lightest, occupies the upper part of the reservoir rocks; water, the lower part; and oil, the intermediate section. In addition to its occurrence as a cap or in solution, gas may accumulate independently of the oil; if so, the reservoir is called as gas reservoir. Associated with the gas, in most instances, are salt water and some oil. Volatile oil reservoirs are exceptional in that during early production they are mostly productive of light oil plus gas, but, as depletion occurs, production can become almost totally completely gas. Volatile oils are usually good candidates for pressure maintenance, which can result in increased reserves. In the typical dry gas reservoir natural gas exists only as a gas and production is only gas plus fresh water that condenses from the flow stream reservoir. In a gas condensate reservoir, the hydrocarbons may exist as a gas, but, when brought to the surface, some of the heavier hydrocarbons condense and become a liquid.

Reservoir rock a permeable rock that may contain oil or gas in appreciable quantity and through which petroleum may migrate.

Resource a concentration of naturally occurring solid, liquid, or gaseous hydrocarbons in or on the crust of the Earth, some of which is currently or potentially economically extractable.

Resources concentrations of naturally occurring liquid or gaseous hydrocarbons in the earth's crust, some part of which are currently or potentially economically extractable.

Rig the drilling equipment used to drill the well that can either be installed on a platform or a MODU.

Riser a pipe or assembly of pipes used to transfer produced fluids from the seabed to the surface facilities or to transfer injection fluids, control fluids or lift gas from the surface facilities and the seabed.

Risk analysis the activity of assigning probabilities to the expected outcomes of drilling venture.

ROV (remotely operated vehicle) an unmanned subsea vehicle remotely controlled from a vessel or an offshore platform. It is equipped with manipulator arms that enable it to perform simple operations.

RSCR (reeled steel catenary riser) installation of the SCR by the reel-lay method which, compared to conventional installation solutions, allows most of the welding to be performed onshore in a controlled environment, thereby reducing offshore welding which brings many benefits, particularly for fatigue sensitive components of the pipeline.

Samples 1. The well cuttings obtained at designated footage intervals during drilling. From an examination of these cuttings, the geologist determines the type of rock and formations being drilled and estimates oil and gas content. 2. Small quantities of well fluids obtained for analysis.

Sandstone a sedimentary rock composed of individual mineral grains of rock fragments between 0.06 and 2 mm (0.002 and 0.079 in.) in diameter and cemented together by silica, calcite, iron oxide, and so forth. Sandstone is commonly porous and permeable and therefore a likely type of rock in which to find a petroleum reservoir.

Satellite well usually a single well drilled offshore by a mobile offshore drilling unit to produce hydrocarbons from the outer fringes of a reservoir.

SCR (steel catenary riser) a deepwater steel riser (see Riser) suspended in a single catenary from a platform (typically a floater) and connected horizontally on the seabed.

Seafloor the bottom of the ocean; the seabed.

Sedimentary rock a rock composed of materials that were transported to their present position by wind or water. Sandstone, shale, and limestone are sedimentary rocks.

Seep the surface appearance of oil or gas that results naturally when a reservoir rock becomes exposed to the surface, thus allowing oil or gas to flow out of fissures in the rock.

Seismic of or relating to an earthquake or earth vibration, including those artificially induced.

Seismic data detailed information obtained from earth vibration produced naturally or artificially (as in geophysical prospecting).

Seismic method a method of geophysical prospecting using the generation, reflection, refraction detection, and analysis of sound waves in the earth.

Seismic survey an exploration method in which strong low-frequency sound waves are generated on the surface or in the water to find subsurface rock structures that may contain hydrocarbons. The sound waves travel through the layers of the earth's crust; however, at formation boundaries some of the waves are reflected back to the surface where sensitive detectors pick them up. Reflections from shallow formations arrive at the surface sooner than reflections from deep formations, and since the reflections are recorded, a record of the depth and configuration of the various formations can be generated. Interpretation of the record can reveal possible hydrocarbon-bearing formations.

Seismic wave the record of an earth tremor by a seismograph.

Semisubmersible drilling rig a floating offshore drilling unit that has pontoons and columns that, when flooded, cause the unit to submerge to a predetermined depth. Living quarters, storage space, and so forth are assembled on the deck. Semisubmersible rigs are self-propelled or towed to a drilling site and anchored or dynamically positioned over the site, or both. In shallow water, some semisubmersibles can be ballasted to rest on the seabed. Semisubmersibles are more stable than drill ships and ship-shaped barges and are used extensively to drill wildcat wells in rough waters such as the North Sea. Two types of semisubmersible rigs are the bottle-type and the column-stabilized.

Shale a fine-grained sedimentary rock composed mostly of consolidated clay or mud. Shale is the most frequently occurring sedimentary rock.

Shallow gas natural gas deposit located near enough to the surface that a conductor or surface hole will penetrate the gas-bearing formations. Shallow gas is potentially dangerous because, if encountered while drilling, the well usually cannot be shut in to control it. Instead, the flow of gas must be diverted.

Show the appearance of oil or gas in cuttings, samples, or cores from a drilling well.

Sour gas gas containing an appreciable quantity of hydrogen sulfide.

SPAR deep draft surface piercing cylinder type of floater, particularly well adapted to deepwater, which accommodates drilling, top tensioned risers and dry completions.

Spinel group a class of minerals of general formulation $A^{2+}B^{3+}_2O^{2-}_4$, which crystallize in the cubic (isometric) crystal system, with the oxide anions arranged in a cubic close-packed lattice and the cations A and B occupying some or all of the octahedral and tetrahedral sites in the lattice; a and b can be divalent, trivalent, or quadrivalent cations, including magnesium, zinc, iron, manganese, aluminum, chromium, titanium, and silicon.

Spool short length pipe connecting a subsea pipeline and a riser, or a pipe and a subsea structure.

Spud to begin operations on a well.

Spud cans cylindrically shaped steel shoes with pointed ends.

Stratigraphic test a borehole drilled primarily to gather information on rock types and sequence.

Stratigraphic trap a petroleum trap that occurs when the top of the reservoir bed is terminated by other beds or by a change of porosity or permeability within the reservoir itself. Compare *structural trap*.

Structural trap a petroleum trap that is formed because of deformation (such as folding or faulting) of the reservoir formation. Compare *stratigraphic trap*.

Structure a geological formation of interest to drillers. For example, if a particular well is on the edge of a structure, the wellbore has penetrated the reservoir (structure) near its periphery.

Subduction zone a deep trench formed in the ocean floor along the line of convergence of oceanic crust with other oceanic or continental crust when one plate (always oceanic) dives beneath the other. The plate that descends into the hot mantle is partially melted. Magma rises through fissures in the heavier, nonliquid (unmelted) crust above, creating a line of plutons and volcanoes that eventually form an island arc parallel to the trench.

Subsea Technology all products and services required to install and operate production installations on the seabed.

Supergiant field recoverable reserves >5 billion barrels of oil equivalents.

SRB sulfate reducing bacteria

SURF subsea umbilicals risers flowlines.

Sweet crude oil oil containing little or no sulfur, especially little or no hydrogen sulfide.

Syncline a trough-shaped configuration of folded rock layers. Compare *anticline*.

Tar sand a sandstone that contains chiefly heavy, tarlike hydrocarbons. Tar sands are difficult to produce by ordinary methods; thus it is costly to obtain usable hydrocarbons from them.

Tectonic of or relating to the deformation of the earth's crust, the forces involved in or producing such deformation and the resulting rock forms.

Teta wire wire with a specific, patented, T-shape used in flexible pipe to resist the radial effect of the internal pressure. Used for high pressure and harsh environments.

Thermal decomposition the breakdown of a compound or substance by temperature into simple substances or into constituent elements.

Tie-back connection of a satellite subsea development to an existing infrastructure.

Tight formation a petroleum- or water-bearing formation of relatively low porosity and permeability.

TLP (tension-leg platform) a floating production unit anchored to the seabed by taut vertical cables, which considerably restrict its heave motion, making it possible to have the wellheads on the platform.

Trap a body of permeable oil-bearing rock surrounded or overlain by an impermeable barrier that prevents oil from escaping. The types of traps are structural, stratigraphic, or a combination of these.

Tsunami a long-period wave produced by a submarine earthquake or explosion. If it strikes land, it can be very destructive. Also called seismic sea wave.

Ulvospinel a mineral from the spinel group of minerals being iron and titanium oxide of the formula Fe_2TiO_4 (orthotitanate iron).

Umbilicals an assembly of hydraulic hoses which can also include electrical cables or optic fibers, used to control subsea structures from a platform or a vessel.

Underbalanced drilling (UBD) a procedure used to drill oil and gas wells where the pressure in the wellbore is kept lower than the formation pore pressure. As the well is being drilled, formation fluid flows into the wellbore and up to the surface. The circulating fluid for underbalanced drilling can be lower density mud, air, nitrogen, mist, or foam.

Undiscovered resources resources inferred from geologic information and theory to exist outside of known oil and gas fields.

VLS (vertical lay system) a Technip proprietary technology for installation of flexible pipes.

Water-producing interval the portion of an oil or gas reservoir from which water or mainly water is produced.

Weathering the breakdown of large rock masses into smaller pieces by physical and chemical climatological processes; the evaporation of liquid by exposing it to the conditions of atmospheric temperatures and pressure.

Well the hole made by the drilling bit, which can be open, cased, or both. Also called borehole, hole, or wellbore.

Wellbore any hole drilled by the bit for crude oil, natural gas, or water.

Well completion the activities and methods of preparing a well for the production of oil and gas or for other purposes, such as injection; the method by which one or more flow paths for hydrocarbons are established between the reservoir and the surface; the system of tubulars, packers, and other tools installed beneath the wellhead in the production casing; that is, the tool assembly that provides the hydrocarbon flow path or paths.

Well control the methods used to control a kick and prevent a well from blowing out. Such techniques include, but are not limited to, keeping the borehole completely filled with drilling mud of the proper weight or density during all operations, exercising reasonable care when tripping pipe out of the hole to prevent swabbing, and keeping careful track of the amount of mud put into the hole to replace the volume of pipe removed from the hole during a grip.

Well Intervention (well work) any operation carried out on a crude oil or natural gas well during or at the end of its productive life, which alters the state of the well and/or well geometry, provides well diagnostics, or manages the production of the well.

Well logging the recording of information about subsurface geologic formations, including records kept by the driller and records of mud and cutting analyses, core analysis, drill stem tests, and electric, acoustic, and radioactivity procedures.

Well servicing intervention in subsea production wells carried out from a floating rig or a dynamically positioned vessel.

Wildcat a well drilled in an area where no oil or gas production exists.

Wireline a cable technology used by operators of crude oil and natural gas wells to lower equipment or measuring devices into the well for the purposes of well intervention, reservoir evaluation, and pipe recovery.

Zone the term applied to reservoirs and used to describe an interval that has one or more distinguishing characteristics, such as lithology, porosity, permeability, and saturation.

INDEX